T0396455

Chinese Handicrafts

Hua Jueming · Li Jinsong · Wang Lianhai
Editors

Chinese Handicrafts

Volume 1
Compiled by Hua Jueming, Li Jinsong, Wang Lianhai,
Guan Xiaowu, Li Hansheng, Luo Xingbo

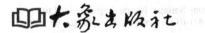  Springer

Editors
Hua Jueming
The Institute for the History of Natural
Sciences, Chinese Academy of Sciences
Beijing, China

Li Jinsong
The Institute for the History of Natural
Sciences, Chinese Academy of Sciences
Beijing, China

Wang Lianhai
Academy of Arts and Design
Tsinghua University
Beijing, China

Translated by
Zhang Weihong
Zhengzhou Business University
Gongyi, China

Jelle Smets
Vancouver, Canada

经典中国国际出版工程
China Classics International

ISBN 978-981-19-5378-1   ISBN 978-981-19-5379-8 (eBook)
https://doi.org/10.1007/978-981-19-5379-8

Jointly published with Elephant Press Co., Ltd
The print edition is not for sale in China (Mainland). Customers from China (Mainland) please order the
print book from: Elephant Press Co., Ltd.
ISBN of the Co-Publisher's edition: 978-7-5347-7329-7

Translation from the Chinese language edition: "Zhongguo Shougong Jiyi" by Hua Jueming et al.,
© Elephant Press Co. Ltd 2014. Published by Elephant Press Co. Ltd. All Rights Reserved.
© Elephant Press Co., Ltd 2022

This Springer imprint is published by the registered company Springer Nature Singapore Pte Ltd.
The registered company address is: 152 Beach Road, #21-01/04 Gateway East, Singapore 189721,
Singapore

Preface: Re-introducing Handicrafts

China is universally renowned for its traditional handicrafts.

China's traditional handicrafts are characterized by their long history. The fact that it goes back to ancient times, is technologically exquisite, and has rich, distinct features is socially and culturally significant. Its handcrafted cultural relics, that have been either unearthed or handed down, its ancient buildings, and its ancient projects were all created using traditional crafts. To this end alone, we can see how important role handicrafts have played in the national economy and people's livelihoods, as well as in the formation and development of Chinese civilization itself. Like traditional Chinese medicine, they are also a treasure trove of Chinese science and technology and have a far-reaching influence worldwide.

For a long time, due to misconceptions and the influence of the ideology of the "extreme left", there has been a prejudice against handicraft techniques and crafts-manship. Handicraft techniques are often considered outdated and obsolete, as if they should all be replaced by modern technology and only exhibited in museums. They are regarded as dispensable and insignificant skills. Such misconceptions are even popular among all levels of the handicraft industry's administration staff. For a long time, traditional handicrafts have been neglected and their protection, inheritance, and discipline have drawn little attention from the responsible authorities, resulting in many precious techniques becoming endangered or dying out. This worrying situation is intrinsically tied to people's misconceptions about handicrafts.

In light of this inheritance crisis, the lack of further development of handicrafts, as well as the people's misunderstanding of them, it is necessary for us to get rid of the confusion and give a clear revision of their connotations, categories, essential characteristics, value, and prospects in the new historical period of reform and opening up. This way, we can remove obstacles for creating a new era of protection and revitalization of traditional crafts.

Categories of Handicrafts

What are Handicrafts?

In the early days, human beings made objects using their hands (and other limbs) and tools (and simple devices). This kind of labor, aimed at making objects, is called handicrafts or handcrafts.

Handicrafts are both material and spiritual, both technical and artistic, both historical and actual. In the social division of labor and economy, it falls in the category of the handicraft industry. It is also known as craft, handcraft, handwork, manual techniques, and traditional handicraft, all of which are used on different occasions, with slight deviations in meaning.

Handed down since ancient times, handicrafts are humanity's most basic form of labor and lifestyle, and thus will continue to be handed down to future generations. Handicrafts are such a part of humanity that it can be seen as the embodiment of the human essence.

Crafts (traditional handicrafts) is an umbrella term that encompasses 14 different categories, namely, (1) making tools and devices, (2) agricultural and mineral processing, (3) construction, (4) weaving, dyeing, and embroidering, (5) ceramics, (6) metallurgy and metalworking, (7) sculpture, (8) weaving and tying, (9) lacquering, (10) furniture making, (11) making calligrapher's tools (12) printing, (13) carving and painting, and (14) special handicrafts and others.

There are different types within each category, and every type contains various kinds. For example, there are different types of lacquering, such as carved lacquering, inlay lacquering, and bodiless lacquering. Then there are different kinds of inlay lacquering, such as mother-of-pearl inlay and metal inlay. In this book, we will use this three-level classification system to sort all types of traditional Chinese handicrafts.

Some techniques are used to solely make artwork, such as carving and painting. However, some others are used to make tools as well as artwork. For example, there are household ceramics and art ceramics; weaving and dyeing are used to produce not only cloth and brocade but also artwork such as *kesi* silk tapestries and double-sided embroidery.

For a long time, industrial arts have received more attention than handicrafts for making useful items, because they enable foreign exchange through export. In the eyes of some people and even some experts and scholars, traditional crafts are equivalent to arts and crafts (such as paper cuttings and New Year pictures), while other more basic techniques for making tools and devices, oil, salt, sauce, and vinegar, which are more closely related to the national economy and people's livelihoods, are neglected. This is, of course, a great misunderstanding, and in order to better understand, rescue, and preserve traditional handicrafts, it is of the utmost importance to rectify it.

Essential Characteristics of Handicrafts

Practicality

As the saying goes: "Necessity is the mother of invention". All handicrafts are invented to meet people's needs in production and their daily life.

"A handy tool makes a handy man". In order to transform and adapt to nature, human beings have to be able to do all kinds of work. This makes it necessary to create and use a great variety of tools, devices and machines. From the most rudimentary sharp-pointed wooden stick to the *lei* (similar to a spade), *si* (similar to a plow), axe, pickaxe, yardstick, marker, divider, carpenter's square, pestle and mortar, well sweep, water mill, windmill, seismograph, compass, astronomical clock, drawloom, and many more, all of these tools and devices have been widely used in people's daily lives. Due to its fundamental significance to human life, their production techniques have become one of the main categories of traditional handicrafts. For example, the large windmill in Sheyang county, Jiangsu province (as shown in Picture 1) is 8 meters high and has 8 fans, which can rotate from all sides. The fans drive the keel of the waterwheel underneath to pump seawater onto the salt field where the salt is extracted. The windmill was restored by Zhang Baichun, director of the Institute for the History of Natural Sciences, Chinese Academy of Sciences. It is now installed at the gate of the Southern Taiwan University of Science and Technology in Tainan city, Taiwan province. There are two major schools of wind-driven machinery, one

Picture 1   Large windmill, located in Sheyang county, Jiangsu province

in the East and one in the West. The shafts of the large windmills in the Netherlands (West) are horizontal, while those in China (East) are vertical.

Such basic necessities as oil, salt, soy sauce, vinegar, and tea are indispensable in the daily lives of Chinese people. The agricultural and mineral processing techniques like flour-grinding, rice-pounding, oil-extracting, sauce-making, vinegar-brewing, tea-making, salt-extracting, and leather-tanning are closely related to people's livelihoods and are still widely used by Chinese people. Luzhou *Laojiao Baijiu* in Sichuan province is going through aging in storage caves. The more *baijiu* ages, the more fragrant it becomes (as shown in Picture 2). Some people say that brewing techniques should not be categorized as intangible cultural heritage. This is incorrect. Fermentation is applied during the alcohol brewing process, which is one of the earliest forms of bioengineering involving microorganisms. It is of great value and has a place in the history of science and technology as well as the national economy.

If people hope to live and work in peace and contentment, they need to build dwellings and houses. Therefore, geotechnical engineers, masons, carpenters, and bricklayers are necessary. The design and production of various building structures, decorations, urban planning, and garden construction are generally called "construction". Picture 3 shows *tulou*—local pieces of architecture characterized by their strong ethnic and regional features, embodying the culture of the Hakka people in Fujian province.

Clothing is a symbol of civilization. Perhaps because of this, clothing takes precedence over food, housing, and transportation. Cotton, hemp, silk, and wool are woven, printed and dyed, and cut and sewn into clothes, shoes, and hats. China is the birthplace of silk-weaving technology. The Silk Road, named after this highly desired commodity, has played a great role in promoting transportation, trade, and cultural exchanges between China and foreign countries. Picture 4 shows a *da'ao,* a women's coat decorated with brocade from the Qing dynasty. The beautiful patterns and colors are what inspired Bai Juyi (772–846), the famous poet of the Tang dynasty, to write

Picture 2 *Baijiu* for aging

Picture 3 *Tulou*

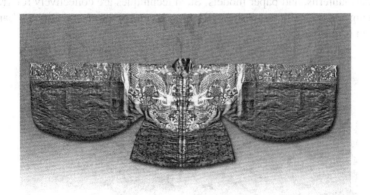

Picture 4 *Da'ao*

in his poem: "The delicate patterns of silk are required by the royal palace… The colors of silk are like those of river water in the south".

Chinese people can't live without pots and jars. The making of porcelain (china) is one of China's great inventions, hence the country's English name. Earthen pots and urns are used for cooking and storing water, while purple clay teapots and porcelain vases are used for drinking tea and arranging flowers, yet having the additional function of being used as an ornament on their own. Colored glazing, glass, and glassware are all categorized as silicate products. Picture 5 shows a woman glazing porcelain in Chenlu town, Shaanxi province.

China had a splendid bronze and iron age and has long been a global frontrunner when it came to the technology of metallurgy and metalworking. Even nowadays, Wang Maizi knives and scissors, Zhang Xiaoquan knives and scissors, gold foil, and iron paintings are still popular among Chinese people. Tibetan, Mongolian, and Husa knives, as well as Bao'an waist knives, are common tools and ornaments among ethnic minorities such as the Tibetan, Mongolian, Hui, Achang, Bao'an, and Salar. Picture 6 shows ironworkers of Jiuxinglu Company in Zhangzi county, Shanxi province, forging gongs with ultra-high-tin bronze (containing 24% tin). The 1-meter-wide gong is the only Chinese musical instrument used in modern symphony orchestras.

Sculpting is one of the oldest artforms. Wood carving, stone carving, jade carving, ivory carving, clay sculpting, and dough modeling are used in the construction and decoration of houses, furniture, bridges, boats, and carriages, but also simply as art, to convey a sense of beauty. Picture 7 is a masterpiece by *Nirenzhang* (Clay Figurine Zhang) of Tianjin city. Two senior farmers are playing chess with pieces of fruit on a self-drawn chessboard in the field.

Grass, hemp, bamboo, rattan, and paper are in themselves quite common materials, but they can be skillfully woven and tied into baskets, mats, tables, chairs, shoes, hats, kites, lanterns, and paper models. Such techniques are collectively referred to as weaving and tying. Picture 8 shows the grand scene of the annual Qinhuai Lantern Market in Nanjin city.

Picture 5 Glazing

Picture 6 Forging of a gong

Picture 7 Game of chess

Ten Defects in *Han Feizi* records that Yu the Great, the first king of the Xia dynasty once "painted black on the outside and red on the inside" of sacrificial vessels. Lacquering originated from the need for the protection and decoration of utensils. It later developed into a variety of types such as engraved cinnabar lacquering, engraved cloud lacquering, and ramee-lacquering. Picture 9 is the work "Radiation" by Professor Qiao Shiguang, a master lacquer artist, who stood on a high place and scattered the paint downwards to get the desired visual effect. The materials used are traditional, while the technique is modern.

Picture 8 Qinhuai lantern market

Picture 9 Radiation

Picture 10 Alcove bedstead

As early as the Warring States period, there were different patterns of mortise and tenon joint structures, which have become a major feature of making Chinese furniture. The design and making of furniture reached its highest peak in the Ming dynasty, and it is now a museum favorite. In recent years, classical furniture has become more and more popular in China. Picture 10 shows an alcove bedstead commonly used in the past. It contains a large bed with a small cabinet and a wooden urinal beside it, thus forming some private space. Picture 11 is an armchair for women. According to an old folk custom in Zhejiang province, women shouldn't lean against the back of the chair when sitting down. Therefore, the front part of the armrest was purposely missing, so women could only sit on the front half of the chair.

Papermaking and printing are two of China's four great inventions and have played a great role in the inheritance and spread of human civilization. *Xuan* paper, or rice paper, has remained a must-have for Chinese calligraphers and painters, and can in no sense be replaced by regular machine-made paper. Picture 12 shows 18 craftsmen lifting *xuan* paper with a width of one *zhang* and two *chi,* that is, approximately 4 meters. Engraved block printing techniques are still used to print ancient books and Buddhist scriptures. The people in Rui'an city, Zhejiang province, still print their genealogical books with wooden movable types. Picture 13 is the Woodblocks Storage Hall of Dege Parkhang Sutra-Printing House in Garzê Tibetan autonomous

Picture 11   Armchair for women

prefecture, Sichuan province, which is the largest among the three Parkhang Sutra-Printing Houses in the Tibetan region, with 220,000 woodblocks in its possession. Picture 14 shows two Tibetan craftsmen printing Tibetan Buddhist scriptures: one coloring and the other printing.

Paper cuttings, fine-grain paper cuttings, shadow play, New Year Pictures, and inner paintings embody both the richness of humanity and the value of artistry, especially since they all originate from people's living habits and aesthetic needs. Picture 15 shows a *Yangliuqing* New Year Picture, from Tianjin, depicting peace, auspiciousness, and abundance, which are the eternal themes of these types of Pictures.

Some techniques, such as how the Li ethnic group in Hainan province drills wood to make fire, are relics of ancient times and extremely rare (as shown in Picture 16). Drilling wood to make fire is the first significant invention of mankind, which, together with the steam engine, electricity, nuclear power, and computers, is a milestone in human progress. Human beings separated themselves from animals and started walking the path of civilization by learning to make fire and preserve kindling. Nowadays, such an ancient technique is only practiced in areas such as

Picture 12 Lifting *xuan* paper with a width of one *zhang* and two *chi*

Hainan, China, and a few countries and areas in the South Pacific. In Picture 17, a native Hawaiian is demonstrating how to drill wood to make fire at the Polynesian Cultural Center in Honolulu, Hawai'i. Some techniques, such as how the Oroqen ethnic group uses fish skin and birch bark in crafting, are so unique that they cannot be classified into the above categories. Therefore, we describe those and other special techniques in Chap. 14, aiming to include as many remaining handicrafts as possible.

Everything mentioned within the 14 categories of traditional handicrafts listed above originates from common people's needs in production and their daily life. Handicrafts are a way of life. Chinese people love them and cannot do without them. Even today with increasing modernization, various kinds of crafts are still quite popular. In recent years, crafting activities such as pottery and knitting have shown signs of revitalization. This is due to its practicality and its place in the modern lifestyle. Handicrafts embody the essence of human culture and labor. Li Bocong, a famous contemporary Chinese philosopher, said, "I create, therefore I am". By imitating "Je pense, donc je suis", the well-known saying of Descartes, French mathematician and thinker, Li's remarks reveal the significance of handicrafts to human culture and labor. The practicality of handicrafts is its most essential and important feature, as Japanese scholar Sōetsu Yanagi (1889–1961) once said, "Practical crafts are the true crafts".

Picture 13 The woodblocks
storage hall of dege parkhang
sutra-printing house

Picture 14 Printing

Picture 15 *Yangliuqing* New Year picture

Picture 16 Drilling wood to make fire in Baoting county, Hainan province

Picture 17 Demonstration of wood drilling to make fire by an aboriginal Pacific Islander

Rationality

Manual work is one of the main ways for people to understand both the natural and artificial world.

Hegel (1770–1831), a German philosopher, once said that what is rational is actual, and what is actual is rational. What is learnt from handicrafts is concrete and rational knowledge. Rationality, a form of high-level cognition, is the soul not only of human nature and science but also of handicrafts. Whether we are making sharp-pointed wooden sticks or drawlooms, we require at least some rational guidance, like knowledge of materials and mechanics. This kind of rationality may be clear or hazy in a craftsman's mind, but craftsmanship is rational in every sense since it is a purposeful activity. Otherwise, even a single rudimentary sharp-pointed wooden stick couldn't be made, let alone other objects.

Let's take the traditional Chinese lost-wax casting as an example. The composite clay-mold casting technique reached its highest peak and yet also encountered its technological bottleneck during the Warring States period. To meet the demand for surface decoration of bronze wares, lost-wax casting was invented: a brand-new casting technique which could produce permeable decoration in three-dimensional space. In doing so, it made possible such fine bronze creations, known from the Shang and Zhou dynasties, as (1) the Moiré Copper Ban, unearthed in Xichuan county, Henan province, (2) Marquis Yi of Zeng's Bronze *Zun* (wine vessel) and *Pan* (plate), and (3) the Chen Zhang Pot. Although there are no records of the precise production process and there is no way to know the terms and formulas used at that time, there is plenty of physical evidence that such technical concepts as the fusibility and cladding of molds were used, as well as such techniques as plastic wax production and wax molds welding. These concepts and techniques were quite

Picture 18 Gating system
of the Moiré Copper Ban
unearthed in Xichuan county,
Henan province

mature during this era, some of which were, in turn, passed on and developed into lost-wax casting and modern precision casting. Picture 18 shows the gating system of the Moiré Copper Ban unearthed from a tomb of the Chu State in the Spring and Autumn period in Xiasi village, Xichuan county, Henan province. It consists of main runners, sub-runners, and branch runners, whose diameters decrease with every turn, totaling five layers. Decorative coiled chi-dragon patterns are formed at the end (on the outermost layer). This system, though quite complex, conforms to the principles of fluid mechanics and foundry engineering.

All forms of handicrafts are based on people's rational cognition of the natural and artificial world and must conform to the objective laws of physics, chemistry, biology, and other scientific disciplines, either by understanding them but not knowing why or understanding them and knowing why. No matter which of the two groups people belong to, rational cognition is essential. That is the rationality of handicrafts.

Aesthetics

Manual work is a combination of wisdom and strength. Even a bone spindle from ancient times (as shown in Picture 19) shows its beauty in its simplicity as well as its harmony between shape and function.

However, not all handicrafts are beautiful. Every period in history inevitably produces some shoddy or overcomplicated handiwork, but these do not represent the majority of all craftsmanship.

In our daily life, the handicraft products we see or use are either handy or pleasing to the eyes, but they are all beautiful. Who is willing to acquire them if they are ugly or inconvenient to use?

Aesthetics is one of the essential characteristics of manual work. As Marina Tsve-taeva (1892–1941), a Russian poet, once said, "Venus was hand made". Great artists use their hands to imbue their creations with great spirit. *"The Thinker"* (as shown in Picture 20) by Auguste Rodin, a French sculptor, is a typical example of this.

Picture 19 Bone spindle

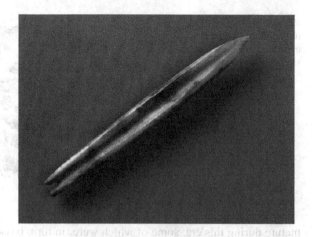

Picture 20 *"The Thinker"*
by Auguste Rodin

Humanity

Just like home-brewed rice wine can invoke homesickness, so can a hand-knitted
sweater still envelop us with the warmth of our loving mother after many years.

Cloth shoes with multi-layered soles may look a bit clumsy, but they are comfortable. Though in order to get these comfortable shoes, they need to be meticulously
made by hand, stitch by stitch, layer by layer.

   The cloth tigers in Picture 21 were made by an elderly woman in Shaanxi province. They are vivid and full of life. She said she made them by imitating the appearance and facial expressions of her grandson. Surprisingly, the cloth tigers actually resemble the boy.

   An old Tibetan man was carefully engraving decorative patterns on a pottery jar. He said that he wanted to make this pottery jar out of his desire to wish his family peace and good fortune ("tashi delek" in Tibetan). He didn't plan to sell it. He did

Picture 21 Cotton tiger

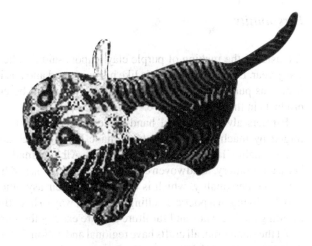

Picture 22 A dressed-up little girl of the Miao ethnic group

not care about how much time he had already spent on the pottery jar, nor how much time he would still spend on it.

Handicrafts embody humanity.

Picture 22 shows a dressed-up little girl of the Miao ethnic group. The Miao people are especially fond of silver. Brides will wear complete sets of silver accessories, such as silver headwear, earrings, neck ornaments, and chest ornaments.

Personality

A master in the making of purple clay teapots said that he only made a few teapots every year. Every one of them is like his own child and has its own distinct character. If he was put in front of a pile of teapots, he would be able to recognize his own products in the blink of an eye.

Farmers always say that hand-made sickles and hoes are different from those forged by machines. The products made by machines are almost the same, with no personality, lacking character and spirit. All handmade products, such as sickles, hoes, embroidery, handwoven cloth, silver ornaments, and hand-made furniture, have their own personality, which is conveyed in their asymmetry and irregularity. It is said that living in a poetic dwelling is the best way to live. But can you imagine such a dwelling having walls and furniture that are exactly the same as any other dwelling?

At the same time, all crafts have regional and national characteristics, which reflect their personality and their roots. *Xuan* paper, or rice paper, is made of blue sandalwood bark, long-stalk straw, and carambola vine juice produced in Jingxian county, Anhui province. All the other kinds of paper made in other areas are not permitted to be called *xuan* paper. The Naxi people in Yunnan province make *dongba* paper with a fixed mold, which is used to print *jiama* (paper charms) and record *dongba* scriptures. Tibetan paper, made of chamaejasme, is mothproof and used to print Tibetan Buddhist scriptures. These kinds of paper are so closely related to the traditional culture, living customs, religious beliefs, and ethnic pride that they cannot, in any way, be replaced by machine-made paper. The essence of their handicrafts is deeply rooted in each culture's personality and location. Shown in Picture 23 are window paper cuttings from Yuxian county, Hebei province, with a style typical of northern China, while shown in Picture 24 is a Dai-style paper cutting, and in Picture 25 a fine-grained one from Yueqing city, Zhejiang province with a style typical of southern China.

Picture 23 Paper cuttings from Yuxian county, Hebei province

Picture 24 Dai-style paper
cutting

Picture 25 Fine-grained
paper cutting from Yueqing
city, Zhejiang province

Combining Variance and Invariance

Handicrafts are dynamic and always changing with time.

There are no handicrafts in the world that do not undergo change and simply remain the same. Porcelain made in the Ming and Qing dynasties are different from those of the Song and Yuan dynasties, which, in turn, are different from those of the Han and Tang dynasties; likewise, celadon made in the Han dynasty is different from the proto-porcelains of the Western Zhou and Shang dynasties. However, all of them are considered porcelain, which exemplifies the fact that variance and invariance are closely related. No one would fault the porcelain of the Song and Yuan dynasties for their differences from those of the Han and Tang dynasties, or those of the Ming and Qing dynasties for their differences from those of the Song and Yuan dynasties. Then who can guarantee that the handicrafts we know today will be the same in 100 years, or even force those handicrafts of today to remain static and unchanged?

The technique of drilling wood to make fire has not changed much since ancient times, although it takes different forms in different parts of the world. If a technique fails to keep pace with the times due to its innate limitations, it can only decline, die out, and be replaced by other more advanced techniques. Isn't it quite natural and reasonable to replace wood with flint, flint with matches, and matches with a lighter to make fire?

It is often said that those traditional handicrafts which have been recognized as intangible cultural heritage should be preserved in their original state. Of course, this kind of advice is positive to some extent, since they should be preserved conditionally and selectively. To meet the needs of mass production and market competition, the Anhui Jingxian Xuan Paper Making Mill has used machinery, instead of manual labor, to pulp paper and steam, instead of fire, to dry paper. No one has raised any objections to this change, as it is, in fact, impossible to pass on the tradition and keep it intact. However, to preserve the original techniques—techniques used about one century ago instead of those in the Ming dynasty—the mill has built a paper-making production line which uses only traditional techniques. The superior *xuan* paper that is made in this traditional way is then sold at a higher price. A wise decision, indeed. Some handicraft techniques are highly complex and cannot be understood in a short time. For example, the brewing of distilled spirits is a kind of biological engineering. Scientific analysis shows that some kinds of *baijiu* contain more than 400 types of beneficial microorganisms. The effect of these microorganisms on the quality of spirits and the way they act and react with the raw materials, yeast, water, and production procedures have not yet been closely examined. In this case, most of the traditional production procedures have been preserved, such as the spreading, airing, and loading of fermented materials, assessing the quality of *baijiu*, and blending, because those techniques heavily depend on the artisans' senses and judgement. Only procedures like the handling of raw material have been replaced by machinery. Also a wise decision.

It is widely recognized that Japan has done the best in preserving their intangible cultural heritage. Moreover, Japan is quite entrepreneurial when it comes to the inheritance and development of traditional handicrafts. As early as the 1980s, or even earlier, Japanese traditional metal crafts and lacquering adapted to the needs of modern life and the changes in people's aesthetic preferences. Since then, there have been many innovative changes when it came to shape, color, and patterns, whereas they maintained the natural materials and traditional techniques used in production. No one can deny that these are still authentic traditional craft products, and no one can accuse such practices of parting with tradition, because keeping pace with the times is in fact the tradition of handicrafts. China lags behind Japan both in theory and practice in this respect, so we still have a long way to go. Nevertheless, it is comforting to see that experts and scholars in some Chinese colleges and institutions are working towards this very goal. For example, outstanding progress has been made in ceramics and lacquer painting. If we use the correct concepts and take appropriate measures, then traditional handicraft techniques will surely become better and more effective in combining both variance and invariance.

Eternality

Handicrafts go hand in hand with human beings, who have changed the world and themselves using their own two hands. Long-term labor makes manual labor an instinctive need in people's lives. There are even famous people who liked handicrafts. For example, among those who loved woodworking, there was the Chinese

Emperor Zhezong of Song (1077–1100), the great Russian writer Lev Tolstoy (1828–1910), and former U.S. President Jimmy Carter. In India, Mahatma Gandhi (1869–1948) (as shown in Picture 26) was a big fan of the spinning wheel. There are also many common people who are fond of handicrafts and love them purely as hobbies.

Handicrafts that aim to create things that are useful are also eternal. Many people say that the pre-modern era was the era of manual labor, the industrial era was the era of machines, and the post-modern era of automation, information, or digitalization. It is not very accurate, but also not very inaccurate. Like Professor He Xuntian, a contemporary Chinese musician, once said, "It is an era where manual and non-manual labor coexist",

Nearly all original work requires a manual touch, especially in manufacturing. At present, there is a shortage of 100,000 senior technicians in China, which is related to our common misunderstanding of what manual labor entails. People who have a bit of common sense when it comes to manufacturing all understand the extreme importance of mold making for machine manufacturing, and mold making is inseparable from manual labor. Similarly, the modification of all cars is based on car models, which are made by hand. The best clothes, hats, leather shoes, and jewelry around the world are all handmade. The handmade leather shoes of the French Massaro family cost $3,000 per pair. How can machine-made ones even begin to compare? Academician Wu Mengchao (1922–2021), a famous Chinese surgeon who won the Highest Science and Technology Award in 2005 (as shown in Picture 27), said that all his operations could be made public. You might know in theory how he did the operation, but you would not be able to replicate it in practice, because you do not possess the same skills. That is why capable surgeons are called "Golden Fingers" and why they can all proudly call themselves craftsmen or artisans.

Picture 26 Mahatma Gandhi is spinning

Picture 27 Dr. Wu
Mengchao during an
operation

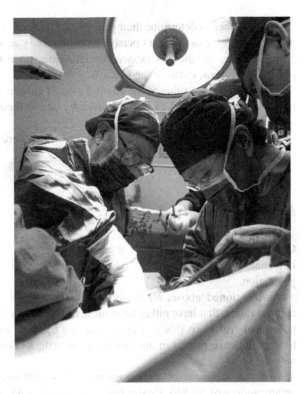

Therefore, it is definitely a great misunderstanding to call the current era "the era of machines or automation".

Where there are humans, there are handicrafts, there is a fondness of participating in craft activities, and there is a love for handmade products. Handicrafts are eternal.

The Value of Handicrafts

Inherent Value

The value of handicrafts is determined by their core characteristics.

Their practicality determines that they are of great value both in people's daily lives (practical value) and on the market (exchange value), both on a personal and national level.

Their rational foundation determines their academic value. There are scientific principles behind many kinds of traditional handicrafts and they deserve to be preserved, just like wildlife.

Their aesthetics determine their general and educational artistic value. Handmade products can give us different perspectives on beauty and mold our temperament.

Meanwhile, handicrafts possess social value, since they are closely linked to people's emotions, customs, and beliefs.

They are of great value to the continuation of culture and the preservation of our national identity as well as cultural diversity due to their intrinsic regional, ethnic, and local characteristics.

Because handicrafts are eternal and dynamic, they are a testimony to humanity's shared universal values and historical continuity.

Historical Value

Chinese people are famous for their ingenuity. This is the natural result of the influence and cultivation of handicrafts being passed down from generation to generation.

As mentioned above, all ancient buildings, ancient projects, and handcrafted cultural relics, that have either been unearthed or handed down, were created using traditional crafts. To this end alone, we can see how important a role traditional handicrafts have played in the formation, growth, and development of the Chinese nation.

Even in the field of historiography, it is still necessary to gain a deeper understanding of the historical value of handicrafts and appraise them properly.

Modern Value

This is a topic that has hardly been talked about or simply avoided, be it intentionally or unintentionally. After understanding more about the essential characteristics and value of handicrafts in this chapter, I believe you, my dear readers, will be able to draw your own conclusions. The main thing that we, as the authors, would like to convey is the following:

Traditional crafts are one of China's treasures as well as a treasure trove of Chinese science and technology comparable to traditional Chinese medicine.

Much of what we eat with, wear, use and play with every day are still handmade products, such as axe, chisel, hoe, pickaxe, black tea, green tea, yellow rice wine, *baijiu*, woven bamboo ware, rattan plaited articles, wood carvings, jade carvings, clay sculpting, dough modeling, tie-dyed cloth, batik cloth, brocade, embroidery, gold leaf ornaments, silver ornaments, celadon ware, purple clay ware, paper cuttings, New Year Pictures, Tongrentang traditional Chinese medicine, Quanjude roast duck, Wang Mazi knives and scissors, Zhang Xiaoquan knives and scissors, cloisonné enamelware, Chinese knots, oil, salt, soy sauce, vinegar, clothes, shoes, hats, fire-works, firecrackers, brush, ink, paper, inkstone, and the list goes on and on. With the improvement of people's living standards, their yearning for returning to nature and their roots, and the new perspective in regards to their aesthetic preferences, there will be more people asking for handmade products.

Today, in China, we can see that traditional handicrafts are still of great value in people's daily lives and of great value to the economy, academic research, artistic appreciation, and culture. Handicrafts will play a vital role in the further development of the central and western regions in China as well as the agricultural, rural areas. Handicrafts will prove their value, not only by improving people's living standards but also by improving the national economy. However, we still need to delve deeper into their academic value. It needs further study and exploration, which will hopefully result in great achievements that will attract worldwide attention. At the same time, their artistic value will steadily garner more and more recognition and interest. While facing the impact of globalization and the dominance of Western culture, the cultural value of Chinese traditional handicrafts is especially important and shall draw ever more attention.

Protecting Handicrafts: Global Consensus

It has taken nearly half a century for the world to reach a consensus on the protection of handicrafts.

Japan formulated and promulgated the *Law on the Protection of Cultural Property* in 1950, which was the first step in protecting intangible cultural heritage in the world. Intangible cultural heritage means non-physical cultural heritage. Traditional handicrafts, such as inlaying gold and silver, Japanese shippo, and the making of Japanese blades, swords, and magic mirrors, account for a considerable proportion of Japan's intangible cultural heritage.

In 1982, UNESCO established the Section for the Non-Physical Heritage, and in 1989, the 25th General Conference adopted the *Recommendation on the Safeguarding of Traditional Culture and Folklore*.

In 1997, UNESCO launched the program of *Proclamation of Masterpieces of the Oral and Intangible Heritage of Humanity*, in which handicrafts are listed as one of the five forms of intangible cultural heritage. This program was officially announced and launched the following year.

In 2001, the *Universal Declaration on Cultural Diversity* was adopted by the 31st General Conference, which emphasized the significance of intangible cultural heritage in maintaining human cultural diversity. In the same year, the first 19 *Masterpieces of the Oral and Intangible Heritage of Humanity* were recorded and Chinese *Kunqu* Opera was one of them.

In 2002, the *Istanbul Declaration* was adopted by the *Third Roundtable of Ministers of Culture of the United Nations*, which emphasized the importance of intangible cultural heritage.

In 2003, UNESCO launched the *Convention for the Safeguarding of the Intangible Cultural Heritage*. China joined this convention the following year. At present, the convention has entered into force, after being ratified by nearly 30 countries.

In 2004 and 2006, UNESCO added more elements to the *Representative List of the Oral and Intangible Heritage of Humanity*. Until then there had been a total of 90 elements listed, among which were the *guqin* (the Chinese zither), the Xinjiang Uyghur *Muqam* (folk melodies), and Mongolian *Urtiin Duu* (long folk songs).

So far, the importance of intangible cultural heritage, including handicrafts, and the necessity and urgency of its preservation has become an international consensus. Especially in developing countries where intangible cultural heritage is still well preserved and relatively complete, it usually receives financial support from both the government and NGOs.

The Protection and Revitalization of Traditional Handicrafts are Part of Chinese Modernization

Chinese handicrafts have long been neglected and improperly protected. The reasons for this are closely related to profound social and ideological misconceptions. For a long time, under the influence of the ideology of the "extreme left", cutting off history, ignoring tradition, and even acting against tradition have become a new kind of "tradition". At the same time, modernization has caused many people to become prejudiced against traditional handicrafts due to a decline in knowledge and understanding, especially in the humanities.

However, since China joined the UNESCO *Convention for the Safeguarding of the Intangible Cultural Heritage*, the preservation and inheritance of intangible cultural heritage have been officially put on the Chinese government's agenda. In 2006 and 2008, the State Council approved and published *List of the First and Second Batch of National Intangible Cultural Heritage* respectively. Among the more than 1,200 pieces of national intangible cultural heritage, accounting for more than a quarter, there are more than 300 traditional handicrafts. The importance of handicrafts and the necessity and urgency of their preservation and inheritance have become the consensus of all sectors of society. With the joint efforts of the government, communities, enterprises, artisans, and experts, a number of precious handicrafts are expected to be protected, inherited, developed, and revitalized. When we have a better and more objective understanding of traditional handicrafts and recognize their value and importance in modern society, we should naturally think of their protection and revitalization as part of China's modernization, and by no means as dispensable or a burden. The protection, revitalization, and development of traditional handicrafts should be put forth as a national decree signed by the executive branch of the central government, in which their short-, medium-, and long-term goals are formulated, worked out, and put into practice. All in all, realizing that goal will require all of our joint efforts (Picture 28).

Picture 28 *Painting of Gusu, the Flourishing City*

How to Regard and Live with Handicrafts?

Handicrafts are a Treasure

Please regard them with respect and develop alongside them.
There is a Chinese TV series by CCTV called *Liuzhu Shouyi*, which means *Retaining Handicrafts*. It is rich in content, practical, and simply made really well. In the series, they interview Japanese scholar Shiono Yonematsu. In fact, the name of the program *Retaining Handicrafts* is exactly the same as Shiono Yonematsu's book. Just like the title of the book and TV series, I agree that we should "retain handicrafts". However, I also think that "developing alongside handicrafts" sounds more positive and reflects reality more accurately, because we do not only retain handicrafts but also develop and revitalize them. Since people will not and cannot live without handicrafts, they are eternal and inseparable from human life. This is why we grow with them, why we develop alongside them, as one.

Beijing, China

Hua Jueming
Li Jinsong
Wang Lianhai

Translators' Preface to "Chinese Handicrafts"

As one of the four ancient civilizations, China has a long history of material and spiritual culture. Its traditional handicrafts have played a crucial role in the formation and development of its civilization, and have also made outstanding contributions to the development of the rest of the world, continuously promoting the exchange and mutual appreciation of Chinese and foreign civilizations. For example, in terms of science and technology, the *Four Great Inventions of China*, namely, paper, the printing press, the compass, and gunpowder, have been an immense contribution to global development. Furthermore, in terms of trade, Chinese silk, tea, and porcelain have been popular and exported in large quantities to Central Asia and Europe for a long time.

Chinese traditional handicraft techniques are the crystallization of thousands of years of hard work and wisdom of folk artisans and represent the cultural heritage and lineage of the Chinese nation. These techniques mainly include (1) making tools and devices, (2) agricultural and mineral processing, (3) construction, (4) weaving, dyeing, and embroidering, (5) ceramics, (6) metallurgy and metalworking, (7) sculpture, (8) weaving and tying, (9) lacquering, (10) furniture making, (11) making calligrapher's tools, (12) printing, (13) carving and painting, and (14) special handicrafts. A detailed introduction to all of the above is provided in this book. The editors are professional scholars in various industry fields, with a solid grasp of the content of each sector, supplemented by credible field research and empirical studies, making this book a professional and readable monograph on traditional Chinese handicraft techniques with rich content, detailed data, and illustrations.

The English translation and publication of Chinese Handicrafts in other countries will be of great value to the world in understanding the history and contemporary protection of Chinese traditional handicrafts, its contribution to Chinese civilization and world civilization, and its contemporary significance.

The two translators of this book are a native Chinese and an English speaker. Zhang Weihong is a senior scholarly translator with a solid background in traditional Chinese culture, as well as profound basic Chinese and English language skills and over twenty years of teaching and practical experience in translation. Jelle Smets has a strong interest in traditional Chinese culture, has lived and worked in China for many years, and is specialized in Chinese-English language and cultural translation. Our collaboration ensured an accurate understanding of the language and issues of the original text, as well as the accuracy and fluency of the translation.

When we saw the nearly 400,000-word Chinese Handicrafts for the first time, we were very proud to be working on its translation and doing our part for the exchange of Chinese and foreign cultures, but at the same time we were under great pressure. In addition to the time constraint and the heavy workload, the most troublesome problem was that there were many technical terms related to the materials, production, and usage of traditional Chinese handicraft techniques in the book, many of which did not have ready English equivalents and needed to be translated by the translator according to the actual situation. Therefore, after reading through the entire book before actually translating it, we developed our own translation principles and strategies for translating terminology.

Our main purpose of translating this book is twofold: first, to preserve the expressions and connotations of traditional Chinese culture as much as possible in its original form; second, to focus on the fluency and acceptability of the translated text for readers. Therefore, when translating the terminology, we have extensively consulted a large number of materials and decided that if there is a corresponding English expression, we will adopt it directly; if not, we will adopt the method of transliteration plus free translation, which means adding the corresponding explanatory translation besides the Chinese *pinyin* to facilitate readers' understanding, and occasionally we will directly adopt the free translation method according to the specific situation. In terms of annotation, in order not to add extra burden to the readers and not to disturb their reading fluency, we have incorporated most of the explanatory text directly into the translation, and only added a table of Chinese dynasties and corresponding Western chronology in the appendix, as well as the English-Chinese comparison of Chinese classical works, so that those readers who are interested can cross-reference these themselves. In addition, all the photos in the original text have been retained in the translation.

We hope that our translation efforts will be well received by our readers. We also sincerely hope that our readers will enjoy reading this book and that it will enhance their understanding of traditional Chinese technology, especially handicrafts, and ultimately promote mutual understanding and communication among world civilizations.

Gongyi, China   Zhang Weihong
Vancouver, Canada   Jelle Smets

Contents

1 **Making Tools and Devices** 1
   Guan Xiaowu and Feng Lisheng

2 **Agricultural and Mineral Processing** 115
   Li Jinsong

3 **Construction** 227
   Luo Xingbo and An Peijun

4 **Spinning, Dyeing, and Embroidering** 297
   Wang Lianhai and Zhao Hansheng

5 **Ceramics** 425
   Hua Jueming and Qiu Gengyu

6 **Metallurgy and Metalworking** 521
   Hua Jueming

7 **Sculpture** 637
   Luo Xingbo

8 **Weaving and Tying** 693
   Wang Lianhai

9 **Lacquering** 757
   Zhou Jianshi and Hua Jueming

10 **Furniture Making** 809
    Hua Jueming

11 **Making Calligrapher's Tools** 875
    Hua Jueming and Guan Xiaowu

12 **Printing** 935
    Guan Xiaowu and Fang Xiaoyan

13 **Carving and Painting** 1007
    Wang Lianhai

14 **Special Handicrafts and Others** 1053
    Yang Yuan and Li Jinsong

15 **Protection, Inheritance, and Revitalization of Traditional
    Crafts** 1119
    Hua Jueming

**Conclusion: Destiny of Traditional Crafts in Contemporary
Times** 1135

Postscript 1141

Appendix I 1143

Appendix II 1145

Chapter 1
Making Tools and Devices

Guan Xiaowu and Feng Lisheng

"A handy tool makes a handy man". Tools and devices are what human beings use to adapt to and tamper with nature. The invention and use of tools, devices, and instruments have helped to fundamentally change the relationship between man and nature, laid the foundation of today's artificial world, and improved the means of production and labor as well as people's quality of life, thus playing a key role throughout the course of human civilization. Of all the different kinds of traditional handicrafts, the making of tools and devices is of fundamental significance. In this chapter, we will cover the making of tools, devices, and instruments.

1.1 Farm Tools, Hand Tools, and Simple Devices

1.1.1 Farm Tools

1.1.1.1 *Lei* and *Si*

The *lei* and *si*, resembling an ancient spade and plow respectively, were the main farm tools during the pre-Qin period. Xu Shen and Zheng Xuan, scholars of the Han dynasties, regarded the *lei* and *si* as the same thing, but according to *King of Sea* in *Guan Zi* and evidence from unearthed objects, they are actually different.

The earliest form of a *lei* was a pointed wooden stick that was used to turn the soil. The rear was fitted with a crossbeam which could be stepped on to drive

G. Xiaowu (✉)
The Institute for the History of Natural Sciences, Chinese Academy of Sciences, Beijing, China
e-mail: gxiaowu@ihns.ac.cn

F. Lisheng (✉)
Tsinghua University, Beijing, China
e-mail: fls@tsinghua.edu.cn

© Elephant Press Co., Ltd 2022
H. Jueming et al. (eds.), *Chinese Handicrafts*,
https://doi.org/10.1007/978-981-19-5379-8_1

1

the wooden stick deeper into the soil. Traces of these *lei* have been found in Late Neolithic archeological sites. If you look closely at the Chinese character "*lei*" (耒), as inscribed on oracle bones from the Shang dynasty, you can see that it depicts the object's general shape at that time. Later, through continuous farming, two-toothed wooden *lei* (as shown in Picture 1.1) and crank *lei* were developed. In Warring States literature, the *lei* was often mentioned alongside the *si*. The *si*, more curved than a spade, was also used for digging. It was recorded in verse "*office of mountain*" in *Office of Earth* in *Rites of Zhou* that "when you make a *si*, you have to cut young trees at the right time".

The *lei* and *si* at that time were made of wood, stone or bone (as shown in Picture 1.2). For example, the handle and teeth of the *lei* of the Warring States period unearthed in Jiangling county, Hubei province were made of wood, with the teeth ends covered in iron edges. It is the same with the wooden *lei* unearthed in the Mawangdui Tombs of the Han dynasty in Changsha city.

Picture 1.1 Stone carving at Wuliang Ancestral Temple built during the Eastern Han dynasty in Jiaxiang county, Shandong province, depicting Shennong, legendary founder of Chinese agriculture and medicine, holding a *lei*, cited from Volume 1 of the *Atlas of Chinese Farm Tools* edited by Song Shuyou

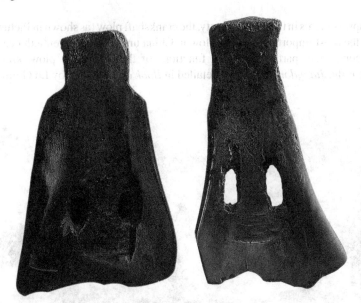

Picture 1.2 Neolithic-era bone *si*, unearthed at the Hemudu site, Yuyao city, Zhejiang province, cited from *Gems of Ancient Chinese Technological Inventions*

The invention and combined application of the *lei* and *si* not only improved farming efficiency, but also shaped the future of plowing and sowing tools. The combination of the *lei* and *si* helped agricultural development enter the next stage.

In addition to the *lei* and *si*, there are many more kinds of hand-made farm tools, such as draw hoes, scuffle hoes, sickles, and other types of shovels, but they are not listed here.

1.1.1.2 Plow

The plow has a long history in China. The first rudimentary form of the plow already existed in the Spring and Autumn period and the Warring States period, but its main structure was perfected during the Qin and Han dynasties. The application and development of plows and plowing were greatly propelled by the prosperity of the metallurgical industry and Zhao Guo of the Western Han dynasty's reform of farm tools.

There are two kinds of traditional Chinese plows: (1) straight shaft and (2) crankshaft. Both of these are usually pulled by animals. There are also two types of straight shaft plows: (1) those with a single straight shaft and (2) those with double straight shafts. As shown in Picture 1.3, in the 1930s, people in Ya'an city in Sichuan province and Guanghua county in Hubei province, as well as the Naxi people in Yunnan province, plowed their fields with a single straight shaft plow pulled by two oxen. In the 1980s, these were still used in Qinghai, Gansu, and Ningxia provinces.

After improvements in the Tang dynasty, the crankshaft plow (as shown in Picture 1.4) became the most important form of plow in China until the late twentieth century.

The component parts, sizes, and functions of the crankshaft plow, known in Chinese as the *Jiangdong* plow, are detailed in *Book of Lei and Si* by Lu Guimeng, a

Picture 1.3 Naxi people in Yunnan province farming with a single straight shaft plow pulled by two oxen, cited from volume *"Farming"* of *the Material Culture of Yunnan*

Picture 1.4 Postcard sent from Shanghai to Belgium on September 29, 1908, depicting farmers plowing in northern China

scholar of the Tang dynasty. It consists of an iron *chan* (share) and *bi* (mouldboard), wooden *di* (slade), *ya chan* (land side), *ce'e* (mouldboard brace), *jian* (strut), *yuan* (beam), *shao* (stilt), *ping* (wedge), *jian* (bolt), and *pan* (whipple-tree).

The iron crankshaft plow appeared in the late Qing dynasty and became popular during the time of the Republic of China (1912–1949). Compared to its previous form, this plow was featured with a simpler frame, longer service life, and better performance, because it has no strut and the iron beam can be turned at will. It was still used in many places in China in the 1970s and 1980s.

Feng Lisheng and Huang Xing, two Chinese scholars investigating Chinese farm tools, went to Dingjiafang village, Xuanhua district, Zhangjiakou city, Hebei province, to interview an old man who had been making and repairing farm tools such as plows, seed plows, carriages, wheel barrows, and wind boxes (as shown in Picture 1.5) for more than 60 years. The old man emphasized that the plow shaft has to be made of a suitably curved tree trunk, otherwise you wouldn't be able to use it and it would break easily. There are four steps to making a plow, namely: (1) sanding the rough trunk, (2) laying out the components of the plow, (3) fashioning all components, and (4) assembling the plow. First, determine the size of each component part according to the bend in the trunk. Then, place all the completed parts in the right position to determine the overall frame of the plow and draw markings on the slade, bend, tail, and other parts that need to be drilled. Once marked and drilled, assemble all the parts and reinforce all the joints with wedges. Finally, meticulously sand and polish the plow, making it light, beautiful, smooth, and applicable.

The shaft, mirror (mouldboard) and head of an iron plow are all made of iron. The complete casting set of the plow mirror and share, as found in Yangcheng county, Shanxi province, was still used until the 1990s, and was listed in *List of the First Batch* of *National Intangible Cultural Heritage* in June, 2006.

1.1.1.3 Seed Plow

The seed plow, a sowing tool used by farmers in the north where the land is mostly dry, has been in use since the Western Han dynasty over 2,000 years ago. There are one-, two- and three-foot seed plows, but the sowing principle and operation method are the same. It usually consists of three parts: (1) the frame, (2) the hopper and seeding regulator, and (3) the share. There are three different kinds of seed regulator, however, the most commonly used one has a discharge hole cut in the middle and lower part of the back of the hopper and a gate clamped on the outside with a wedge. When the wedge is pulled up, the gate can be moved up and down to regulate the size of the hole, thus controlling the outflow of seeds. The seeds flow through the hole into the tubes and are planted into the soil via the legs. A thin bamboo strip goes through the hole, with one end fixed on the inner upper part of the hopper and the other tied to a hanging rope outside. The strip can swing freely around the hole. When the seed plow is in use, the heavy object tied on one end of the rope will swing with the plow's movement, making the thin bamboo strip vibrate, and thus helping the seeds flow out.

Picture 1.5 Seed plow used by farmers in Xuanhua district, Zhangjiakou city, Hebei province (Photo by Feng Lisheng)

Just like plows, the choice of material for seed plows is of great importance. The various parts of the frame are mostly made of locust wood, and the hopper is made of catalpa, locust, ailanthus or willow wood. Since the row spacing varies, such as one *chi* and two *cun* (about 40 cm), one *chi* (about 33 cm), eight *cun* (about 27 cm), or five *cun* (about 17 cm), the sizes and specifications of the seed plows are also different.

To construct a seed plow, firstly, the top and feet of the frame, the upper and lower horizontal braces, and the shafts are made. On both the top and feet of the legs, tenons, and holes need to be made: tenons to hold the horizontal braces and the pallet, oblique tenons fit for the tubes, and the discharge hole should be cut at the right angle. Then tenons are made on the shafts to match the mortises on the outer end of the front beams of the pallet. The length and width between the two shafts should easily accommodate livestock.

Secondly, the pallet frame and the hopper are made. The pallet frame is connected to the legs by mortise and tenon joints, and the hopper is fixed on the pallet. The hollow tubes are fixed between the pallet and the legs, with the upper part connected with the holes on the bottom of the seed-separating box and the lower part with the hollow part of the legs. Several bars are nailed on the front, left, and right sides of the shafts, and their lower ends are connected with the pallet and the front beams of the pallet. Holes of the same size and number as on the seed-separating box are drilled into the bottom of the pallet, which connects the seed-separating box and the tubes.

Finally, the shafts are installed. The tenons of the shafts are joined with the mortises at the outer ends of the front beams of the pallet. The shafts are fixed onto the pallet with ropes and their rear parts are fixed to the legs.

Only highly skilled carpenters are capable of making seed plows and only skillful farmers can properly handle them. The service life of a proper seed plow can be as long as 10 or even 30 years.

1.1.2 Hand Tools

1.1.2.1 Axe

The axe is a kind of wedge-shaped cutter, which, together with the chisel, is one of the most basic and widely used hand tools. The history of the stone axe (as shown in Picture 1.6) can be traced back to hundreds of thousands of years ago, and it once symbolized a clan leader's power. Later, there was the jade axe, which evolved into a symbol of kingship, chiefs, and tribal alliance leaders. It was also called a *yue* (tomahawk).

The bronze axe (as shown in Picture 1.7) was first seen during the time of the Qijia culture (about 2,200–1,600 B.C.). The primitive form of a socketed axe later developed into a shaft-hole axe during the Eastern Zhou dynasty. During the Warring States period and Han dynasties, tools such as axes and adzes were mostly cast from pig iron and then forged. However, since the Tang and Song dynasties, they were forged directly.

After the Jin dynasty, the blade of the axe was widened and the handle was shortened to make it easier to handle and operate.

During the Tang and Song dynasties, poleaxes were often used on the battlefield. In the *General Principle to the Art of War*, by Zeng Gongliang of the Song dynasty, records were found of big axes and phoenix-headed axes being used during the Sui and Tang dynasties.

1.1.2.2 Chisel

The earliest chisel, made of bronze, was first seen during the time of the Qijia culture. There were at least five different types of chisel blades used up to the Shang and Zhou dynasties, namely, (1) single-sided, (2) double-sided, (3) round, (4) wide, and (5) narrow. The round-bladed chisel was used to drill round holes. This chisel can be traced back to the Warring States period at the latest. The rear part of the chisel was made of steel and had a conical hole in which the wooden handle was inserted (as shown in Picture 1.8). The chisel is an essential carpentry tool.

Picture 1.6 Perforated
stone axe, belonging to the
Dawenkou culture and
unearthed in Tai'an city,
Shandong province, cited
from Volume 1 of the *Atlas
of Chinese Farm Tools* edited
by Song Shuyou

1.1.2.3   Saw

The first saws were made of stone or clam shells, and later of bronze and steel. Stone
and clam shell saws were first seen during the Neolithic Age. Bronze saws were first
used halfway through the Shang dynasty and became popular during the spring and
Autumn period and the Warring States period. Some of them were unearthed in the
Lijiazui area of the Panlongcheng site, Huangpi district, Wuhan city, Hubei province,
and in Taixi village, Mancheng district, Baoding city, Hebei province. Legend has
it that the saw was invented by Lu Ban, the master carpenter. However, this legend
is unfounded. Bronze saws can be classified under four or five different categories
according to their shape and holding position: (1) tenon saws, (2) compass saws, (3)
gent's saws, (4) double-bladed saws, and (5) frame saws. The first four kinds can only
be used to cut shallow grooves and objects that are not too thick, but the performance
of the old frame saws is similar to that of more recent kinds of saws. A saw blade of
the Warring States period is currently exhibited in the Sichuan Museum. There is a
small hole on one end of it, in which a handle can be fixed. If there were small holes
on both ends, it may have been a frame saw. Bronze saws are weaker and softer than
steel saws, so the blades cannot be made too long or too wide.

Steel saws, known as hacksaws, were first seen in the late Warring States period
and gradually became popular after the Han dynasty. I-shaped frame saws similar to

Picture 1.7 Copper axe, unearthed at the Tonglushan mining site in Daye city, Hubei province, cited from Volume 1 of the *Atlas of Chinese Farm Tools* edited by Song Shuyou

Picture 1.8 Chisel, cited from *China at Work: An Illustrated Record of the Primitive Industries of China's Masses, Whose Life is Toil, and Thus an Account of Chinese Civilization* written by Rudolf P. Hommel and translated by Dai Wusan, et al.

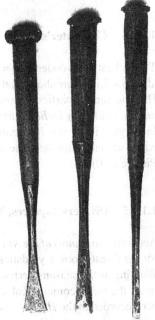

Picture 1.9 Frame saw, cited from *China at Work: An Illustrated Record of the Primitive Industries of China's Masses, Whose Life is Toil, and Thus an Account of Chinese Civilization* written by Rudolf P. Hommel and translated by Dai Wusan, et al.

those used nowadays (as shown in Picture 1.9) were first seen in the Song dynasty painting *Along the River During the Qingming Festival*. The structure and manufacturing process of I-shaped frame saws are briefly described in *Forging* in *Heavenly Creations*. The main features of this kind of saw are "a crossbar in the middle, which serves as a girder, and a thin bamboo strip, which straightens the saw". When the saw blade is pulled straight and tight, the frame can be made larger. This way, the saw blade transmits less resistance to the back of the saw when objects are sawn and it can be used to process thicker and longer pieces of wood.

1.1.2.4 Carpenter's Plane

The earliest flat wooden tool was a single-bladed adze. The hand plane was invented after that. Literature shows that the plane may have appeared in the period between the Han and Tang dynasties. The earliest record can be found in *Mirror of Craftsmanship and Guidelines by Lu Ban* during the Wanli period of the Ming dynasty (1573–1620). In *Forging* in *Heavenly Creations,* the making and usage of planes are described, which are basically the same as the carpenter's planes we know today (as shown in Picture 1.10).

1.1.2.5 Dividers, Squares, Yardsticks, and Markers

According to *Annals of the Xia* in *Records of the Historian*, when he tamed the water, Yu the Great "took a yardstick in one hand and a divider and square in the other. Working as the seasons permitted and with a vision of opening up the kingdom, he made the roads communicable, banked up the marshes, and surveyed the hills". It was recorded in the *Huai Nan Zi* that Yu the Great ordered Taizhang and Shuhai, his ministers, to measure distances in steps, draw circles with the divider, make squares

Picture 1.10 Carpenter's planes (Photo by Guan Xiaowu)

with the square, and record lengths and heights with the yardstick. During the Shang and Zhou dynasties, the divider, square, and yardstick were widely used in the making of bronze wares and vehicles as well as structures. They were also often mentioned in the books of Pre-Qin philosophers, for example, in *Li Lou I* in *Meng Zi*, it says: "Without the divider and square, one could not make squares and circles".

The earliest dimensional measuring tools discovered so far are the bone and ivory rulers of the Shang dynasty, which were unearthed in the Yin Ruins. After Qin Shi Huang, the first emperor of the Qin dynasty, unified China, he standardized units of measurement nationwide, including the length of the ruler (one *chi*, approximately 23.2 cm), based on the measuring system of the Qin Kingdom and the bronzeware *sheng* made by Shang Yang (*Shang Yang Fangsheng*). The measuring system of the Qin was passed on to the Han dynasty, where one *chi* equaled 23–23.7 cm. The ten *cun* on the ruler of the Eastern Han dynasty were separated by geometric lines or bird and animal patterns. There were different types of rulers, such as gold-plated copper rulers, painted bone rulers, copper rulers with dragon and phoenix patterns, bamboo rulers, and wooden rulers. The length of the ruler in the Wei and Jin dynasties increased to about 24.5 cm. The Southern Dynasties period adopted the old dimensional measuring system of the Qin and Han dynasties, with a slight increase of one *chi* equaling about 25 cm, while in the Northern Dynasties period, the length of one *chi* increased to about 29 cm. The Sui dynasty unified weights and measures using the old system of the Northern Dynasties period. Then the Tang dynasty adopted the system of the Sui, with one *chi* equaling about 30 cm. Exquisitely carved ivory and rosewood rulers were often presented as gifts to nobles and envoys of various countries. After that, in the Northern Song dynasty, one *chi* was lengthened to 31.6 cm. Finally, in the Ming and Qing dynasties, one *chi* was officially set at 32 cm, yet in practice it was 33.3 cm among the common people.

A divider is a tool for making circles and measuring roundness. Its working principle is the same as that of the modern divider, but its structure is slightly different. The shape of the early dividers is like the tool held by Nuwa as carved on a stone carving from the Eastern Han dynasty in Wuliang Ancestral Temple, Jining city, Shandong province. The measurement units and tools for measuring angles were

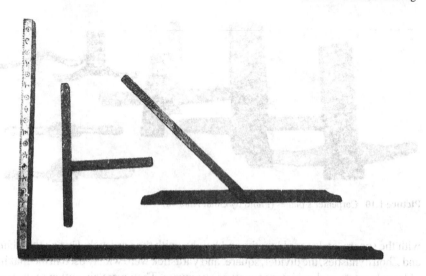

Picture 1.11 L-squares, cited from *China at Work: An Illustrated Record of the Primitive Industries of China's Masses, Whose Life is Toil, and Thus an Account of Chinese Civilization* written by Rudolf P. Hommel and translated by Dai Wusan, et al.

invented because of the use of perspective and graduation in the making of spoke wheels and the *qing* (chime). A square is a tool for measuring right angles. The ones from the distant past are exactly the same as the L-squares (as shown in Picture 1.11) used by carpenters today.

Mo dou, known as a carpenter's ink marker (as shown in Picture 1.12), is a snap line tool in traditional carpentry. The ink-soaked cotton thread in the ink box is usually used for level measuring, but can also be used for vertical measuring. In the past, the most important skill for carpenter apprentices to learn was to make tools. Once they knew how to make, use and "clean up" tools, it meant they had mastered the fundamentals of carpentry. The way to assess whether or not an apprentice was qualified was not by having them make furniture, but by having them design and make a woodworking tool by themselves, usually a carpenters' ink marker. Those who created tools with a sound structure, beautiful shape, and excellent workmanship were considered the best. An apprentice who could not make a carpenter's ink marker that was up to his master's standard, did not qualify as a "fine carpenter", but only as a "rough carpenter".

The divider, square, yardstick, and marker are the most primitive and basic measuring tools. The axe, chisel, drill, and saw are the most primitive and basic hand tools. The *lei*, *si*, sickle and plow are the most primitive and basic farm tools. The invention and use of all of these tools marked the beginning of human civilization and they are still widely used in modern production and life. In light of this, we should recognize and respect their value and importance.

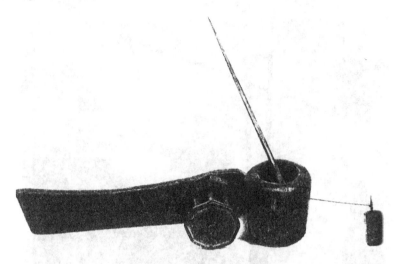

Picture 1.12 *Mo dou*, cited from *China at Work: An Illustrated Record of the Primitive Industries of China's Masses, Whose Life is Toil, and Thus an Account of Chinese Civilization* written by Rudolf P. Hommel and translated by Dai Wusan, et al.

1.1.3 Simple Devices

1.1.3.1 Well Sweep

The well sweep, known in Chinese as the *"jie gao"* in *Fortification of the City Gate* in *Mo Zi*, is a water intake device that uses the principle of leverage. It was widely used as early as the Spring and Autumn period. Along with the well windlass, it was a common water-lifting and irrigation device in ancient rural China. There is a specific description of its use in *Heaven and Earth* in *Zhuang Zi*. The image of the well sweep (as shown in Picture 1.13) was carved on a stone relief (in 147 A.D.) at Wuliang Ancestral Temple, which was built during the Eastern Han dynasty in Jiaxiang county, Shandong province. In *Heavenly Creations* there is also a Figure of it with a stone as the balancing weight.

1.1.3.2 Well Windlass

The well windlass was said to be invented by Shi Yi, an official historian in the early Western Zhou dynasty. It became quite popular for irrigation during the Spring and Autumn period, especially for irrigating small plots of land in water-deficient areas in the north. Even nowadays, it is still used for drawing water from deep wells in mountainous areas with deep groundwater.

Two wooden shafts of the well windlass were excavated at an ancient copper mine site in Tonglushan, Hubei province. They serve as one of the best examples to

Picture 1.13 Copy of the ston ecarving of a well sweep from the Eastern Han dynasty in Wuliang Ancestral Temple in Jiaxiang county, Shandong province

show how a well windlass was used to lift copper ore from a vertical shaft. The well windlass (as shown in Picture 1.14) was depicted in the poem *Tune: Endless as the Sky* by Emperor Li Jing (916–961) as early as the Southern Tang dynasty. It writes "Dreaming of willow banks green with sweet grass; she wakes to find no golden well with its windlass". Its pictures are drawn both in *Book of Agriculture* by Wang Zhen of the Yuan dynasty and in *Heavenly Creations* by Song Yingxing of the Ming dynasty. The former book also described a more complex well windlass, where a container was attached to each end of the rope wound on the shaft barrel. Its usage was described as "turning forward and backward, the suspended device goes down when it is empty and goes up when it is full. Thus, the two containers take turns to fetch water, making it more efficient". This kind of device can save the time it takes for each container to either travel down or back up by using the other container's weight as counterbalance. The early form of the well windlass was actually a fixed pulley. It is still uncertain when the well-known hand-held well windlass was invented. What we do know is that its earliest image can be found on murals from the Jin dynasty at Peijiabao Tomb in Jiangxian county, Shanxi province.

1.1.3.3 Pulley

The pulley is a simple lever device, which is recorded in *Mo Zi* or *Mohist canon*. With the fixed pulley, we cannot conserve energy, but we can change the direction of

Picture 1.14 Model of a double well windlass, unearthed in the Eastern Han dynasty site in Yanshi county, Henan province

force, while with the movable pulley we can conserve half the energy, but we cannot change the direction of force. In practice, a number of movable pulleys and fixed pulleys are often combined into various forms of pulley blocks, as to achieve both the conservation of energy and the ability to change the direction of force.

Since the Warring States period, pulleys have been widely used for combat and for extracting water and brine. The images of pulley devices can be seen in the brick and stone reliefs of the Han dynasty. What's more, there are many illustrations in *Heavenly Creations* depicting pulleys being used to draw water from wells and extract brine.

1.1.3.4 Bow

The bow, an elastic long-ranged weapon, is light and convenient to carry and use. Early bows, called slingshots (as shown in Picture 1.15), launched pellets, which were later substituted with arrows. A stone arrowhead made of flint was unearthed at the Paleolithic site in Shiyu village, Shuozhou city, Shanxi province, indicating that the use of bows and arrows can be traced back to at least 28,000 years ago, predating written records. The original bow was made of a single piece of bent elastic wood or bamboo. At the Yin Ruins in the Shang tombs, bronze bow nocks as well as arcs were discovered. From the retained dust traces as well as the images of the Chinese character "bow" (弓) in oracle bone inscriptions and bronze inscriptions, it can be inferred that the handle of the bow was incurvated towards the archer, which in

Picture 1.15 Yang Fuxi is making bows in his workshop

turn shows that the shape of the bow has undergone great change and improvement. Bows and arrows were widely used during the Spring and Autumn period. The composite bows unearthed from the Chu tombs in Changsha city, Hunan province are consistent with the bow-making techniques recorded in *Bow-making Craftsman* in *Book of Diverse Crafts*. After the Warring States period, the shape of the bow did not change much, but the material and manufacturing process were improved.

Traditional composite bow-making techniques are still in use today, such as the Beijing Juyuanhao Workshop, the bow-making technique of the Xibo people in Xinjiang, and the bow-making technique of the Mongolian horn sinew bows, all of which are listed in *List of National Intangible Cultural Heritage*.

The Juyuanhao Workshop, located in the Dongsi Bow and Arrow Courtyard in Beijing, was once the Qing dynasty's royal bow workshop. Yang Wentong, its ninth-generation successor, resumed his old career in 1998 and the next in line is his son Yang Fuxi. Their bow- and arrow-making technique follows the traditional ways of making the recurve Chinese traditional composite bow. They first make a core for the bow from bamboo, with horns attached to the outside, sinew to the inside, and wooden nocks at both ends of the bow. The making process mainly consists of "*bai huo* (woodwork)" and "*hua huo* (decoration)". The bowstring is usually made of cotton thread. To make an arrow, the craftsman should first adjust, peel and scrape the shaft, and then attach the head and the feathers. The materials, tools, and techniques used by the Juyuanhao Workshop to make bows and arrows are similar to those recorded in *Book of Diverse Crafts, Dream Pool Essays,* and *Heavenly Creations*. In 2006, it was listed in *List of the First Batch of National Intangible Cultural Heritage*.

1.1.3.5 Crossbow

The crossbow is an important ranged weapon from ancient times. Based on such evidence as the bone daggers of the Yangshao culture in Miaodigou and the shell ornaments unearthed in the Gaohuang Temple site in Xuzhou city, Jiangsu province, its origin can be traced back to late primitive society. To shoot a crossbow, the archer should first cock and load the bow, and then aim and shoot. The crossbow can be cocked with the help of stirrups or winches so that the ejection force becomes greater and the range longer, which increases its accuracy and lethality.

The crossbow consists of a bow and a bow-cocking mechanism (the stock and the trigger mechanism) that makes it possible to launch at a deferred time. The trigger mechanism (as shown in Picture 1.16) consists of a sight (*wang shan*), a trigger (*xuan dao*), a sear (*gou xin* or *cheng niu*), a linkage mechanism, two rods (*jian*) and a few other connectors. The earliest crossbow, a device from the early Warring States period, was unearthed in the ruins of the capital of Lu Stage in Qufu city, Shandong province.

During the Han dynasty, the crossbow was improved. Firstly, the *guo* (housing box) was added, which meant that the trigger mechanism could be assembled more tightly and that the stock was able to bear greater tension; the *wang shan* (sight) was raised higher, its curved edges were replaced by right ones and the scales were carved, which caused aiming to become more accurate. After the Han dynasty, one-foot or two-foot crossbows and belt crossbows with stirrups or belt hooks to help cock the bows were invented, and the shooting force increased continuously.

Picture 1.16 Bronze trigger mechanism of a crossbow from the Warring States period, currently exhibited in the National Museum of China

Crossbows, similar to those in the Han dynasty, were frequently used during the Three Kingdoms period and the Western Jin dynasty. However, after the Western Jin dynasty, central China was ruled by nomadic people from the north, among whom horse-riding and archery were quite popular, which reduced the use of crossbows. Crossbows regained their popularity during the subsequent Eastern Jin and Southern Dynasties period: large crossbows such as the "divine crossbows" (*Shen nu*) and the "super-large divine crossbows" (*Wanjun Shen nu*) came into being. The copper trigger mechanism of the Southern Dynasties period discovered in the Qinhuai River in Nanjing is 39 cm long. It is estimated that the entire bow was about 4 m long and the limb about 2 m long. Therefore, it could not be anything but a mounted crossbow installed on a frame.

During the Tang and Song dynasties, there were many new types of powerful one person and large crossbows. However, because the efficiency of iron armor kept getting better and better, the long-range killing advantage of one person crossbows became less and less. During the Ming and Qing dynasties, under the influence of the ethnic minorities in south and southeast China, crossbows were mainly used for self-defense or hunting. At that time, troops were mostly equipped with firearms.

1.2 Devices

1.2.1 Cutting and Processing Devices

1.2.1.1 Emery Wheel

The axle mechanism used for treating jade (as shown in Picture 1.17) appeared in the early Neolithic Age at the latest. Jade wares were unearthed at the type sites of Xinglongwa along the Liaohe River, Peiligang and Dawenkou along the middle and lower reaches of the Yellow River, Hemudu along the lower reaches of the Yangtze River, and at Majiabang. By the time the Hongshan, Longshan, and Liangzhu cultures came along, jade wares were already quite popular. Early techniques of jade treatment mainly consisted of grinding and engraving simple decorative patterns on jade wares. During the period of the Longshan culture (2500–2000 B.C.) and the Liangzhu culture (3300–2300 B.C.), such techniques as cutting, the incised carving of string patterns, drilling, hollowing, and embossing were widely employed with superb craftsmanship. Since then, many improvements have been made at different points in time. In later times, a rotating emery wheel and emery were used to process raw jade by means of abrasion, such as polishing. It is thought that various kinds of emery wheels were invented in the Shang dynasty or even as early as the Liangzhu culture, but the original emery wheels and staving press were first recorded much later, in the Ming dynasty. It is recorded in *Jade* in *Heavenly Creations* that "When jade is first cut, a round iron disc is made and mounted on a frame connected with pedals underneath and a basin of emery is placed beside it. The disc is turned by

Picture 1.17 Jade polishing

pedals, and at the same time emery is sprinkled on it so as to cut through the jade. Finally, the jade will be cut open as emery is gradually added into the basin". The disc mentioned in the passage is the same as the emery wheel. The basic structure of a jade polishing bed in the Ming dynasty was like this: The rectangular wooden frame is supported by a horizontal axle, through which an iron disk is fixed in the middle. Leather strips rotate around both ends of the axle counter-clockwise and are connected to pedals underneath. When the left and right feet step on the pedals respectively, the disc will rotate back and forth touching the emery from the basin beside it to the surface of the raw material being polished. There were different types of emery wheels for different kinds of work, such as, for engraving fine lines, for embossing, for polishing, and for hollowing.

1.2.1.2 Lathe

The lathe, which is used to carve curved surfaces, was invented in the pre-Qin era and was used first in the processing of wood and later of metal. On the gold and silver wares, such as plates, bowls, and boxes, unearthed in Hejia village in the southern suburbs of Xi'an, obvious turning lines can be seen on their surfaces, and the threads, the starting points and cutting points are clearly distinguishable. The concentric thread lines on the unearthed golden boxes were carved very finely and close together. The conical surface of the male latches is also made to fit perfectly inside the female latches. Each part is processed so carefully that, once fitted together, they create a tightly fixed whole. The lacquer woodwares unearthed from the Chu Tombs at Yutai Mountain, Jiangling county, Hubei province, were made using three techniques: (1) chopping, (2) lathing, and (3) carving. Of those woodwares, the round wooden boxes, as well as *zhi* and *zun*, which are two kinds of wine vessels from ancient China, were lathed on the outside and chopped on the inside. Lathing-made woodwares were recorded in *Essential Techniques for the Peasantry,* which states that elms "can be used as rafters when they are five years old [...] and lathed to make a *dule*", which is a gyroscope. In *Essentials of Fire Attacks,* Jiao Xu of the Ming dynasty describes the method of making molds for *huo tong,* a kind of tubular metal firearm, by saying that we should "make the exterior mold by lathing the dried camphor wood or fir wood according to the desired size of the *huo tong*".

The technique of lathing metal wares began in the Han dynasty. For example, in Nanyang city, Henan province, bronze boats from the Han dynasty were unearthed that have fine and uniform spiral patterns on the outer surface. Similarly, some of the gold and silver wares of the Tang dynasty unearthed in Hejia village, Xi'an city, Shaanxi province, show signs of being made with lathing techniques. In *Civilian Officials* in *Compilation of Regulations in the Song Dynasty*, it was recorded that there were altogether 81 kinds of craftsmen in the manufacturing department of the Imperial Palace, which belonged to the Imperial Office Bureau. One of these 81 kinds of craftsmen was a "lathing craftsman". According to *Treaties of Foods and Commodities* in *History of Ming,* "There were three kinds of coins at that time, namely, those with a golden back, those with a black and shiny back, and those with lathed edges. It was difficult to cast coins as it required the hard work of craftsmen, so the lathing bed, used for lathing coins, was later replaced by the furnace".

In some places in China lathing beds are used to make wooden toys nowadays. Wooden toys in Xunxian county, Henan province, are mainly made in Yangpitun village of Liyang town. They are generally made of waste materials and scraps, which keeps the manufacturing costs and the prices low. Wang Minde, an old craftsman, inherited the handicraft from his ancestors. His family has been making lathed wooden toys for generations. They first lathe the surface of the wood on the lathing bed to make it smooth, then hollow it out, and finally assemble and fix all the parts together.

1.2.2 Agricultural Devices

1.2.2.1 Water Ladder

Fanche, also known as the dragon spine, dragon wheel, or water ladder, is used for irrigation (as shown in Picture 1.18). According to ancient literature, in the third year of the Zhongping period in the Eastern Han dynasty (186), Bi Lan "invented the water ladder". Ma Jun also "made the water ladder" during the Three Kingdoms period. It was recorded in Japanese literature that in the third year of the Dahe period during the reign of Emperor Wenzong (829) in the Tang dynasty, there were three types of water ladders: (1) hand-cranked, (2) foot-pedaled, and (3) ox-drawn. *Painting of Planting and Harvesting* by Yang Wei in the Northern Song dynasty depicts a water ladder pedaled by four people. *Book of Agriculture* by Wang Zhen in the Yuan dynasty described the structure of the water ladder, which is still in

Picture 1.18 A water ladder pedaled by four people

use in some villages in China to help distribute water throughout the farmland. The contemporary Chinese scholar Zhang Baichun made a close study of a water ladder in the rural part of Zhejiang province, currently exhibited in the China Agricultural Museum, and produced an engineering drawing of it in 1992.

All of its components are made of wood. According to Qing dynasty literature, "The tub (of the water ladder) is made of fir and chestnut wood and the axles are made of sandalwood". The wooden dowels and dovetail grooves, used as links between components, become firm due to expansion in water, and do no rust as easily as iron nails.

The chains are used to achieve the purpose of lifting water to a greater height. Differently from modern chains, its chains (dragon bones) and sprockets (a big and a small wheel) are linked together by means of meshing the cogs of the wheels against the paddles on the chain links (dragon bone links). This kind of meshing method simplifies the structure and making of the water ladder without reducing the strength of the cogs, for which wooden components are perfect.

Based on the analysis of his own example, Rudolf P. Hommel thinks that the best inclination angle for square grooves of the cogs is 24°, but the angle is even smaller in practice. Sometimes, several water ladders are combined in sections so as to improve the overall lift.

Water-powered Trip-hammer under the Willow, a painting of the Tang dynasty, which is exhibited at the Palace Museum in Beijing, depicts a water ladder driven by an ox. It is also recorded in *Book of Agriculture* by Wang Zhen that, "The ox-driven

Picture 1.19 The ox-driven water ladder in Wuxian city, Jiangsu province, cited from Volume I of *Atlas of Chinese Farm Tools* edited by Song Shuyou

water ladder, used on river banks, is a modification of the water-driven one. The small horizontal wheel is placed beside it instead of under it, and the ox on the bank drags the wheel shaft, which will make it turn up and down. It is many times faster than the one pedaled by man". Its structure is the same as that of the "ox water lift" (ox-driven water ladder, as shown in Picture 1.19) depicted in *Heavenly Creations*.

1.2.2.2 Winnower

The winnower, known in China as *shanche, fengshanche* or *fengche*, is a device that uses the wind to separate grains and divide them according to their size and weight.

The winnower was first recorded in the *Quick Approaches of Chinese Characters* by Shi You of the Western Han dynasty, "*Dui wei shan tui chong bo yang*". [The mortar and the millstone pound the grain, the winnower winnows it.] According to the note of Yan Shigu of the Tang dynasty, the word "*shan*" refers to "the winnower". The pottery winnower model of the Western Han dynasty unearthed in Jiyuan city, Henan province and those of the Eastern Han dynasty unearthed in Ruicheng county, Shanxi province, all confirm the information found in literature.

The winnower described in *Book of Agriculture* by Wang Zhen (1271–1368) was quite mature. They could be classified as hand-cranked and foot-pedaled based on their different driving modes. They could also be classified as vertical or horizontal ones according to the placement of the fan. The cylindrical blower casing at that time made more sense than the cuboid in one of the Han dynasties. The air outlet is located in front of the winnower and the fan is placed inside the casing. On one end of the shaft is a crank handle. When in operation, the grains with chaff are first loaded into the hopper. As the cranking of the handle drives the fan, the air will blow out the non-grain materials, which are lighter, from the outlet, while the grains, which are heavier, will flow out along the grain outlet.

The structure of the winnower with a single grain outlet in Huamin village, Kaihua county, Zhejiang province is similar to the ones recorded in *Book of Agriculture* and *Heavenly Creations*. The little door underneath the hopper of the winnower in rural Yunnan province can be adjusted to control the flow of the grain. However, the structure of the winnower in Wuxian city, Jiangsu province (as shown in Picture 1.20) is slightly different in that it has a winnower and an extra grain outlet, which can be used separately or combined to work simultaneously.

That is to say, there are two-grain outlets for this kind of winnower, with the two either located on the same side of the handle or on both sides. By adjusting the rotating speed of the fan, the strength of the flow of air will, for example, make the good grains flow out of the first grain outlet and the secondary ones out of the second grain outlet, while the non-grain materials, such as chaff, will fly out of the chaff outlet. This kind of winnower, commonly used in northern China, is said to have originated in the south a long time ago.

1.2.2.3 Mills

Since ancient times, people have used millstones and mill rods to process grain. After the period when "the wood was broken to be used as the pestle, and the ground was dug to be used as the mortar", the pestle and mortar simply became more efficient to process grain. The earliest archeologically discovered stone mills could be traced back to the Warring States period, while the earliest written records were found in *Quick Approaches of Chinese Characters* by Shi You of the Western Han dynasty.

There are many kinds of mills, which fall into the following categories based on the driving power used: manual, animal-driven, hydraulic, and wind mills. They can also be categorized as hand, small, medium, large, extra large, and hydraulic mills according to their size. They can also be categorized according to the materials and methods used for processing, such as gristmills (as shown in Picture 1.21), colloid, and oil mills.

Picture 1.20 Winnower in Wuxian city, Jiangsu province, cited from Volume I of *Atlas of Chinese Farm Tools* edited by Song Shuyou

Picture 1.21 Millstones from the upland gristmills in Heitu'ao village in the eastern suburb of Hohhot city (Photo by Zhang Zhizhong)

A mill usually consists of a hopper, a damsel (iron spindle), vertebra, rod, disc, millrind, and millstones. Millstones come in pairs: the upper stone (runner stone) and the lower stone (bedstone). The upper stone has a spindle hole, convex grinding chamber, furrows, and grinding mouth, and its top also has one or two eyes (i.e., feeding mouth). The lower stone, which is fixed onto the grinding disc, has a concave grinding chamber, large furrows, and grinding mouth. On the side of the upper stone's furrows, there is a round spindle hole that matches the damsel of the lower stone. The hopper is fixed above the eyes on the upper stone. The millstones are made of iron, stone, and wood, or built of adobe, stone, and brick. The millrind or mill bed is also usually made of adobe, brick and stone, or sometimes wood.

The millstones are divided into several harps, and the number of harps depends on the millstones' diameter and the level of grinding. There are three common types, namely, those with 6, 8 and 10 harps, those with 8 being the most popular. The density of the furrows depends on the particle size of the abrasive and the quality of the millstones. If the particles of the abrasive are small and the quality of the millstone is high, each harp will usually have 8 to 10 furrows, including 3 to 4 auxiliary ones. However, when the quality of a millstone is poor, each harp will only have 5 furrows, including 1 to 2 auxiliary ones.

Each mill is built with local materials, for example, the mills in Inner Mongolia, as well as Shanxi and Shandong provinces are made of granite or basalt. How is a stone mill made?

Make a stone slab. A big rough stone from a quarry is cut into a vertical slab, only a skilled craftsman can do it horizontally.

Make a workblank. Find the center of the slab by making a cross with string and drill a small hole in the center. Then draw a circle around the hole, cut the slab into the desired shape and size, and trim it into a workblank. Next, draw lines on the stone and trim it further to round and flatten it, shaping the workblank into the required shape.

Chisel furrows and drill holes. Divide the circumference of the stone into eight equal harps, draw furrow lines in each harp, and chisel out long and short furrows, grooves, half slopes, and vertical slopes.

Assembly. Put the damsel into the spindle hole of the lower stone and add an iron wedge to lock it in place. Put a round iron sleeve in the spindle hole of the upper stone, insert the wooden adjustment handle into the side hole of the upper stone, wedge it tightly, and then pour alum or rosin to lock it in place as well.

Adjust and correct. Turn the upper stone and further trim the uneven parts according to how well the two millstones fit together when grinding until it runs smoothly.

After a long time, the millstones will need to be trimmed again, and the furrows and grooves chiseled once more for maintenance (as shown in Picture 1.21).

1.2.2.4 Tilt Hammer

The *Dui*, known in English as a treadle-operated tilt hammer, is used for processing grain. Evolved out of the use of the pestle and mortar, it made pounding more labor-saving and efficient. The pestle's body and mallet are made of wood, its hoop and head are made of iron, while the mortar is made of stone. The earliest tilt hammer ever unearthed in China was a kind of burial object and dated back to the Western Han dynasty. The earliest record in literature was found in *New Theories of Huan Zi* by Huan Tan, dating back to the end of the Western Han and the beginning of the Eastern Han dynasties.

The tilt hammer can be divided into four kinds according to the power used, namely (1) hand-, (2) pedal- (as shown in Picture 1.22), and water-driven, more specifically, by means of (3) water troughs and (4) waterwheels. The hand-driven tilt hammer, operated by one's hands, is more laborious than that driven by pedals. The principle of the water-driven tilt hammer that uses troughs, as recorded in *Book of Agriculture* by Wang Zhen, is the same as that of the pedal-driven one. Water from a natural stream flows into the trough that rests on the one end of the pestle, which causes the mallet at the other end to be lifted when the trough is full. The mallet will fall down again when the water pours out.

1.2.2.5 Edge Runner

The *nian*, known in English as an edge mill or edge runner, dates back to the Northern and Southern dynasties period and was once widely used as one of the most important machinery for the peeling and crushing of materials, like grain, oil plants, porcelain clay, and paper pulp. It can be categorized as man-, animal-, or water-driven according to the power used. It can also be categorized as a roller or trough edge runner according to its structure, both of which are described in *Book of Agriculture*, *A Complete Book on Agriculture* and *Heavenly Creations*.

The roller edge runner in Shanxi province and Inner Mongolia autonomous region (as shown in Picture 1.23) are mainly used for husking grain, as well as rolling flour, oil, and paper pulp.

It consists of a plate, a damsel (iron spindle), a roller, a rod, an iron chain, and an iron wedge. The center of the plate is equipped with a damsel, which is either fixed on the bed or supported by one or more stones. The periphery of the plate is made larger in order to hold the rolled materials and the roller is mounted on top of it. The roller will rotate with the moving of the rod.

The middle part of the plate is convex and gradually lowers outwards, thus making the material move from the center to the periphery. The grooves on the plate play a role in pressing and cutting grains, which is helpful for grain peeling. The middle part of the roller is a little bit thinner than the two ends, with the diameter of the middle part being about 4 mm thinner than that of the two ends. The grooves can be carved in different patterns, such as straight lines, diagonal lines, or fish scale lines. When there are no materials on the plate, both ends of the roller are in contact with

Picture 1.22 A tilt hammer
for hulling rice in Lijiang
city, Yunnan province, cited
from Volume I of *Atlas of
Chinese Farm Tools* edited
by Song Shuyou

Picture 1.23 A roller edge
runner in Xiaojing village in
the eastern suburb of Hohhot
city, Inner Mongolia, China
(Photo by Zhang Zhizhong)

the plate's surface while there is a gap of about 2 mm between the middle part and the plate, so that the rice grains won't be crushed or damaged.

The trough edge runner mainly consists of a trough, a wheel, and a rod. It can be divided into different types, such as a single wheel, double wheel, and bogie type. Both sides of the wheel have decorative patterns and one iron chain on each end of its spindle hole. The trough is spliced by multiple stone troughs, of which the rod is pulled by animals or hydraulic power that causes the wheel to move. The outer circular surface and both sides of the wheel are designed to husk or crush the materials.

The plate and the roller are made of granite, the rod is made of elm or locust wood, and the damsel, the chains, the sleeve, and the wedge are forged iron. How is an edge runner made?

Make the workblank. Trim a big stone into the workblank for the plate with the help of a carpenter's ink marker (*mo dou*) and wedge.

Make the plate. Round off the workblank, and trim the surface and circumference by drawing cross lines and a circle. Drill the damsel hole, and then carve small holes or mill grooves on the plate.

Make the roller. Cut the stone in the shape of a squat tube. Draw a cross and circles at both ends of the workblank, and connect the crosses. Roughly trim the circumference and carve holes and grooves on the outer surface of the roller.

Assembly. Put the damsel into the spindle hole on the plate, add a wedge to tightly lock it in place, and then pour alum or rosin for solidification.

Adjust and correct. Test the device after assembly and trim the imperfections on the plate and the outer surface of the roller. It needs to be trimmed again, for maintenance, after 2–3 years of use.

People in Fusheng village of Zhuozi county in Inner Mongolia usually put seven or eight copper coins in the middle of the circular grinding course on the plate and then push the roller. If the copper coins are not crushed, it means that the gap between the roller and the plate is about 2 mm, which is perfect. If the gap is too small, the grains are likely to be crushed, if the gap is too big, the grains are likely to be left unhusked.

1.2.3   Hydraulic Devices

1.2.3.1   Water Mill

A water mill is used for processing grain. It uses two types of water wheels: vertical and horizontal. A water mill with a horizontal wheel consists of millstones, a water wheel, and a vertical shaft. The more complicated water mill, one with a vertical wheel, consists of millstones, a water wheel, a propeller shaft, gears, and a control mechanism. The vertical water wheel can be further categorized into the overshot and the undershot water wheel.

Picture 1.24 The hydraulic tilt hammer and mill in Changxi town, Shexian county, Anhui province (Photo by Guan Xiaowu)

Records show that both Ma Jun and Han Ji of the Three Kingdoms period were already making vertical water wheels. There are more and more records of water mills and water edge runners since the Western and Eastern Jin dynasties. *Painting of the Panche at the Water Gate* in the Northern Song dynasty depicts the processing of grain with a water mill. The water flowing from the chute pushes the paddles of the horizontal water wheel into motion and thus turns the millstones on the upper floor. The connecting water mills described in *Book of Agriculture* by Wang Zhen use a vertical water wheel to drive the several connecting mills with a gear propeller mechanism. The gears at the end of the horizontal shaft can also drive multiple tilt hammers. In case of drought, barrels are attached to the water wheel, thus converting it into a water-lifting irrigation device.

Water mills with vertical and horizontal wheels are still used in southern Anhui, Yunnan and Tibet. Changxi town, Shexian county, Anhui province, has a scenic spot with a water mill (as shown in Picture 1.24), which was reconstructed on the original site according to the excavated remains. It consists of an undershot vertical water wheel, a horizontal shaft, transmission gears, a tilt hammer, and millstones. The tilt hammer is used to process rice grains or glutinous rice flour, and the mill is used to grind corn flour. The water wheel has a diameter of about 5 m and consists of a spoke plate, a rim plate, blades, wedges, connecting pins, and a reinforcing block.

According to the craftsmen who took part in the reconstruction process, the water wheel was made with the wood from the abundant trees from the surrounding area.

The hydraulic tilt hammer and mill in Shuiduixia village, Jixi county, Anhui province uses an overshot water wheel with a diameter of 2.2 m, above which is an inlet groove with a bucket-like structure.

According to Guan Xiaowu and Huang Xing's field investigation in Shannan and Shigatse, Tibet, water mills are still widely used there. The combination of planting highland barley as the main crop and the abundant mountains and water resources means water mills remain the main tool for processing Zanba, a kind of roasted barley flour. Niangre township, Chengguan district, Lhasa, Tibet, also has many water mills for processing Zanba. The 62 water mills contracted with Gurong Langzi Zanba Company in Duilong Deqing county, Lhasa all use rimless horizontal water wheels.

As shown in Picture 1.25, there are two mills in the well-known Jiami Water Mill area, which is a scenic spot in the Niangre Folk Customs Garden in the northern suburbs of Lhasa city, about 6 km outside downtown. The mill once offered Zanba to the 5th Dalai Lama in the seventeenth century. Afterward, it was named "Jiami Water Mill" by the 7th Dalai Lama, meaning "delicious Zanba water mill" in Tibetan, and was designated as the court's Zanba supplier. The water wheel has 16 rectangular holes along its shaft for installing paddle blades. The number, width, and length of the openings and blades on the shaft are determined by the diameter of the vertical shaft and the power needed. The water wheels in greater Lhasa are generally installed with 16 blades but with differing widths and lengths.

1.2.3.2   Hydraulic Tilt Hammer

The hydraulic tilt hammer, used for processing food, consists of a vertical water wheel, a propeller mechanism, and a tilt hammer. According to *New Arguments of Huan Zi* and *Slang Dictionary*, the hydraulic tilt hammer was in use as early as the Han dynasty and its structure already became more advanced and complex in the Eastern Jin dynasty. The structure of the hydraulic tilt hammer with a vertical water wheel is described in *Notes on Comprehensive Mirror to Aid in Government* by Hu Sansheng, a historian from the end of the Southern Song and the start of the Yuan dynasty. *Book of Agriculture* by Wang Zhen depicts two types of hydraulic tilt hammers using a vertical water wheel. The (1) *liaoche dui* described in the book uses an undershot vertical water wheel while (2) the *gu dui* uses an overshot one with bucket-like blades. They are also both described in the engraved edition of *Hydraulic Engineering* in *the Annals of Shaoxing* from the Qing dynasty's Kangxi period (1662–1722). The *liaoche dui* is used when the water flow is flat and runs through the lower part of the wheel while the *gu dui* is used when the water flow is rapid and falls down on the upper part.

Empirical evidence from Songpingzi village, Lijiang county, Yunnan province, shows that the structure of the rimless water wheel of the hydraulic tilt hammer is relatively simple: with blades directly tenoned on the shaft, and the speed of the water wheel adjusted by a sluice.

Picture 1.25 Jiami water mill 1. Water Millstone and Grain Hopper. 2. Water Wheel (Photo by Guan Xiaowu)

Meanwhile, the hydraulic tilt hammer in Tongcun town, Kaihua county, Zhejiang province, which uses wood blocks to crush flour, has the same structure and assembly method as those described in *The Book of Agriculture* by Wang Zhen. The water wheel consists of 8 spokes, 8 rim plates, 32 blades, wedge rods, connecting plates, and reinforcing bars. Each blade is composed of an outer and an inner blade plate, and a bottom plate, which is wedged into the groove of the rim plate, thus forming 32 tubular blade buckets. Iron sleeves are installed at both ends of the water wheel shaft to increase the strength and wear resistance of its surface. At the same time, bamboo troughs are used to carry water to the bearings and the pins of the brackets to cool and lubricate them. The hydraulic tilt hammers in Dali city's Xizhou town, Yunnan province, and Nanning city's Taishou township of Binyang county, and Kunlun town of Yongning district, Guangxi, all work the same way.

1.2.3.3   Water Edge Runner

The water edge runner consists of a roller and a horizontal or vertical water wheel (as shown in Picture 1.26). The water edge runner driven by a horizontal water wheel is equipped with a gear propeller mechanism. According to Zhang Baichun and others, these two types of water edge runner were still in use in Guangxi and Yunnan in the 1990s. For example, in the Miao people's Gaoxiantun, Rongshui county in Guangxi, there are 18 water edge runners for grinding rice, corn, and other grains. The driving device is a horizontal water wheel with a special spoke structure. The gear teeth of the propeller gear were installed on one side of the wheel, which is different from those in other Chinese regions, but similar to those in Europe.

Most parts of the water edge runner are made of Chinese fir except the stone roller and trough, the iron shaft sleeve, hoop ring of the horizontal gear and part of the lower end of the vertical shaft. Hardwood is preferred to make gears, especially the teeth. The whole water edge runner is assembled, not with nails, but by using tenons, pins and wedges. It needs maintenance every 5–10 years. If well maintained, it can last for a hundred years.

1.2.3.4   Composite Hydraulic Devices

There are different types of composite hydraulic devices, which means any combination of tilt hammer and mill, or tilt hammer, mill and edge runner, or tilt hammer, mill, edge runner, and *long* (huller). With different machinery combined, it is more powerful than a single hydraulic tilt hammer, water mill, or water edge runner on its own. The connecting water mills recorded in *Book of Agriculture* by Wang Zhen (1271–1368) can drive nine mills to work at the same time and the gear at the shaft head can also drive several tilt hammers. This kind of mechanical system became more and more advanced and complex during the Ming and Qing dynasties. Even nowadays it is still in use in some rural areas in China. According to Zhang Baichun and Feng Lisheng, farmers in Huamin village, Kaihua county, Zhejiang province

Picture 1.26 Vertical water edge runner in Yuanyang county, Yunnan province, cited from Volume I of *Atlas of Chinese Farm Tools* edited by Song Shuyou

used composite machinery (as shown in Picture 1.27) that combined the tilt hammer, mill, edge runner, and huller for processing food. It was driven by an undershot water wheel, which, at one point, had also been equipped with bamboo containers for irrigation.

1.2.3.5 Chinese Noria

The Chinese noria, called a *tongche* in Chinese, is a machine used for lifting water from rivers for irrigation. With man, animal, or hydraulic power, it can rotate containers filled with water along the rim of a water wheel until it automatically spills into an aqueduct at the top of the wheel.

It came into being no later than the Tang dynasty. What Chen Tingzhang, a scholar of the Tang dynasty, described exactly, in *Ode to the Hydraulic Wheel* in volume 948 of *Full Literature of the Tang Dynasty*, was, in fact, the noria. Since the Song dynasty, records about it appeared more frequently. It was widely used in Jiangxi, Zhejiang, Hubei, Guangdong, Guangxi, Yunnan, Gansu, and Ningxia in the Ming and Qing dynasties, and even in the 1950s and 1960s. Even nowadays, it is still used in some areas in China.

The Chinese noria can be divided into four types based on its structure and power source: (1) the animal-driven noria, (2) the double still water noria (with two water

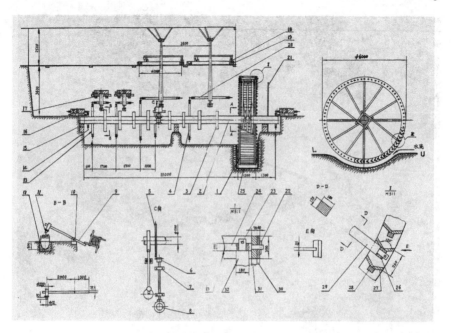

Picture 1.27 A detailed structure of the water tilt hammer in Huamin village, Quzhou city, Zhejiang province 1. undershot vertical water wheel; 2. main axle (φ500); 3. protrusive board (8 sets); 4. gear A (φ2800, 2 pieces); 5. gear B (φ2500, 1 piece); 6. gear B' (φ580, 2 pieces); 7. axle seat (2 pieces); 8. huller (φ500, gear φ580, 1 set); 9. rod (16 pieces); 10. holder (16 pieces); 11. stone hammer (16 pieces); 12. stone mortar (16 pieces); 13. gear C (φ2500, 2 pieces); 14. gear D (φ1300, 2 pieces); 15. main axle seat (3 pieces); 16. lotus-shaped mill (φ1000; 2 pieces); 17. high-placed mill (φ1000; 2 pieces); 18. roller (2 sets of 6 pieces); 19. trough (2 sets); 20. gear A' (φ3000, 2 pieces); 21. water baffle; 22. dowel; 23. dowel; 24.dragon bone; 25. grating plate; 26. dragon bone; 27. blade; 28. dragon bone; 29. wheel spoke; 30. outer rim plate; 31. inner rim plate; 32. wheel spoke; 33. blade (Note: The dimensions in the Figure are in millimeters). (Measured and Drawn by Zhang Baichun and Feng Lisheng)

wheels, one low and one high), (3) the single running water noria, and (4) the double running water noria. The first two are used in still water, such as ponds and lakes, while the latter two use running water.

Norias made of bamboo or wood are usually small in size and are mainly used China's southern regions, such as Guangxi and Yunnan. The noria (as shown in Picture 1.28) near Shuiyuan cave in Fengshan county, Guangxi province is described in *Investigation and Research on Chinese Traditional Machinery* by Zhang Baichun et al. The larger wheel is 9 m wide and the smaller one 6.4 m. All the components are connected by mortise and tenon joints, bamboo knitting, and rattan binding. How to build a noria: (1) Lay the foundation, erect stakes, and make the frame for the wheel. (2) Construct and place the water wheel onto the frame and install the smaller parts. (3) Attach the bamboo buckets to the wheel, build the water basin at the top of the wheel and connect it to the aqueduct.

Picture 1.28 The noria in Poxin village, Paoli township, Fengshan county, Guangxi province (Photo by Zhang Baichun)

Another kind of noria is mainly made of wood, with the largest wheel measuring over 20 m in diameter. It consists of a shaft, large and small spokes, piercing braces, *wangxian* (network of planks), *denggunzi* (connecting rods), scraper blades, and water buckets. The Lanzhou waterwheel (as shown in Pictures 1.29 and 1.30) falls into this category of noria, and the steps on how to construct it are as follows:

(1) Dig the water channel, erect the foundation, and lay the groundwork, i.e., the frame. The upper part of the frame is a joist (a column supporting the water wheel), which is used to adjust the height of the water wheel to adapt to changing water levels. (2) Install the shaft, both ends of which are wrapped in cast iron rings and are supported by the basin-shaped "*yangyu*" on the joists. (3) Install the large spokes and small wheels. (4) Install the smaller spokes and the *chuancheng,* i.e., the braces supporting them. (5) Install the *wangxian* and *denggunzi*. The outer wheel of the noria is called the *wangxian*, the network of planks, which is tenoned with the large spokes, while the inner wheel is called the *denggunzi*, which consists of rods that lock the large spokes in the middle in place. (6) Install the scraper boards and buckets. The scraper boards, where the buckets are hung, are nailed between the large and small spokes as well as the *wangxian* and *denggunzi* and are used to help direct water into the buckets. The number of buckets can be adjusted according to the water flow.

When the ice on the Yellow River thaws in spring and the water level rises, the water will flow from the stone dam against the wheel, which causes the water wheel

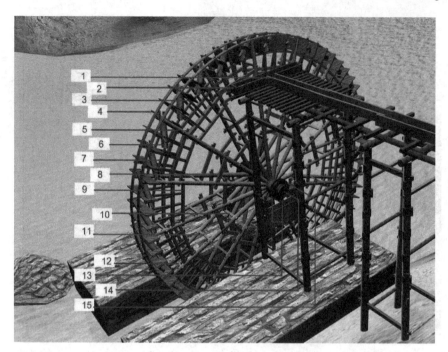

Picture 1.29 The Yellow River water wheel in Lanzhou city, Gansu province (Drawn by Huang Xing). 1. aqueduct; 2. elevated water basin; 3. small spokes; 4. scraper boards; 5. large spokes; 6. buckets; 7. *denggunzi*; 8. small wheel; 9. *chuancheng*; 10. *kaidangguang* (the connecting bars at one end); 11. *zhongguang* (the connecting bars in the middle); 12. *shoutouguang* (the connecting bars at the other end); 13. *yangyu*; 14. axle shaft; 15. axle bearing

to move and the buckets to rotate with it. The buckets fill with water at the bottom, rotate upwards, then spill the water into the water basin at the top, which finally guides the water to the fields via the aqueduct.

1.2.4 Wind Devices

There are two types of windmills in China: vertical and horizontal types, which were mainly used to turn water wheels. The cloth vanes are similar to the cloth sails of sailboats. Written records show that the Chinese wind-powered water ladder was invented in the Southern Song dynasty at the latest, which is later than the one in Persia (now Iran).

Picture 1.30 The water wheel in Xiachuan village, Xincheng township, Xigu district, Lanzhou city, Gansu province (Photo by Guan Xiaowu)

1.2.4.1 Horizontal Windmill (Vertical-Axle Windmill)

In the poem *Ode to the Water Wheel* by Tong Ji of the Ming dynasty, he describes the operation of a wind-powered water wheel: "The wind drives the water wheel in Lingling county. The sound is loud in the night air; The wheel the size of three *zhang*, lets the wind carry the sound of water [...]" The kind of windmill Tong Ji is referring to must be a horizontal one, one with a vertical axle, especially when taking the Qing dynasty records into account. The most important of these records are volume 36 of *Salt Industry and Trade* by Zhou Qingyun (1866–1934), in which the theory behind the horizontal windmill was published, and volume 13 of *Records of Inkstone, Lei and Jade Tablets* by Lin Changyi from the Late Qing, in which its installation method was described.

In the early 1950s, there were 600 horizontal windmills (as shown in Picture 1.31) in the two districts of Hangu Zhaishang and Tangda alone, which is along the coast of the Bohai sea, an inland sea in eastern China. They are made up of eight spokes fixed on the upper part of the vertical axle; the masts are connected to the upper horizontal spokes, the lower horizontal spokes, the ropes, and the links. The sails are then hung on the frame to form the wind wheels. The vertical axle is fitted inside the iron ring at the top, and the needle, which means the lower end of the tapered axle shaft, is fitted inside the axle support (the iron bowl), thus forming two pairs

Picture 1.31 The horizontal windmill near Dagu port in Tianjin city

of sliding bearings. The spur gear is fixed at the lower part of the vertical axle and meshed with the vertical gear. The vertical gear is installed on a big axle and can move left and right to accommodate the meshing and separating of gears. The big axle is equipped with *shuitou,* a driving sprocket with twelve teeth to drive the water ladder.

When the windmill is set in motion, the spur gear is meshed with the vertical gear, and the sails are raised to the optimal height depending on the wind. Once the ropes tied to the lower horizontal spokes are released, the windmill will cause the water ladder to start turning. The height of the sails and the length of the ropes can be adjusted to change the rotating speed. When the wind is too strong, the windmill should not be used at all, because once the rotating speed exceeds the limit, the whole device will be damaged.

One of the great advantages of the sails on a horizontal windmill is that their direction can be adjusted automatically. When they turn downwind, they tend to be perpendicular to the wind and receive the greatest wind force; when the sails turn upwind, they turn parallel to the wind direction and suffer the least wind resistance.

There were more than 200,000 horizontal and vertical windmills in Jiangsu province in 1959. In the 1960s, windmills were used in several places, such as Yancheng, Jianhu, Dafeng, Funing, Xiangshui, Sheyang, and Ganyu in Jiangsu province. In 1993, vertical windmills were still used in northern Jiangsu. In 2006, Institute for the History of Natural Sciences of Chinese Academy of Sciences and the Southern Taiwan University of Science and Technology jointly invited Chen Ya, a craftsman from Qinghe village, Sheyang county, Jiangsu, to build a large horizontal windmill. It took Chen and his son eight months to complete the windmill and the whole construction process was recorded and filmed on the spot by Sun Lie.

Picture 1.32 Restored horizontal windmill, installed at the gate of the Southern Taiwan University of Science and Technology in Tainan city, Taiwan

According to *feng shui* tradition, an auspicious day was chosen to hold a "ground breaking ceremony" before construction of the restored windmill could officially begin. As Picture 1.32 shows, it is now a scenic spot at the gate of the Southern Taiwan University of Science and Technology in Tainan city, Taiwa.

1.2.4.2 Vertical Windmill (Horizontal-Axis Windmill)

Both *Notes on Physics* by Fang Yizhi (1611–1671) and *Discrimination on Sound and Meaning* by Zeng Tingmei (1734–1816) contain records of the vertical windmill able to hold three to six sails. It is also called the horizontal-axis windmill because the vanes are attached to a horizontally inclined axle. In China, this kind of windmill is mainly used in the southeast coastal areas.

Zhang Baichun and Feng Lisheng measured and drew the vertical windmill in Xilinzi village, Ganyu county, Jiangsu province (as shown in Picture 1.33). The horizontal axle is linked with the driving gear, which, in turn, is connected to a vertical axle's driving gear by horizontal coupling. The diameters and numbers of teeth of both driving gears, which are both fixed on a long hub, are the same. Except for a few iron pieces, the whole windmill is made of wood, mainly Chinese fir. All the wooden components require to be polished and coated with tung oil in order to prolong their service life. The sails used are also typical Chinese ones, and the

Picture 1.33 A vertical
windmill in Wuxian county,
Jiangsu province, cited from
Volume I of *Atlas of Chinese
Farm Tools* edited by Song
Shuyou

method of using the sails is similar to that of the horizontal windmill. Although the
vertical windmill cannot automatically adapt to the changing directions of the wind,
it has the advantages of being simple to construct, easy to use, and that it occupies a
minimum amount of space.

1.2.5 Chinese Bellows

Chinese bellows were widely used in cooking and some crafting industries requiring
high temperatures such as ceramics, smelting, and casting. As the Chinese saying
goes: "Where there is wind there is iron". Researchers believe that a significant reason
for China to be able to smelt iron and fire porcelain so quickly and early in history
is due to the development of the Chinese bellows.

The earliest bellow was an air bag made of animal skins, which was called *tuo*
or *nangtuo* by the ancient Chinese. *Record on the Subject of Education* in *Classic
of Rites* says: "The son of a good smelter is sure to learn how to make a *qiu*". That
is to say, the children of smelters had to learn to make a *qiu* (the air bag used as a

bellow) from an early age, otherwise they couldn't inherit their ancestral handicrafts, which were usually passed down from generation to generation. *Preparation against Tunnelling* in *Mo Zi*, states: "Prepare a furnace and bellows, the bellows being made of ox hide. For the furnace use two pots. Use well-sweeps weighing 100 *jin* to blow the bellows". Multi-bellows, meaning several bellows connected by *fou*-shaped pots, were used as early as the Spring and Autumn period and the Warring States period. The image of the ox hide bellows is engraved on the Eastern Han dynasty stone relief in Hongdao Temple in Tengxian county, Shangdong province. Wang Zhenduo (1911–1992), the Chinese scholar restored the bellows based on a careful study, which is shown in Picture 1.34. A major innovation in bellows came when the one-step human-powered bellows evolved into the horse-driven and water-driven bellows respectively utilizing animal and hydraulic power. *Biography of Du Shi* in *History of Later Han* reads that when Du Shi was the prefect of Nanyang, he was "good at planning, loved the common people and wished to save their labor. While serving as governor, he invented hydraulic bellows for the casting of agricultural tools, which allowed people to enjoy great benefit for little labor". *Book of Agriculture* by Wang Zhen (1271–1368) records two types of hydraulic bellows: the vertical wheel type and the horizontal wheel type. Both are still used in Sichuan, Yunnan, Hunan, and Zhejiang in modern times. Joseph Needham (1900–1995), a well-known British biochemist and sinologist, pointed out that the way the hydraulic bellows convert rotational motion into reciprocating motion through crank-connecting rods was a brilliant mechanical creation. Similar mechanisms were discovered in European water pumps from the fifteenth century, which may have been influenced by the ones used in China.

In the Tang and Song dynasties, the ox hide bellows were replaced by the wooden box bellows, which were first recorded in the Northern Song dynasty in *Part I* of *General Principle to the Art of War* by Zeng Gongliang (999–1078). The door of the wooden box actually acts as a piston. If it uses double wooden doors, air can be supplied continuously. However, since this kind of bellows requires the opening and closing of wooden doors, it easily leaks air which is not very efficient. As a result, it was later replaced by the famous double-acting piston bellows, the earliest image of which was found in *Book of Anthroposcopy according to Animals and Numbers* from the end of the Southern Song and the start of the Yuan dynasty. A more detailed description can be found in *Mirror of Craftsmanship and Guidelines by Lu Ban* of the Ming dynasty. It is characterized by its ability to draw in air from both sides of the handle. The bellow box is separated into two parts in the middle by a plate. A two-way valve is arranged at the joint of the two nozzles on both sides of the box and the piston, separate from the bottom of the box, is covered with feathers. When the piston moves to the left, the valve on the right opens to let air enter the box, while the one on the left closes to discharge air into the bottom layer, forcing the two-way valve to swing to the right, covering the air inlet on the right, thus letting the air flow out through the nozzle on the left. When the piston moves to the right, air enters from the air inlet on the left, and the air on the right side of the piston flows out from the nozzle on the right. In this way, as the piston pushes and pulls, air can be constantly supplied (as shown in Picture 1.35).

Picture 1.34 A model of ox hide bellows, cited from *Gems of Ancient Chinese Technological Inventions*

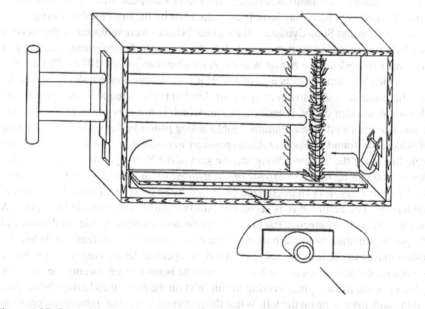

Picture 1.35 Structural map of double-acting piston bellow (Measured and drawn by Feng Lisheng and Ren Jinggang)

The shape of this kind of bellows is either rectangular or cylindrical, with the latter being more efficient than the former. *A Brief Map of Mines in Southern Yunnan* by Wu Qijun of the Qing dynasty reads: "The bellows used in furnaces are made of a whole piece of wood, which is cylinder-shaped and hollow. The diameter is about one *chi* and four *cun* (47 cm) and it is two or three *chi* (66.7–100 cm) long. It needs three people to operate. If a single piece of wood is not available in that size, wooden planks can be fixed together, however, the force of the blown air will diminish". There is also the single-acting piston bellows, which has only one air inlet opposite the handle, and can only let air enter and flow out once with each push-and-pull combination. It is the kind of manually-powered bellows that is most suitable to be used by a single person, which is why it is still being used by regular people today.

The double-acting piston bellows (as shown in drawing 1–35), which are ingeniously designed, convenient to use, and highly efficient, are one of ancient China's great inventions in mechanical engineering, just as Rudolf P. Hommel writes in *China at Work*: "It surpassed all air pumps in terms of blowing effect until the invention of modern machines".

1.2.6 Well Drilling

In China, obtaining salt from brine wells has a history of over 2,000 years. In the Northern Song dynasty, deep well drilling was invented for creating shaft wells, called *zhuotong* wells in Chinese, which refers to the wells being as straight as bamboo shoots. The depth for extracting underground brine in these wells was reached through percussion drilling. This type of technology developed to a very high level in the Ming dynasty, when devices such as the *zhuangzi* hammer drill were invented. Both the article *"On Salt in Sichuan"* by Su Shi (1037–1101) and the missive to the throne *"Plea to Send Officials to Jiangyan County"* by Wen Tong (1018–1079) recorded the time, place, drill technique, depth, structure, and brine obtaining tools for creating a shaft well. The extraction pipe is encased with bamboo, which is fastened with hemp ropes on the outside and coated with putty on the inside, thus preventing fresh water from flowing in from the sides, but ensuring natural brine flowing in from below. During the Wanli period in the Ming dynasty (1573–1620), the bamboo casing was replaced by wooden ones due to the increased depth of the wells. Deep well drilling originated in Jingyan, Rongzhou, and other counties in Sichuan province, the most renowned shaft well being the one in Daying county, which was drilled during the Qingli period (1041–1048) in the Northern Song dynasty. However, the brine well industry in Zigong county, Sichuan province, was even larger in scale and more famous.

It had 1,711 shaft wells with, at its peak, an annual output of more than 4,000 tons of salt. At that time, during the Song dynasty, the process of obtaining salt from wells included drilling the well, extracting and drying (filtering) brine, and crystalizing the salt. The cable tool drilling technology experienced the following three developing stages: (1) the germination stage featured by shallow wells with large caliber from the Eastern Han to the early Song dynasty, (2) the transition stage featured by *zhuotong* wells in the Song dynasty, and (3) the maturity stage featured by deep wells with small caliber in the Ming and Qing dynasties. In 1835, the world's first deep well exceeding 1,000 m, the sea well (1,001.42 m), was drilled. It was an important milestone, which served as a testimony to the sophistication of Chinese drilling technology and promoted the development of drilling technology in the whole world. Percussion drilling includes steps such as drilling and measuring the well, correcting deviation, filling cavities, lifting tools, and repairing wooden pillars. The process and tools used (as shown in Picture 1.36) have been completely preserved and are included in *List of the National Intangible Cultural Heritage*.

Textile machines and tools are also a major category of machinery in ancient China. See Chap. 4 of this book, which focusses on the creation of fabrics, for further details, since they have been omitted here.

Picture 1.36 The tilt hammer, cited from *Study on Tools Obtaining Well Salt in Ancient China*

1.3 Transportation

1.3.1 Land Vehicles

All the carriages that archeologists found at the Yin Ruins of Anyang, Henan province, the Laoniupo site of Xi'an, Shaanxi province, and the Qianzhangda site of Tengzhou, Shandong province, date back to the late Shang dynasty. The construction methods and overall shape of these carriages were quite advanced at that time. *Book of Diverse Crafts* reads: "Many craftsmen are required to construct a carriage". After a carriage has been constructed, other steps are necessary as well, such as varnishing, painting, and attaching the harness and reins. Vehicles in the pre-Qin era were mainly single-shaft carriages, which were used for hunting or for waging war. Common people with ox carts were relatively uncommon. Later, during the Qin and Han dynasties, double-shaft carriages replaced the single-shaft ones, and four-wheeled carriages and wheelbarrows appeared as well. Since the Han dynasty, the number of passenger and freight vehicles has never stopped increasing. Even all the way to modern times, traditional freight carts and passenger carriages have played an important role in China.

1.3.1.1 Cart

A traditional freight cart (as shown in Picture 1.37), fit for either short or long distances, is pulled by mules or horses, which carries heavier loads and goes faster than a cart pulled by oxen or donkeys.

It consists of a body, shafts, wheels, and axles. The body of a cart is made of wood and the shafts are placed in front of it. The most common wheeled cart has 16 or 18 spokes on its wheels, and the rim of each wheel is equipped with an iron hoop. The outer side of each wheel is reinforced with rivets, which makes the cart strong and durable. In northern China, they also used a kind of cart with H-shaped spokes and a rather thick axle, which made it more suitable for heavy loads.

Since the 1940s, due to the fact that the iron hoops around their cart's wheels damage the roads' surface, people fitted them with rubber and added bearings to the wheel hub. These carts became commonly known as "rubber carts". This kind of cart was used in China until very recently, which shows its usefulness and longevity.

The wheel hubs of the cart are made of hardwood such as locust wood or elm wood. A square hole is drilled in the center of the wood and the hub is planed and cut into shape. Then the outer circumference is divided according to the number of spokes and the spoke holes are cut equidistantly. Next, the round shaft hole is drilled and a cast iron bearing sleeve is embedded onto the two ends of the hub. Four round iron hoops are installed at both ends of the hub and the outer side near the spoke holes, so that the hub is firm and does not easily deform. The rim is formed by combining several spliced circular arc-shaped boards and two layers of riveted boards. Spokes are installed into the spoke holes of the riveted rim. After the hub, spokes and rim

Picture 1.37 A freight cart shown on a postcard sent from Beijing to Canada in 1909, cited from *Old Dreams Reappear*

are assembled, the protective iron hoop is embedded on the outer edge of the rim with rivets.

The axle is made of hard wood such as birch, locust, or elm wood, and an iron "gourd head" (*hu lu tou*) is installed at the joint with the bearing sleeve to increase wear resistance.

The body is usually made of hard wood with a straight grain such as Chinese oak, maple, locust, or elm wood. Only some secondary components, such as planking, can be made of slightly softer wood.

1.3.1.2 Wheelbarrow

A wheelbarrow (as shown in Pictures 1.38 and 1.39) is a handy vehicle for carrying goods and people. It consists of a wheel, a tray, and a frame.

The wheelbarrow is also referred to as a small cart, handcart, and single-wheeled cart. It can be easily pushed by one person on narrow and rugged roads, and even easily carried across more extreme terrain. What's more, it costs little to make. Therefore, it was widely used in China, especially in northern rural areas, where almost every family owned one.

In the Eastern Han dynasty, images of the wheelbarrow were carved on brick and stone reliefs in Sichuan and Jiangsu. The *muniu* and *liuma*, which literally translates to wooden ox and wooden horse cart, were two kinds of wheelbarrows invented by Zhuge Liang and Pu Yuan of the Three Kingdoms period. The wooden ox had shafts

Picture 1.38 Drivers crossing the street with their wheelbarrows

both in the front and in the back, and a small wheel. Both sides of the cart could be loaded with grain. Due to the heavy loads, it had to be pulled and pushed at the same time, making it move slowly. The wooden horse cart could not carry as much grain as the ox cart. However, the lighter load and its larger wheel made it possible to be pushed by just one person as well as move faster.

Wheelbarrows are sometimes designed to move using wind or animal power. *Heavenly Creations* depicts two draught animals pulling a wheelbarrow. The book of Van Braam Houckgeest, a Dutch-American merchant, from 1795 contains pictures of wheelbarrows equipped with a sail. The sail, made of cloth or woven bamboo, was equipped with a mainsheet, a boom vang, and a halyard. *The Hong-Sueh Sketches* (1839) by Lin Qing of the Qing dynasty contains records of a wheelbarrow equipped with a sail, with one person supporting the shaft and draught animals pulling in front.

When the Eight-Nation Alliance occupied Beijing in 1900, a foreigner drew a cartoon satirizing Empress Dowager Cixi's escape with her pulling a wheelbarrow containing her belongings. It was probably around that time that the use of the wheelbarrow began to spread abroad. A photo taken in a British port city on Christmas Day in 1935 shows a mother carrying two children in a wheelbarrow. The wheel was at the front of the shaft, similar to the wheelbarrow of the Chinese Han dynasty. Today, the wheelbarrow, despite its changed shape, is still a necessary tool in many rural households.

Picture 1.39 A nurse
visiting a pregnant woman
by wheelbarrow in Shanghai
in the early years of the
Republic of China

民国初年，坐着独轮车的护士下乡视诊产妇。

1.3.1.3 *Lele* Cart

The *Lele* cart, also known as the jigger cart, *lolo* cart, and ox wagon, was a common means of transportation on northern Chinese prairies. The axle and wheels of a *lele* cart are mostly made of hard wood such as birch and elm. They are simple in structure, easy to make and repair, and suitable for moving on grasslands, snow, swamps, and sandy beaches. "When you are on a trip, the *lele* cart will be the room, when you stop somewhere, the roofed cart will be the house". This proverb shows that various kinds of *lele* cart are indispensable tools for herders' daily lives, in ancient as well as modern times, including food, clothing, housing, transportation, weddings, and funerals.

The *lele* cart (as shown in Pictures 1.40 and 1.41) is composed of an upper part, which is the two shafts and frame, and a lower part, which is the wheels and axles. The wheels, which are the core components, are composed of hubs, spokes, and rims.

The wheel usually has 9 rims with 18 spokes, but the size of the rim depends on the wheel's diameter. Two spokes are installed symmetrically into each rim, with each rim riveted with mortise and tenon joints. There is usually a gap left between

Picture 1.40 *Lele* cart (Photo by Guan Xiaowu)

Picture 1.41 A train of *lele* carts

the inner edges of the rims and the tips of the spokes, so that through years of use, the worn rim will slowly move along the spokes towards the center of the wheel. Generally speaking, it takes 8–10 years for the rim to fit tightly over the tip of the spokes. This kind of design causes the connection between the rim and the spoke to continuously remain firm which prolongs the service life of the wheel. After the hubs, spokes and rims are fitted together, circular spindle holes are made at the hubs' ends. To finish the wheel, a cast iron bearing sleeve is embedded into each hole.

The body of the cart consists of shafts, boards for the bottom and sides, cross bars, vertical bars, supports, plugs, clasps, and other fasteners.

The wheels and the body form a complete *lele* cart. Some parts of it can be painted in bright colors, which not only serves esthetic purposes but makes it more resistant to corrosion as well. The main difference between various *lele* carts lies in the upper part. The sedan cart, which is equipped with an arc hood made of fine wool felt, is an advanced means of transportation for the elderly, children, and honored guests. The box cart is basically a wooden box on wheels. It is similar to the warehouse of the Han people and is used to store clothes and food. The truck cart is used to carry yurts, foraged items, fuel, and livestock products.

1.3.2 Water Vehicles

1.3.2.1 Dugout Canoes

Great Treatise in *Book of Changes* says: "They hollowed out trees to make canoes". In ancient times, people made a canoe by using wet mud to paint the parts of a tree trunk that did not need to be dug out, then setting fire to the part that needed to be dug out and finally digging it out with a stone axe. In 1958, three dugout canoes of the Spring and Autumn period and the Warring States period were unearthed in Wujin county, Jiangsu province. They are now exhibited in the History Museum of China (nowadays National Museum of China).

There were three types of canoes in ancient China: (1) those with a flat bottom, (2) those with a flat bottom, pointed bow, and square stern, such as the canoes unearthed in the river running through Yancheng village, Wujin county, Jiangsu province, and (3) those with a round bottom and a pointed bow and stern.

In both Manchu and Hezhe languages, the dugout canoe (as shown in Picture 1.42) is called a "*weihu*", while in some other places it is known as a "*kuaimazi*". It is made of a whole tree trunk, which is cut and carved. It is more than two *zhang* (6.7 m) long and wide enough for one person to sit in. The dugout canoe, open on top with a round bottom, has a pointed and slightly upturned bow and stern. The paddle is nearly one *zhang* (3.3 m) long and is used to alternatingly row left and right. The smallest canoe is big enough to hold a single person, while the largest one can hold five to six people. In addition to rowing in a single canoe, two "*weihu*" can be connected to each other, which is called "canoe pairing" and is used to transport vehicles and goods during floods.

Picture 1.42 Dugout canoe

Hunters in the northeastern forests of China use some very distinctive water vehicles. The animal skin boats used by the Oroqen people, for example, are made of an entire moose's or deer's skin, which can hold two to three hundred *jin* (100–150 kg). However, it cannot be used for too long, because when the leather softens in the water it loses some of its buoyancy.

In Hunan, Hubei, and Sichuan, a popular water vehicle is a small boat called a "sampan" or "*huazi*". Its shape is similar to that of a canoe. It is mostly used by fishermen who use cormorants for fishing. Seeing someone poling a "*huazi*" on a river or lake in southern China is a scene that is quintessentially Chinese.

1.3.2.2 Sheepskin Rafts

Skin rafts, also known as leather boats, are traditional water vehicles for the Salar, Hui, Dongxiang, Bao'an, and Tu people. They are popular along the Yellow River in Qinghai, Gansu, and Ningxia. These rafts can be categorized based on which animal's skin is used: sheepskin or cowhide. Even though they're called sheepskin rafts, they are mostly made of goatskin (as shown in Picture 1.43). To produce these, a high proficiency in slaughtering and skinning animals is necessary. Starting from the neck of the goat, the whole skin is skinned in one piece without any tears or breaks. Once skinned, air is blown into the skin to inflate it like a bag, a small amount of boiled oil, salt, and water is poured into it, and then the head, tail, and limbs are tied closed. After days of airing, the dried bags become yellowish-brown and transparent. Then, wooden strips of a northeast China ash tree are tied into a square frame with hemp ropes and several wooden strips are also tied horizontally onto the frame. Finally

Picture 1.43 Sheepskin rafts along the Yellow River in Lanzhou city (Photo by Guan Xiaowu)

the skin bags are tied underneath, thus finishing the sheepskin raft. The production of cowhide rafts is roughly the same as that of sheepskin rafts. When a raft need to be used, the frame is simply placed on top of the skin bags. It can carry people and cargo, with small ones able to support a weight of two to three tons and the large ones able to support over 10 tons. With its light weight and it laying shallowly in the water, the sheepskin raft doesn't easily strand or strike rocks. It is also flexible and easy to operate.

Skin rafts are not only easy to make, strong and durable, but also very light. Therefore, they can be carried by a single person. The Salar people of Dahejia, Gansu province, often use cowhide rafts to cross the Yellow River. There are two main types: large ones and small ones. The large ones are made of six to eight cowhide bags connected side by side, while the small ones only use four bags, all of which are connected into squares with crossbars attached to them. Large rafts can carry more than 20 people, while small rafts can carry seven to eight people. When crossing the river, passengers squat on the middle part of the raft, and three to four sailors stand at the bow and stern, moving their oars in unison to the rhythm of their work songs.

1.3.2.3 Inland Water and Sea-Going Vessels

Annals of Xia in *Records of the Historian* states that the tributes from Yangzhou were transported north to the Central Plains via inland rivers. In the Western Zhou dynasty, there were already designated officials in charge of boats and ships. During the Spring and Autumn period, there were different kinds of boats and ships, such as king warships, light boats, small boats, and grain ships. The different states at that time fought not only on rivers but also at sea. More than 20 wars on a small and large scale were recorded in the *Wu and Yue States during the Spring and Autumn Period*. There were different types of warships in the Wu State such as *dayi*, *zhongyi*, and *xiaoyi*, respectively meaning warships with large, medium, and small wings, as well as tower ships, charge ships, and bridge ships. During the Warring States period, both shipbuilding and water transport were well-developed. The Golden Tally to Monarch Qi of the Er, which was unearthed in Qiujia Garden village in Shouxian county, Anhui province, served as a duty-free pass issued by King Huai of the State of Chu. The inscriptions on it specified the exact number of permitted boats and ships as well as the navigation route. The burial ship of the Warring States period excavated in Pingshan county, Hebei province, is the oldest ship relic discovered in China so far. The boards for the deck of the ship were connected by iron hoops, it had a hull length of 13.1 m and a displacement of 13.28 tons.

Shipbuilding greatly progressed in the Qin and Han dynasties. In the Han dynasty, sea routes were set up from Guangdong to the Indian Ocean and from the Shandong Peninsula to Japan via the Korean Peninsula. The most famous type of ship of the Han dynasty was the multistoried tower ship, and there were about a thousand of such warships in its fleet. Liu Xi of the Eastern Han dynasty recorded the shipbuilding technology and achievements of that time in great detail in *Interpretation of Ships* in *Interpretation of Names*, including the appurtenances, structure, and classification of ships as well as their theories on stability.

The wooden and pottery ship models of the Han dynasty unearthed in ancient tombs in Changsha, Guangzhou and Jiangling, Hubei province (as shown in Picture 1.44), are equipped with decks and superstructures. Overlooking plates on the starboard and port sides are used as passages, which could be bound with bamboo or wood, thus providing partial buoyancy when overloaded, increasing stability and reducing swing when the ship tilts. From this, we can see that the ships of the Han dynasty were equipped with appurtenances such as oars, poles, sculls, tow ropes, rudders, and sterns. The sailing techniques were quite advanced for their time with the sails working in unison with the rudder and the stern. During the Three Kingdoms period, sea-going vessels were built in the Qing, Yan, You, and Ji states, as well as in Panyu, Nanhai prefecture, which had been an important shipbuilding town ever since the Warring States period.

In the eighth year of the Sui dynasty's Kaihuang period (588), the navy led by Yang Su fought fiercely on the Yangtze River using *wuya* warships against the defenders of the Chen dynasty. A *wuya* warship (as shown in Picture 1.45) has five floors and is powered mainly by rowing with large-tailed sculls on both sides. In rapids, all the oars, sculls, and rudders are used at the same time. A small attic is built above

Picture 1.44 Pottery boat of the Eastern Han dynasty, unearthed from Han tombs in the eastern suburb of Guangzhou province in 1955, cited from *Gems of Ancient Chinese Technological Inventions*

the fifth deck for observation and command. During the Sui dynasty, the expansion and excavation of the North–South Grand Canal promoted the development of grain transportation and the prosperity of the shipbuilding industry. Emperor Yangdi of Sui led a huge fleet to patrol the canal all the way to Jiangdu (present-day Yangzhou) three times, and tens of thousands of dragon boats and cruise ships were built for this purpose.

In the Tang dynasty, inland water shipping played an important role in accruing both individual and national wealth. The wooden boat discovered in Puxi township, Rugao county, Jiangsu province, in 1973 was most probably built after the year of 649, during the Tang dynasty. The slender hull is made of three sections of wood, tenoned together. It is 18 m long and is divided into 9 compartments with watertight bulkheads. The bottom consists of a whole piece of wood and is tenoned into the hull. The compartment plate and cover plate are nailed together with iron nails, and seven long boards of timber are piled, nailed, and filled with lime and tung oil to make the starboard and port sides.

The sea vessels of the Tang dynasty were as long as 20 *zhang* and could carry six to seven hundred people or hundreds of tons of goods. When sailing in the Persian Gulf, it could only berth in the lower reaches of the Shatt al-Arab. If you sailed westward to the mouth of the Euphrates, you had to transfer the goods to smaller ships.

The silk and porcelain trade in the Song dynasty relied on sea ships, which started from Guangzhou or Quanzhou in the south to Southeast Asia, South Asia, West

Picture 1.45 Model of a *wuya* warship, cited from *Five Thousand Years of Chinese Science and Technology*

Asia, North Africa, and even East Africa via the South China Sea. Around that time, Guangzhou, Hangzhou, and Quanzhou's foreign trade departments were established. Shipbuilding in the Northern Song dynasty mainly revolved around grain ships as well as passenger ships, warships, and ships for carrying horses. However, in the Southern Song dynasty, the number of grain ships that were built gradually decreased and the number of warships increased. Shipbuilding workshops, which were either government- or private-run, were spread all over inland states and major coastal ports. The sea-going vessels unearthed at Houzhu port, Quanzhou, in 1974 were built no later than the seventh year of the Song dynasty's Xianchun period (1271). The hulls are multi-plate structures and are divided into 13 compartments with 12 watertight bulkheads. The ships started with the help of a winch shaft. Another Song dynasty sea-going vessel discovered in Ningbo in 1979 is equipped with an anti-rolling keel, which reduces the rolling of the ship in the wind and waves.

The ocean-going vessels from the Yuan dynasty became world famous because of *The Travels of Marco Polo*. There are two types of sea-going grain ships, namely, the larger *Zheyang* and the smaller *Zuanfeng*. The former can carry 24–30 tons of goods and the latter 12 tons. Both types are slightly larger than the canal grain ships with ceylon ironwood rudderstocks, which are very sturdy. The hull of a canal grain ship is narrow and long, with an aspect ratio of 7:6. The sea-going cargo ship of the Yuan dynasty salvaged in Gwangju, Jeollanam-do, South Korea, in 1976 is a pointed ship with a curved keel. According to the shipbuilding customs in Fujian, copper mirrors and coins were embedded at the mortice and tenon joints of the keel, which is called "seven stars accompanying the moon". Seven bulkheads separate the ship into eight compartments of which the ribs are connected to the outer plate. There is a hole near the lowest point of each bulkhead, which is used for draining dirty water after the compartment has been cleaned. The shape of the outer plate is like fish scales and is connected to the bulkheads with tongue-shaped tenons.

In 1984, three ancient sunken ships were discovered at the Penglai Water City scenic spot in Shandong province. They were built during the Yuan dynasty and deployed no later than the ninth year of the Hongwu period (1376) in the Ming dynasty. The ship, built with castanopsis wood, is comprised of 14 compartments, each of which is separated by the ship's bulkheads. The keel is formed by two sections of square timber connected by a hook structure, and mortise and tenon joints. The outer plate is made with Chinese fir. The strakes are connected in squares with butt joints and further connected with the shell plates with piercing nails and shovel nails.

The grain sent as tribute to Jinling, the capital of the early Ming dynasty and present-day Nanjing, was mainly transported by river. In the beginning or the early Ming dynasty's Hongwu period (1368–1398), there were frequent wars in Liaodong in the north. The grain used to feed the army fighting up there was transported from the south via established south-to-north-sea routes departing from Liujiagang, Taicang, or Suzhou city. In the first year of the Yongle period (1403), Emperor Chengzu of the Ming dynasty ordered mariners, under the command of officers like Wang Xuan, to ship provisions and funds for the army over sea routes to Liaodong and Beijing. In addition to the transportation of grain, other kinds of bulk cargo were also transported. For example, Jingdezhen porcelain was mainly transported and sold all over China and the rest of the world by means of water transportation.

The long-distance ocean voyages of the early Ming dynasty used a direct navigation channel far away from the coast. Zheng He's seven voyages to the West occurred nearly one hundred years earlier than the "Age of Discovery" in the West. *The Overall Survey of the Ocean's Shores* records Zheng He's fourth voyage to the West and describes his Treasure Ships (as shown in Picture 1.46), the larger of which are 44 *zhang* and 4 *chi* (148 m) long and 18 *zhang* (60 m) wide, while the smaller ones are 37 *zhang* (123 m) long and 15 *zhang* (50 m) wide. It can be seen from the size of the ships that they are *Fu* ships, a kind of sharp-bottomed ship popular in Fujian and Zhejiang provinces. After Zheng He's seventh voyage in the sixth year of the Xuande period (1431), Emperor Xuanzong of the Ming dynasty banned sea transportation. As a result, the shipbuilding industry stagnated.

In the Qing dynasty, due to the shortage of timber and the rise of ship prices, many Chinese people began to build ships in Southeast Asia, where materials such as tung oil, hemp ropes, oyster ash, and iron could be transported from China, while Chinese crew members and shipbuilders living overseas could shoulder the task of shipbuilding. Most of the ships built there were *Fu* ships, but the price was only 40–60% of that in China. From the late Ming dynasty to the early Qing dynasty, cargo transportation between China and Japan was done by Chinese merchant ships, which were called Tang ships in Japan. The ships went from ports in Shandong and Nanjing to Nagasaki port in Japan.

Picture 1.46 Model of one of Zheng He's Treasure Ships, cited from *Atlas of Science and Technology through the History of China*

As for inland water transportation, shipping on the Yangtze River flourished from the Yongzheng to the Qianlong period (1723–1796) in the Qing dynasty, when a large volume of rice, salt, lead, tea, brocade, silk, and timber was coming in from Sichuan. There were more than 50 types of ships moored in Jiujiang, a major shipbuilding town in Jiangxi province. In the Qing dynasty, grain transportation on the Grand Canal was the same as in the Ming dynasty. Every year 624 twin type grain ships were built. This type of ship is formed by connecting two ships with iron hinges, which are convenient to connect and disconnect.

After the 1760s, the use of traditional Chinese sailing boats was challenged by the emergence of Western double-decked sailing boats in the Southeast Asian marine trade, and their share rapidly declined. Generally speaking, the shipbuilding industry in the Qing dynasty made only slow progress and ocean voyaging almost stopped. Traditional Chinese ships were eventually overpowered by the more advanced ships and armament of the Western powers. Thus the Westernization Movement emerged in the late Qing dynasty, which was the prelude to the modern Chinese shipbuilding industry.

1.3.2.4 Watertight Compartments

The watertight compartment is a great invention in Chinese shipbuilding, which first emerged in the Tang dynasty and was widely used in sea-going ships and some inland water vessels from the Song dynasty onward.

In ancient times, Quanzhou was famed for its developed shipbuilding industry. In *Miscellaneous Records at Xishan Mountain*, written during the Jiaqing period (1796–1820) in the Qing dynasty, Cai Yongjian writes: "During the Tianbao period (742–759), Wang Yao built boats for Lin Luan with timber transported from Brunei.

The hull of the boat is eighteen *zhang* long [...] The silver inlaid hull is divided into 15 squares, which can store as much as 15,000 to 20,000 tons of goods". The "15 squares" refer to 15 separate compartments. The wooden boat that was unearthed in Yangzhou, Jiangsu province, in 1960 and dated back to the Tang dynasty, is equipped with the oldest watertight compartments to date.

There were several or even dozens of them inside large ships during the Song and Yuan dynasties. Watertight compartments were famous both at home and abroad for their performance in preventing ships from sinking but were not seen in the West until the eighteenth century. The watertight bulkheads separate the compartments from each other, so even if one or two of them are damaged and flood, the whole ship still retains sufficient buoyancy to remain afloat. Even if there is too much water in the flooded compartment, it will not sink quickly as long as the cargo in it is abandoned and the load is reduced. If the damage is not too serious, there is not much water, and the cargo in the flooded compartment is removed, the damage can be repaired without affecting the ship's continued navigation. Even if there is a lot of leakage, the ship will be able to pull into the nearest shore for repair due to the protection of the other watertight compartments. What's more, the separation of the compartments makes it convenient to load, unload and manage cargo, since the cargo can be loaded and unloaded across different compartments at the same time. In addition, the connection of the deck with the hull can increase the strength of the ship, which not only increases the overall transverse strength, but since it also replaces a ship's ribs, it simplifies the shipbuilding process.

Samuel Bentham (1757–1831), a noted English mechanical engineer and naval architect, studied the structure of Chinese ships and was able to improve the existing shipbuilding techniques because of it. In 1795, he was commissioned by the British Royal Navy to design and manufacture six new types of ships. He wrote in his paper that these ships would "have bulkheads that increased the ships' strength and protect them from sinking due to leakage, just as the Chinese do now". Later, in his biography written by his wife, it is noted: "This was not the invention of General Bentham, who himself once publicly said: 'This is what the Chinese today practice, just like they did in ancient times.'". Since then, China's advanced watertight compartment structures were gradually adopted by the shipbuilding industry all over the world. Even now, they are still an important structural part of ship design.

Traditional Chinese wooden sailboat building techniques are still used in Shenhu town, Quanzhou city, Fujian province. The "Princess Taiping" was built completely according to traditional ship design, materials, building, and decoration conventions, as well as ceremonies. The 14 watertight bulkheads divide the ship into 15 compartments. There are two water-passing holes near the keel on each of the compartment plates, and the gaps between the plates are sealed with tung oil, oyster ash, and hemp rope. This technique has been listed in *List of National Intangible Cultural Heritage*.

1.4 Instruments and Meters

1.4.1 Li-*Recording Drum Carriage*

The *li*-recording drum carriage, also known as the *dazhang* carriage, is an ancient machine that automatically records the distance in *li* by gear transmission device. It was used for imperial ceremonies. The carriage has two wooden mannequins that simultaneously beat a drum after every one *li*, or 500 m, traveled. The total traveled distance can be calculated by counting the number of drum beats.

There are different records about its origin, placing it in the Eastern Han, the Western Han or the Jin dynasty. Based on the records in *Notes on the Antiquity and Present Days* by Cui Bao of the Jin dynasty and *Records of Carriages and Costumes* in *Book of Jin*, it is generally considered a ceremonial vehicle used as a guide car and as a symbol for the mighty imperial power. It is usually accompanied by the south-pointing carriage. *Records of Carriage and Costumes* in *History of Song* contains the principle and structure of the drum carriage made by Lu Daolong, and also detailed descriptions of the one improved by Wu Deren. Based on these records, Zhang Yinlin (1905–1942) was able to make reasonable assumptions and estimates about the structure of the drum carriage, which made it possible for Wang Zhenduo (1911–1992) to make a restoration model of it (as shown in Picture 1.47).

1.4.2 *Water Clock*

The water clock is an ancient time device used in several ancient civilizations, such as Egypt, Babylon, and China. It consists of a water container and a gauge. There are two types of water clock: In an outflow one, water fills the container and is drained, while in an inflow one, water fills the marked container. The gauge marks the time. When in use, it is placed in the container and goes up and down with the changing water level. The earliest record of a water clock can be found in *Rites of Zhou*, while the earliest unearthed one dates back to the Western Han dynasty. These copper containers are cylindrical, with a pipe near the bottom and a beam on the lid. Those with opposing rectangular holes on the beam and the lid are called single-container outflow sinking-arrow water clocks. There are only two inflow water clocks left from ancient times. The one exhibited in the National Museum of China was made in the third year of the Yanyou period (1316) in the Yuan dynasty and the one exhibited

Picture 1.47 A restoration model of the *li*-recording drum carriage

in the Palace Museum in Beijing was made in the tenth year of the Qianlong period (1745) in the Qing dynasty. The water clock in the Song dynasty was only several seconds faster or slower than the actual time per day, the accuracy of which was not surpassed until the invention of the pendulum used as the controller of the mechanical clock.

The water clocks in the Han dynasty were mainly floating-arrow types. During the Northern Wei period, the scales water clock was invented by Li Lan, a Taoist priest. It weighs the water in the container with scales and measures the time with its changes. It was recorded in a few books, such as *Sleeve Records* by Shen Yue of the Southern Dynasties period. The most obvious difference between the two types lies in how time was displayed. The scales water clock is more sensitive to water flow and shows a more detailed division of time, so it spread soon after its invention and quickly became the main astronomical timing instrument from the Sui and Tang dynasties to the Northern Song dynasty. According to *Classification of State Offices* by Sun Fengji of the Southern Song dynasty, the steady water flow of the scales water clock keeps such a stable water level that its timing error ranges between one minute or 20 s at most. It could meet the needs of astronomical observation at that time and is considered one of the greatest water clock inventions.

However, it was later replaced by the lotus water clock since its time display could not continuously display the passage of time. In the eighth year of the Tiansheng period during Emperor Renzong's reign (1030), Yan Su invented the lotus water clock, which used an overflow system. It looks like a next-level floating-arrow type. It consists of "a pot, a bamboo tube and a copper water-saving tube" on one side pointing toward the lower water container, which in turn is connected with the arrow pot. In other words, the water flow from the upper container to the lower one is stronger than that going from the lower container to the arrow pot. The water overflows from the copper water-saving tube in the small hole in the upper part of the lower container to the arrow pot via the bamboo tube, thus stabilizing the water level in the lower container and minimizing the influence of the water level to its flow. After repeated tests, the lotus water clock was officially adopted by the Astronomy Department in 1039.

The water level in the arrow pot will still change a little when water is added to the upper container, which will affect the timing accuracy. Therefore, people around the late Northern and early Southern Song dynasties made the mixed overflow compensation floating-arrow water clock (as shown in Picture 1.48) by combining the overflow type and multi-stage compensation type. It became the standard form of the water clock after the Southern Song dynasty and was considered one of the greatest inventions of automatic control systems.

Picture 1.48 Water clock of the Yuanyou period in the Yuan dynasty

1.4.3   Houfeng Seismograph

Houfeng Seismograph was invented by Zhang Heng, a scientist in the Han dynasty. According to *Treatise of Five Elements* in *History of Later Han*, there were 26 major earthquakes in more than 30 years from the fourth year of the Yongyuan period during Emperor Hedi's reign (92) to the fourth year of the Yanguang period during Emperor Andi's reign (125). They sometimes affected dozens of counties, causing huge losses, with earth fissures and landslides, flooding, and collapsing houses. After years of research, in the first year of the Yangjia period (132), Zhang Heng invented the Houfeng Seismograph. It is recorded in *Biography of Zhang Heng* in *History of Later Han* that the seismograph is "made of fine copper, with a diameter of eight *chi* (about 2.7 m)" and "shaped like a wine bottle", with a round cover on it. The surface is carved with seal script as well as patterns of mountains, turtles, birds, and animals. In the center of the instrument is a copper column (*duzhu*), surrounded by eight channels (*badao*) with ingenious mechanisms. Eight dragons, which are arranged in eight directions: north, northeast, east, southeast, south, southwest, west, and northwest are fixed on its surface. The heads of the animals, each with a copper ball in their mouths, are connected to the mechanisms in the channels on the inside. Eight copper toads squat on the ground facing the dragons, holding their heads high with mouths open and ready to receive the copper balls. When an earthquake occurs in a certain place, the body of the instrument vibrates and the mechanisms inside are triggered, so that the dragon in the direction of the earthquake opens its mouth and spits out the copper ball, which falls into the mouth of the toad and makes a sound. This way people will be able to judge the direction of the earthquake.

Its blueprint and production method are said to both have been recorded in *Collections of Instruments* by Xin Dufang of the Northern Qi dynasty and *Record of Copper Seismograph* by Lin Xiaogong of the early Sui dynasty. These records have unfortunately long been lost. In modern times, the most influential research was attributed to Wang Zhenduo, who concluded that the column of the Houfeng Seismograph, with eight channels around it, worked much like the inverted pendulum of the modern seismograph. The column has a high center of gravity so that it easily loses its balance when affected by seismic waves. When it falls down into the channel in the same direction as the earthquake, the lever (*yaji*) is hit, thus causing the upper jaw of the dragon to lift up and spit out the copper ball, which is similar to raising an alarm. The Houfeng Seismograph (as shown in Picture 1.49) exhibited in the History Museum of China (nowadays the National Museum of China) was made based on Wang's restoration.

There have been many arguments over the years, both at home and abroad, about the working principles and structure as well as some doubts about the reliability and sensitivity of the instrument. Various assumptions and designs have also been put forward. However, no recognized conclusion has been reached yet.

Picture 1.49 A model of the Houfeng Seismograph

1.4.4 Astronomical Instruments

1.4.4.1 Armillary Sphere

The armillary sphere is an observation instrument that was used to measure the spherical coordinates of celestial bodies in ancient China. It was created in the early Warring States period and specifically recorded in the documents of the Western Han dynasty. The instrument was created based on the "Theory of the Celestial Globe", which states that "the celestial globe looks like an egg shell, the celestial bodies egg balls, and the earth the yolk". The primitive armillary sphere might have consisted of two rings, one of which was a fixed equatorial ring, and the other a movable *siyou* ring rotating around the polar pivot, with a sighting tube attached to the ring for observation. Later, in order to better observe the celestial bodies such as the sun, the planets, and the moon, more rings were added to the armillary sphere, which evolved into a multi-purpose astronomical observation instrument. For example, in the Eastern Han dynasty, Fu An and Jia Kui added the ecliptic ring, and Zhang Heng added the horizon and meridian rings.

Li Chunfeng, an astronomer in the Tang dynasty, designed a more precise one with three layers. The outer component of the six cardinal points includes horizon, meridian, and equatorial circles; the middle component of the three arrangers of time consists of the moon's path, ecliptic, and equatorial rings; the inner component of the movable *siyou* device includes the *siyou* ring and a sighting tube. The Northern Song dynasty produced the highest number of armillary spheres, including the copper one made by Han Xianfu in the first year of the Zhidao period (995), the one made by Shu Yijian, Yu Yuan, and Zhou Cong in the third year of the Huangyou period (1051), the one by Shen Kuo in the seventh year of the Xining period (1074), and the copper one on the astronomical clock-tower built by Su Song. There was another such instrument during the Shaoxing period (1131–1162) in the Southern Song dynasty. Guo Shoujing also made one in the Yuan dynasty. However, all of those armillary spheres have been lost. Only one made of bronze from the second year of the Zhengtong period (1437) in the Ming dynasty is left. It is exhibited at the Purple Mountain Observatory in Nanjing (as shown in Picture 1.50). In the 26th year of the Qing dynasty's Guangxu period (1900), it was looted and transported to Potsdam, Germany during the invasion of the Eight-Nation Alliance in Beijing, and was returned to China after the First World War. It was sent to Beijing in April 1921, and was transported to the present location after the September 18th Incident in 1931. The armillary sphere was damaged by the Japanese troops after their occupation of Nanjing, and was repaired and properly protected after the founding of the People's Republic of China.

Picture 1.50 The armillary sphere made in the Ming dynasty (Photo by Zhang Baichun)

1.4.4.2 Abridged Armilla

The multiple rings of the armillary sphere are staggered, which covers the sky area and limits the observation range. So Shen Kuo (1031–1095) got rid of the moon's path ring and changed the position of some of the other rings so that they did not block anyone's sight. Later, in the 13th year of Zhiyuan period during the reign of Emperor Shizu (1276) in the Yuan dynasty, Guo Shoujing made the abridged armilla by separately installing the ring sets. The moon's path and ecliptic rings of the armillary sphere were removed and the equatorial device (consisting of the equatorial and right ascension circles) and the altazimuth mount (consisting of horizontal and azimuth circles) are separated, so that none of the rings impede observation. The equatorial ring and the hundred-scale ring are overlapped, and the former can rotate freely along the latter due to the four short cylindrical copper rods fixed horizontally between them. This is the earliest use of roller bearings in the world.

The abridged armilla made by Guo was destroyed by Jesuit missionaries in the 54th year of the Kangxi period (1715). The one exhibited at the Purple Mountain Observatory in Nanjing is the only one left (as shown in Picture 1.51). This was a copy made during the Ming dynasty's Zhengtong period (1436–1439). It was originally kept at the Ancient Observatory in Beijing and was transported to Nanjing by Academia Sinica in 1932. However, it was seriously damaged during the Second Sino-Japanese War (1937–1945).

The rings representing different coordinate systems of the armillary sphere are simplified into two independent devices of the abridged armilla, namely, the vertical

Picture 1.51 The abridged armilla made in the Ming dynasty (Photo by Zhang Baichun)

revolving circle (altazimuth mount) and the equatorial armillary sphere (equatorial device). The vertical revolving circle consists of the horizon and vertical revolving rings, while the equatorial armillary sphere consists of equatorial, hundred-scale, and movable *siyou* rings. Both sides of the *siyou* ring are engraved with the full 365.25 degrees of a circle, according to ancient Chinese mathematicians, and the sighting bar in the middle can rotate around its center. Both ends of the sighting bar have crosshairs, which was the first of its kind. Celestial bodies in any azimuth could be observed by rotating the *siyou* ring and the sighting bar, and their declination number could be read out from the scale on ring. If you multiply the number by 360/365.25 and subtract the product by 90 degrees, you'll get the modern declination value.

The torus of the equatorial ring is engraved with the degrees of 28 lunar mansions and two boundary bars, both ends of which are connected with the north polar pivot by thin lines, thus forming two triangles. The angle between the planes of the two triangles is the difference of right ascension. To get the determinative star distance of a certain lunar mansion, you only need to aim one boundary bar at its determinative star and the other at the lunar mansion itself, and then you can read out the number on the ring. Add it to the distance from the west side of the celestial body to the corresponding place where the vernal equinox is located, then subtract the distance of the vernal equinox, and multiply it by 360/360.25 to get the modern right ascension value.

The hundred-scale ring is fixed to the equatorial ring. When you aim the boundary bar at the sun, the reading obtained on the ring is the true solar time. The altazimuth mount consists of a horizontal ring and a vertical revolving ring. The sighting tube between them can rotate around the center of the vertical revolving ring. The azimuthal angle and altitude of any celestial body can be measured by rotating the vertical revolving ring and the sighting tube.

The abridged armilla is installed on a rectangular copper frame and supported by dragon columns and cloud columns. The lower parts of the columns are connected by flumes, which can help to determine the horizontal position of the instrument by observing the water's surface. The complex patterns of clouds and dragons on the columns show the sheer majesty of the instrument.

The design and creation of the abridged armilla was the best in the world at that time. It was not until 1598 that Tycho Brahe (1546–1601), a Danish astronomer, invented Western astronomical instruments that came close to these Chinese counterparts. The equatorial devices of large modern telescopes used at observatories, especially the British types, were derived from the Chinese abridged armilla. The altazimuth mounting structure of theodolites used in engineering, and topographic and astronomical surveying is also similar to the armilla's longitude and latitude rings and vertical revolving ring. The structure of the astronomical compass used in airborne navigation is also similar. It can be said that the Chinese abridged armilla is the original form of all the above-mentioned modern instruments.

1.4.4.3 Water-Powered Astronomical Clock Tower

The original water-powered astronomical clock tower (as shown in Picture 1.52) was a large-scale astronomical instrument in ancient China. It was built by Su Song and Han Gonglian in the seventh year of the Yuanyou period (1092) and destroyed in the second year of the Jingkang period (1127) in the Northern Song dynasty. The wooden tower, 12 m high and 7 m wide, includes functional, power, control, and transmission systems.

The functional system consists of an armillary sphere, a celestial globe, and a time-keeping mechanism, which are placed on the upper, middle, and lower levels, respectively. The armillary sphere is at the top, with a roof that can be opened, pioneering the operable skylights seen with modern telescopes. The celestial globe is installed on the middle level, demonstrating the celestial movements. One half is hidden under the floor, and the other half is above it. Both are driven by the

Picture 1.52 Model of water-powered astronomical clock tower

gear wheel. The globe completes one revolution in 24 h, showing the astronomical phenomena such as the rise and fall of stars. The time-keeping mechanism is on the lower level, the same level as the power system. The power system consists of a pool and a central wheel. Under the joint influence of the moving water and the control machinery, the central wheel rotates at a uniform speed and drives the functional system to run through the gear system. To ensure its uniform rotation, an escapement mechanism is applied and it works like this. The outer ends of the central wheel spokes are equipped with receiving scoops, with a balancing lever underneath. One end of the lever near the scoops is called a checking fork, and the other end is called the main counterweight, under which a heavy weight is hung. When a scoop is filled with water, the moment of the arrested spoke and its fulcrum will be larger than that of the weight, and the checking fork will be tripped by a projection pin on the scoop, making the central wheel rotate. At the same time, when the balancing lever is pulled, and the upper stop and the left upper lock are lifted, the right upper lock will be pushed up by the wheel. Hence the wheel rotates the amount equal to the angle of a spoke. And then the checking fork bounces up, and the left and right upper locks fall down to lock the next spoke (the right lock is a stop clip that prevents reversal). Water will be filled into the next scoop through the siphon, and the water in the previous scoop will be poured into the container. This way, the central wheel will rotate intermittently at a uniform rate, which is used to divide time.

The transmission system has two sets of toothed gears, which drive the armillary sphere, the celestial globe, and the time-keeping mechanism, respectively.

Su Song wrote the book *Essentials of a New Method for Mechanizing the Rotation of an Armillary Sphere and a Celestial Globe* after the completion of the water-powered astronomical clock, recording the inner workings of the clock tower in detail. Later in the early years of the Southern Song dynasty's Shaoxing period (1131–1162) when an armillary sphere was made, the book, which was kept by Su's descendants, was an important reference.

The water-powered astronomical clock tower represents the highest mechanical level of design in the world at that time. It integrated the armillary sphere for observing stars, the celestial globe for demonstrating astronomical phenomena, the water clock for time-keeping, and the mechanical device for announcing time. It employs various kinds of devices and machines, such as the water wheel, the Chinese noria, the siphon, the well sweep, the cam, the balance, and the scale rod. The linkage control mechanism adopted functions the same as the escapement mechanism of modern mechanical clocks. The large-scale astronomical instrument was built in the Song and Yuan dynasties, when science and technology in ancient China reached their peak. It embodies the scientific and technological achievements in ancient Chinese mechanical design and manufacturing, astronomical observation, metallurgical casting, and architectural technology, and shows the intelligence and creativity of the ancient Chinese people.

In the 1950s, Liu Xianzhou and Wang Zhenduo, two Chinese scholars, did research on the instrument. In 1956, a 1:5 scale model was completed under the supervision of Wang and displayed in the History Museum of China (currently the National Museum of China). In recent years, several restoration models have been built in Taiwan and

Japan, which employ involute gears and other modern techniques. However, we hold that they aren't a true restoration, since the techniques they used were beyond the technical level of the Song dynasty. Some scholars also have doubts about whether the water-powered astronomical clock tower built by Su Song can operate consistently, which will require further research.

1.4.5 Vernier Caliper

The vernier caliper is a precision device for measuring the length of an object, which consists of a main scale, fixed claws, a vernier slide, movable claws, and a vernier scale. When measuring an object, the claws of the main scale and the vernier scale are closed around the object so that it is convenient to read out the dimension and eliminate the error caused by inaccurate calibration. The device is widely used to measure the inside and outside width and depth of round objects and special-shaped objects.

In the first century, during the Xinmang period at the latest, China invented the copper caliper, which has the same shape and structure as the modern vernier caliper. It consists of a fixed scale, fixed claws, a fish-shaped handle, a guide slot, a guide pin, a combined sleeve, a movable scale, movable claws, and a thumb lever. One *cun* on the copper caliper exhibited in the National Museum of China (as shown in Picture 1.53) equals 2.48 cm. There are scales on it, with the minimum reading "*fen*", which equals one-tenth of a *cun*. If the estimated reading value is one-tenth of a "*fen*", it is a "*li*", which equals 1/3 of a millimeter. It can be used to measure the diameter, thickness, and depth of an object. When measuring the diameter or thickness, we first pull away the slider and put the workpiece between the fixed measuring claw and the movable measuring claw. Then we move the slider to clamp the workpiece tightly and read the measured value directly on the fixed scale. When measuring the depth, we move the thump screw gradually to extend the depth rod until the end of the rod makes contact with the bottom of the hole. Then we can take the reading.

The principle of isochromatic amplification was not applied to the Xinmang copper caliper, which makes it inferior to the modern one in accuracy. Yet for a device

Picture 1.53 Xinmang copper caliper

that was invented 2,000 years ago, it gives a good idea of the level of mechanical advancement in ancient China.

1.4.6 Compass and Luopan

It is necessary for people to know directions in their daily lives, at work and during military operations. The earliest directional instrument in China is the south-pointing carriage, which was in the form of a vehicle. A wooden mannequin on the carriage points the direction with the help of differential gears, which is described in ancient books as "although the carriage turns away, the finger of the wooden mannequin always points to the south". Zhang Heng in the Eastern Han dynasty, Ma Jun in the Three Kingdoms period, and Zu Chongzhi in the Southern Qi dynasty all made south-pointing carriages. During the Tang dynasty's Yuanhe period (806–820), Jin Gongli once proposed making a south-pointing carriage to Emperor Xianzong. Yan Su actually built such a carriage in the fifth year of the Song dynasty's Tiansheng period (1027) and Wu Delong also offered the method of making one in the first year of the Song dynasty's Daguan period (1107). *Notes of Kui Tan* by Yue Ke and *Records of Carriages and Costumes* in *History of Song* both detail the making and technical specifications of the south-pointing carriage. After the Jin dynasty, it was commonly installed in the front of the imperial chariot. Since the Yuan dynasty, there has been no further development in the creation of south-pointing carriages. Based on ancient books, Wang Zhenduo reconstructed a model (as shown in Picture 1.54), which is now on display in the National Museum of China. However, Wang pointed out that such a carriage was usually part of the imperial chariot and was probably not workable due to the limited technology at that time.

The compass, or the "south-pointing needle" in Chinese, is one of the Four Great Inventions of ancient China. It is, simply put, a direction finder that uses magnetism. The basic principle of ancient Chinese directional instruments, such as the south-pointing turtle (as shown in Picture 1.55), south-pointing fish, compasses (as shown in Pictures 1.56 and 1.57), and *luopan* is to keep the magnetic needle pointing in the tangent direction of the magnetic meridian under the influence of the geomagnetic field, which pulls the north end of the needle toward the south geomagnetic pole, and pulls the other toward the north geomagnetic pole. They are used to give directions in such fields as navigation, geodesy, travel, and warfare.

The invention of the compass is based on the ancient Chinese understanding of the lodestone, which is highly magnetic. The ancients found out that lodestones attract iron and attract and repel each other. In the Song dynasty, people mastered the method of magnetizing iron pieces and invented various kinds of directional instruments based on many experiments in which they discovered the principles of geomagnetic fields, geomagnetic declination, and geomagnetic inclination. According to literature, there were two ways to magnetize iron pieces in ancient China: One way was to rub iron needles on lodestone in one direction, the other to place iron pieces in the north–south direction after heating them to a high temperature and the geomagnetic field

Picture 1.54 Model of the south-pointing carriage

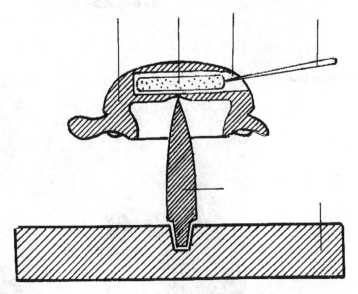

Picture 1.55 South-pointing turtle wooden turtlemagnetyellow waxneedle bamboo nailwooden board

Picture 1.56 Model of silk
thread suspension compass

Picture 1.57 Water float
compass

will magnetize them as they lay there. *Dream Pool Essays* by Shen Kuo (1031–1095) records four methods of installing the south-pointing needle: (1) putting it on the rim of a bowl, (2) putting it on a fingernail, (3) putting it on one end of a suspended silk thread, and (4) having it float on water in a bowl. Ancient literature also records the making of the south-pointing turtle and fish, which are the precursors of the land and water *luopan*, respectively.

It is generally believed that the compass was introduced to Europe from China by the Arabs. The use and spread of the compass have greatly promoted the development of human society since it has promoted land and marine traffic and expanded the scope of people's activities. The gimbal compass was invented by westerners and introduced to China by missionaries. This kind of structure completely counteracts the external interference and the pointer can point stably to the north, so it is widely used in transportation and military activities.

Luopan (as shown in Pictures 1.58 and 1.59), is also known as *"diluojing"* or *"luojingpan"* in Chinese. It is a combination of a magnetized needle and a dial disk and is widely used in navigation and geomancy in ancient China. The dial disk can be traced back as early as the Warring States period and the Qin and Han dynasties, when fortune tellers used the *shipan* in divination. The 24 directions on the *luopan* after the Song dynasty originated from this. However, in the Ming dynasty, due to the different needs in geomancy and navigation, the disk system of the *luopan* underwent some changes. The *luopan* fall into two categories depending on how the magnetized needles are installed, namely, the water *luopan* and the dry *luopan*. The *luopan* before the Ming dynasty were water *luopan*, which were also called "water needles", while the dry *luopan* were called "dry needles". The main difference between the two lies in that one uses a shaft to support the needle. Due to the limited friction of the supporting point, the magnetized needle can rotate freely, which makes it more suitable for navigation than the water *luopan*.

There were also two types of water navigation *luopan* in ancient China: inland and coastal. The latter, centered in Zhangzhou, Fujian province, and Xingning, Guangdong province, were mainly used for navigation. The former, centered on Wan'an town, Xiuning county, Anhui province, were mainly used for geomancy. Therefore, they were also called *"Huipan"*. The *luopan* made in Wan'an requires traditional crafting skills, with strict specifications, steps, and techniques. It generally has to go through the seven processes below.

Select material and make workblank. Select tough, fine, and untextured kurogane holly or ginkgo biloba wood, and saw it into a workblank according to the required diameter and thickness.

Lathe the blank into a round disk, polish it with fine sandpaper and equisetum grass, and dig a round hole in the middle for the magnetized needle. For decorative *lupan*, its edge is engraved with various patterns.

Carve grids on the disk. Carve horizontal grids with radii of different lengths from the center of the disk according to the atlases of different models and disks, and then carve straight grids according to *yin* and Yang, the Eight Diagrams, and the Heavenly Stems and Earthly Branches. Different schools use different grids, none of them allow any mistakes.

Picture 1.58 Pottery figurine holding a *luopan*

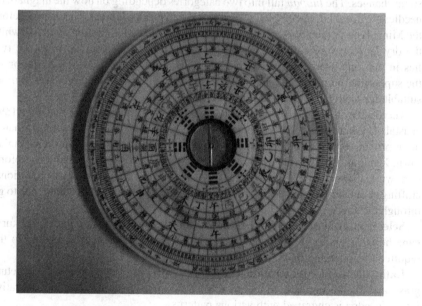

Picture 1.59 Chinese *luopan*

Clean the disk. Get rid of the residual sawdust and smoothen the rough edges, so that the disk surface is clean and tidy, and the grids are clear.

Write on the disk. Fill out the contents of the grids with a fine brush according to the private atlas.

Apply oil. Refine tung oil and apply it to the disk in layers. A well-oiled *luopan* is still bright, clean, and clear a 100 years later, and its color remains the same.

Install the magnetized needle. This process includes magnetizing the steel needle, measuring its center of gravity and installing it on the disk. This is the core of *luopan*-making technology, which is usually done by the shopkeeper himself in a secret room. Finally, cover the disk with a glass sheet.

Wan'an *luopan* dates back to the late Yuan and the early Ming dynasties, but really flourished in the middle of the Qing dynasty. It saw a resurgence of its glory in the early Republic of China, which lasted until the early 1960s. Wu Shuisen, inheritor of the techniques, follows the family tradition and continues the "Wu Lu Heng" brand, reviving the Wan'an *luopan*, which was once forgotten by many. What's more, he also developed dozens of other types such as Lotus Eight Diagrams and Double Dragons Playing with a Ball. In 2006, the technique of making Wan'an *luopan* was approved by the State Council and was listed in *List of the First Batch of National Intangible Cultural Heritage.*

1.5 Musical Instruments

The oldest known musical instrument in China are bone flutes unearthed from the Jiahu Relics in Wuyang county, Henan province in 1987. There were altogether 18 bone flutes gathered from many tombs, but the two similarly shaped ones from tomb No. 82 were the most representative. We hold that they proved the tradition of using a pair of *yin* and *yang* flutes in ancient China. The bone flutes were made from the wing bones of a crane by sawing, drilling, and grinding them. With a small finely processed hole that is made to adjust pitches, the flute can play semitones in two octaves. The Jiahu Relics belong to the early stages of the Peiligang culture in the Neolithic Age, about 9,000 years ago. The discovery of bone flutes promotes our better understanding of the long and brilliant music culture and the making of musical instruments in China.

In the Zhou dynasty, rites and music were used to stipulate the rank of fame and position. So, making rites and music was of great importance in the implementation of politics and religion. In *Chunguan* in *Rites of Zhou*, musical instruments are divided into eight categories, collectively referred to as eight tones, namely gold, stone, clay, leather, string, wood, gourd, and bamboo. Among them, gold refers to the bell cymbal, stone refers to the *teqing* and the *bianqing*, clay refers to the ceramic ocarina, leather refers to drums, string refers to stringed instruments, such as the *qin* and the *se*, wood refers to the *zhu* and the *yu*, gourd refers to the *sheng* and the *yu*, and bamboo refers to wind instruments, such as the *xiao*, the flute, the *guan*, and the *chi*.

After going through a long period of development and evolution, a perfected series of national musical instruments, that integrate many kinds of features from all ethnic groups, took shape in ancient China. There were hundreds of them, including those from minorities. The following is a brief description of the production techniques of some existing musical instruments.

1.5.1 Guqin *(the Chinese Zither)*

The *guqin* (*qin*, or Chinese zither) is a plucked seven-string musical instrument, which appeared in the Western Zhou dynasty at the latest and prevailed among the *shi* (hangers-on of aristocrat) during the Spring and Autumn period. Yu Boya was a famous player at that time. In the Wei and Jin dynasties, the *guqin* already had the *hui*, the markers, on it and its shape and structure were almost the same as the present *guqin*. The Tang Qin, which is a *guqin* made in the Tang dynasty, is the most precious type handed down from ancient times. Those made by the Lei family in Sichuan at that time were especially popular and were named "Lei Qin" after their family name. The "Jiu Xiao Huan Pei" (the highest heaven jade pendant), "Da Sheng Yi Yin" (musical legacy of the sage) (as shown in Picture 1.60), and "Fei Quan" (the flying spring) are exhibited in the Palace Museum in Beijing are all Lei Qin. In the Song and Ming dynasties, there were many well-known skilled *guqin* craftsmen in the inner Palace who made *guqin* for the imperial family. There are many *guqin* forms recorded in the *qin* scores of the Ming and Qing dynasties, such as the *Fuxi* form, *Shennong* form, *Lianzhu* form, and the *Zhongni* form, the latter being the most common and named after Confucius' style-name.

The *guqin* is composed of the top board (as shown in Picture 1.61), the bottom board, and the sound chamber. Its production process includes the steps of selecting material, drying, hollowing, lacquering, and shine removal.

The *guqin* is made of tung wood and catalpa, which has remained the same since ancient times. Lei's family in the Tang dynasty summed it up as "selected material and fine craftsmanship guarantee an excellent instrument lasting more than 500 years". Tung wood, which has smooth grain lines, even width, moderate hardness, and doesn't have knots or insect damage, should be used for the top board of the *qin*. It is best to make the bottom board with catalpa older than 100 years. The selected wood should be soaked in water for a long time, and then hung above the stove or exposed to the wind and sun until it is dry.

The radian of the top board, the height of the *yueshan*, or the bridge, and the dragon's gums have an impact on the timbre of the *guqin* and the playing of the player. There has always been a saying "the front is one-finger high and the back is a piece of paper high", that is, the height of the *yueshan* in the front shouldn't be higher than one finger, and the height of the dragon's gums should only be as high as the thickness of a piece of paper. The sound chamber (as shown in Picture 1.62) of a *guqin* is the key to determine its timbre. There is no uniform standard for the size of the sound chamber, which is related to the thickness and length of the top

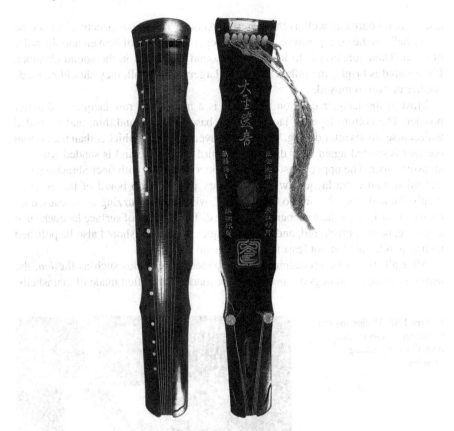

Picture 1.60 "Da Sheng Yi Yin" (musical legacy of the sage), now exhibited in the Palace Museum in Beijing

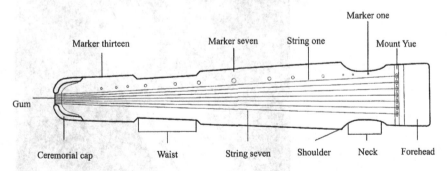

Picture 1.61 Schematic diagram of the top board of a *guqin*, cited from *Guqin* by Zhang Huaying

and bottom boards as well as the material. Another important structural par in the cavity belly is the size, position, and placement of the pillar of heaven and the pillar of earth. Their function is to improve the sound vibrations in the sound chamber. If the sound is bright, the pillars should be larger; if it is dull, they should be made smaller or even removed.

Most of the lacquer used on the *guqin* is a mixture of raw lacquer and antler powder. The bottom layer of lacquer at the base is coarse and thin, and is sanded with coarse stones after drying. The middle layer is finer and thicker than the bottom one and is sanded again after drying. The third one is fine and is sanded with wet abrasive paper. The upper finest layer of lacquer is sanded with finer abrasive paper and filled with extra lacquer where necessary until the top board of the *guqin* is evenly flat and smooth with no "*sha* sound", which is the buzzing noise caused by the rough surface on the top board. There are different kinds of surface lacquer, such as purple, brown, black, red, and yellow lacquer. Usually, it should also be polished with tung oil. Last but not least is shine removal.

After all the above procedures are completed, accessories such as the *hui*, the markers, strings, and pegs are installed. The markers are often made of clamshells,

Picture 1.62 Hollowing out the sound chamber, cited from *Guqin* by Zhang Huaying

Picture 1.63 Fitting the top and bottom boards (Photo by Wei Yong and provided by Jun Tian Fang Guqin Workshop)

sometimes gold or jade. The goose feet and pegs should be made of hard mahogany or rosewood, as well as jade or ox horns. Strings used to be made of silk, but since the 1950s, metal-nylon strings have been used instead. Steel strings are strong, durable, and loud, while silk strings are low and simple. They may be able to embody the essential charm of a *guqin*, but they also easily come loose or break. A *guqin* must be played for a long time for the tone to get finer and finer.

In recent years, the *guqin* has become more and more popular. There are also many famous *guqin* makers. Wang Peng, inheritor of Beijing Jun Tian Fang Guqin Workshop, is one of them. The skill of *guqin*-making has been listed in *List of the Third Batch of Intangible Cultural Heritage List* of Beijing (as shown in Picture 1.63).

1.5.2 Sheng-Guan Musical Instruments Made at the Beijing Hong Yin Zhai Workshop

The production skills for *sheng-guan* musical instruments originated from those made at the imperial palace of the Qing dynasty. From then to now, there have been four generations inheriting the skills at the Hong Yin Zhai Workshop owned by Wu's family, namely, Wu Qirui, Wu Wenming, Wu Zhongfu (as shown in Picture 1.64), and Wu Jingxin. At present, the skill has been passed down to the fifth generation in the traditional way, in which the master teaches his apprentice face-to-face.

Picture 1.64 Wu Zhongfu, the third generation inheritor, at work (Provided by Hong Yin Zhai)

The *sheng-guan* musical instruments made by the workshop (as shown in Picture 1.65) include the *sheng*, the *guan*, the flute, the *xiao*, the *suona,* and the *xun*. The first step of making those instruments is to treat the wood and bamboo to ensure that the finished instruments are not deformed or cracked, and are durable. Silver and tin are added to copper and bell metal in a specific ratio to ensure that the musical instruments can create a loud sound with a long tone. To make a *xun*, soil from five to six different places is selected and mixed in a set ratio to ensure a low and heavy tone.

How is a *sheng* made? There are basically two types of *sheng*: a square one and a round one. The base of a *sheng* is made of a 3-mm copper plate with a varying number of holes punched through it. To make a *sheng*, first, weld the base to the windchest, the outer part of the bucket, and the mouthpiece, to form the *shengdou*, or the bucket. Then, saw bamboo into the required size and file the both ends are flat. This is called the bamboo pipe. Break the bamboo joints with a poker. Lathe mahogany into a cone, hollow it out, treat the surface, and stick it on each pipe with an adhesive. Insert the feet of the pipes into the bucket. Drill the keys on the pipes according to their different pitch and the finger holes according to the hand shape. Make a reed blade out of the bell metal with a knife, grind it until it produces the appropriate note, and smoothen the edges to ensure a consistent, pure sound. Stick the reed on the pipe foot with beeswax, cover it with green paste, which is made by grinding red copper with five flowers stone, and tune it up after it has dried. Put an appropriate amount of cinnabar and yellow wax on the tip of the reed blade according to the required pitch, and then make more fine adjustments. Once that is done, the *sheng* is completed.

How is a *suona* made? To make a *suona*, a piece of square wood is whittled into a cone with a thin head and a thick tail, which is then shaped into a wavy arc. The next

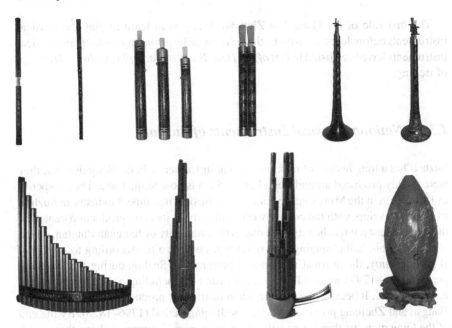

Picture 1.65 *Sheng-guan* musical instruments made by the Hong Yin Zhai Workshop (Provided by the workshop)

steps, such as hollowing, drilling, grinding, and polishing, are the same as making a pipe. The bowl, which is the resonance chamber of a *suona*, is made of a copper sheet pressed into the shape of a bowl with a crowbar on a bowl pressing machine. The whistle is the part that produces the sound. Its core component is the part supporting the connection between the whistle and the pole. The core is inserted into the tube of the pole and then the whistle is inserted into the upper end of the core, which makes it possible to blow into the *suona*. The thick end of the pole is attached to the bowl, which resonates and amplifies the musical sounds. After tuning, the *suona* is completed.

How is a *xun* made? Mixing clay is the key process of making a *xun*. Break different types of clay, gathered from different places, into pieces and mix it together. Then soak it in water for three days and filter it with cloth three times. When the slurry is purified, pour out the water. Let the remaining water in the clay evaporate before mixing it. It must be repeatedly beaten and kneaded before it can be used. Form the shape of the body of a *xun* with clay and gypsum. Press the mixed clay on the body to make the initial shape. Cut the shape in equal halves, hollow out the body, and stick the two halves back together. Then cut the blow and finger holes. After being fired at a high temperature, it will be tuned.

The last and most difficult process is to adjust the pitch, intonation, and operation of a *xun*. It should be done by a tuner under the supervision of senior technicians, and then tried and fine-tuned by performers to ensure its accuracy.

The first rule of the Hong Yin Zhai Workshop is to learn to play the musical instruments before learning to make them. Their skills in making *sheng-guan* musical instruments have been listed in *List of the Third Batch of Intangible Cultural Heritage* of Beijing.

1.5.3 National Musical Instruments of Suzhou

Suzhou has a long history of making musical instruments. In the Song dynasty, they were mainly produced around Legu Lane, which is now Shijia Lane. The prosperity of Kun Opera in the Ming dynasty caused the musical instrument industry in Suzhou to greatly develop, with the county west of the city as its main production center. In the Qing dynasty, it reached its peak due to the prosperity of Jiangnan Pingtan, a kind of Chinese folk ballad singing, and other types of folk arts. According to *Annals of Wuxian County*, the musical instruments produced in Suzhou during the Qianlong period (1736–1793) were "all kinds of instruments including gold, stone, silk and bamboo ones". It became the main musical instrument production and sales hub for Jiangsu and Zhejiang provinces during the Jiaqing period (1796–1820). By the end of the Qing dynasty, there were still 6 famous musical instrument shops there, such as Yang Wanxing, and more than 10 string instrument workshops, such as Jiang Qiu Ji. In the 1930s, there were more than 20 musical instrument workshops around Changmen Gate and Guanqian Street. Since 1951, Suzhou's musical instrument industry has undergone many reforms, with many ups and downs. In 2,000, the Suzhou National Musical Instrument Factory was established, and its production techniques were listed in *List of the Second Batch of National Intangible Cultural Heritage* in 2008.

Suzhou's national musical instruments include stringed, wind, and percussion instruments (drums and bell metal instruments). Bowed string instruments include the *erhu* (as shown in Picture 1.66), *jinghu, jing erhu, gaohu, banghu,* and *qinhu.* Plucked string instruments include the *guqin, konghou* (as shown in Picture 1.67), *pipa, ruan, zheng, sanxian, yueqin, liuyeqin,* and *qinqin.* Wind instruments include the flute, *xiao, sheng,* and *guan.* Percussion instruments include kettledrums, *yugu* drums, chimes, *suluo* gongs, *yunluo* gongs, *shimian* gongs, and cymbals.

The ways to produce these musical instruments vary greatly. Let's take the *erhu* as an example. A total of 90 steps are required, including 32 steps for making the sound box, 20 for the neck, 16 for the tuning pegs, and 12 for the bow. To make a bow, you should go through the processes of selecting bamboo, baking, and bending the top and tail of the bamboo, breaking the top and tail, straightening and grinding the bow, drilling and ironing the hole, drilling the air hole, lacquering, knotting, fixing and mounting the bow. Another 10 steps are for stretching the snake skin, including skin treatment, sticking liner cloth on the back of the skin, stretching the skin on a device and ironing it, and repeat the process for a second time, adjusting and cutting the extra skin, and finally adding color.

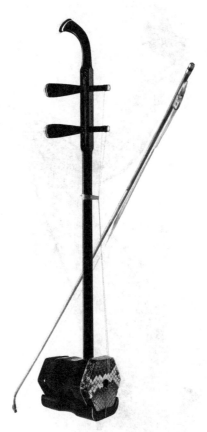

Picture 1.66 *Erhu* made of mahogany and silver silk (Provided by Shen Bowen)

1.5.4 The Flute and Xiao Made in Yuping

The Yuping flute and *xiao* are well-known traditional wind instruments in China, which are so named because of the bamboo produced in Yuping county, Guizhou province. They are famous for their clear and beautiful timbre and exquisite carving. They are of high historical, cultural, and technological value. They are the cultural pearls of ethnic groups such as the Dong, Han, Miao, and Tujia.

The Yuping flute and *xiao* are also known as "*Ping xiao* and Yu flute" because the county was originally named Pingxi. It is recorded that the *Ping xiao* was created by Zheng Weifan during the Ming dynasty's Wanli period (1573–1620), while the Yu flute was first made in the fifth year of the Yongzheng period (1727) in the Qing dynasty. The production of the flute and *xiao* goes through several processes such as selecting materials, making blanks, carving flowers, and polishing. They are created with different beautiful styles and patterns depending on the owner being male or female. The Zhengs, who love music, regard the skill of making the flute and *xiao*

Picture 1.67 *Konghou* (Provided by Shen Bowen)

as a family heirloom and persist in passing it on from generation to generation. Their instruments were once listed as a tribute to the Imperial Palace in the Ming dynasty. During the Qing dynasty's Xianfeng period (1851–1861), the Zhengs had to sell their *xiao* to make a living due to changing circumstances. And since then, the family has been engaged in making and selling *Ping xiao* (as shown in Picture 1.68). Later, they broke the rule of keeping it in the family due to a shortage of products to sell. So, they recruited apprentices to expand the scale of their production. During the Second Sino-Japanese War (1937–1945), the production of the Yuping flute and *xiao* developed greatly, with more than 30 shops in the urban area of Yuping county alone.

Since the foundation of the People's Republic of China in 1949, the production techniques of the flute and *xiao* have been well protected. From the 1980s until the early 1990s is considered the heyday of the development of these instruments, with the highest annual output at more than 500,000. However, modernization has a great impact on the national musical instruments. The protection of the Yuping flute

Picture 1.68 Replica of Zhengs' *xiao*, which won the silver prize at the London International Industrial Exhibition in 1913

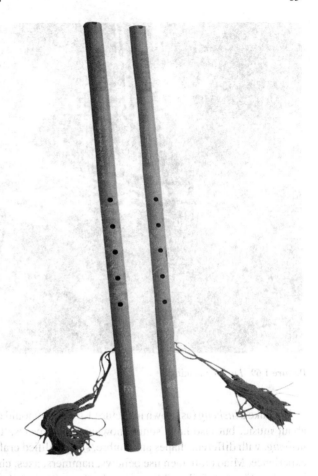

and *xiao* production techniques were ignored. Now there are less than ten senior craftsmen left. Fortunately, the skills were listed in *List of the First Batch of National Intangible Cultural Heritage* for protection and inheritance, and the situation has improved.

1.5.5 *Lusheng of the Miao People*

An important area where *lusheng* are made is Leishan county, Guizhou province in southwestern China. The craftsmen mainly live in Paika village in Danjiang town, Pingxiang village and Queniao village in Fangxiang town, and Taoliang village and Nianxie village in Taojiang town. These Miao villages are located in the mountains and are very difficult to reach.

Picture 1.69 *Lusheng* making

To make a *lusheng* (as shown in Picture 1.69), one should not only know something about music, but also have some knowledge of physics. There are many kinds of *lusheng*, with different shapes and timbre, so a qualified craftsman must have a lot of experience. Miao craftsmen use bellows, hammers, axes, chisels, saws, drills, brass, bitter bamboo, tung oil, and lime to make their exquisite *lusheng*, famed for its pure sound, and bright and beautiful appearance.

The instrument is a symbol of the Miao culture. The Miao people combine lyrics, music, and dance when playing it (as shown in Picture 1.70), adhering to the core traditions and simplicity of the Miao culture and art. The techniques for making *lusheng* have always been passed on from to apprentice. No written materials are kept. Due to its complexity, it is not easy to learn either.

The *lusheng* is made of bitter bamboo, birch bark, Chinese fir, and copper sheets in Tianxing town, Daguan county, Zhaotong city, and Yunnan province. It consists of three parts including the pipes, the bucket, and reeds, and generally has six sound pipes. Based on the techniques inherited from his forefathers, Wang Jiefeng made some reforms on how to make *lusheng* by changing the six pipes into eight or ten and adding lead to the brass to enhance its elasticity and toughness. His instrument is louder and more pleasant to the ear, and he was, therefore, nicknamed "Wang Lusheng" by the Miao villages on the border of Yunnan and Guizhou. Wang's

Picture 1.70 *Lusheng* performance

lusheng-making skill became representative of the Tianxing *lusheng*. The skill was listed in *List of the First Batch of National Intangible Cultural Heritage* in 2006.

1.5.6 Morin Khuur

The predecessors of morin khuur, also known as the horse head fiddle, were the "*quboz*", "*chor*", and "*huqin* with horse tail" and "*huqin*", all of which were made by the common people with no uniform shape or pitch standard. The morin khuur nowadays evolved gradually from the *chor* since 1950s. By the mid-1980s, there was a unified shape and pitch system, and with the efforts of Duan Tingjun, an expert in morin khuur making, the descant, tenor, bass, and contrabass morin khuur were created, making it a national musical instrument that can play multi-voice music.

The descant, alto, and tenor morin khuur share the same shape and components, which consist of the head, the neck, the fingerboard, the sound box, the strings, and the bow (as shown in Picture 1.71). The horsehead consists of a horse head, a copper tuning peg, and a wooden one; the sound box consists of a front, a back, four side plates, a sound bar, a soundpost, purfling, and a saddle; the strings are made of hairs from nylon, which are supported by and connected with the upper and lower bridges, the tail pieces and the tailgut; the bow is composed of the tip, the stick, the frog, the screw and the bow hair. For the bass and contrabass morin khuur, there is no purfling

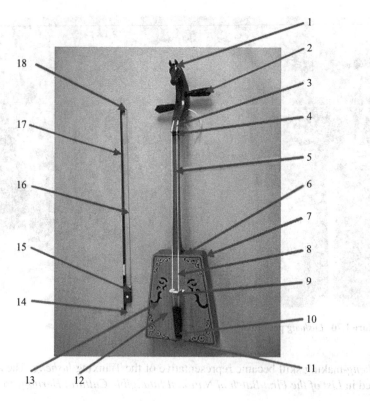

Picture 1.71 Schematic diagram of the components of a morin khuur (Photo by Guan Xiaowu).
1. horsehead, 2. tuning pegs, 3. neck, 4. bridge (upper), 5. fingerboard, 6. side, 7. back, 8. strings,
9. bridge (lower), 10. tailpiece, 11. saddle, 12. front, 13. f-hole, 14. screw, 15. frog, 16. bow hair,
17. stick, 18. tip

on the side plates, but there is fluting in the middle waist just like the back of the violin, and four metal strings according to the violin system.

To make such an instrument, apart from common woodworking tools and general electric tools, special tools such as the caliper, thread saw, soundpost fixing tool, bass sound bar fixture, shaft fixture, G-shaped fixture, sound calibrator, and tuning bracket are also needed, as well as special molds for making the horse head, the neck, the fingerboard, the f-hole and the purfling, and patterns for drawing ornamentation.

The back of the morin khuur is made of plywood, the neck, back of the sound box, and sides are made of painted maple wood, the bass sound bar and the descant soundpost are made of syzygium buxifolium wood (*yulinmu*), the front is made of tung wood and syzygium buxifolium wood, and the fingerboard is made of ebony or mahogany.

The production process of the head and the neck are as follows. First, make the head, the workblank for the neck, and the fingerboard according to the molds and the size of the morin khuur to be made. Process the bonding surface between the neck and the fingerboard and stick the latter on the former. Then process the semi-circular cambered surface on the back of the neck and drill the string hole at the upper end of the fingerboard. After the neck of the horse head is processed, a groove is opened on its back for the copper peg. Secondly, carve the horse head, whose shape depends on the craftsman's inspiration, endowment, and carving skills. Next, saw an inverted V-shaped opening at the protruding part of the lower end of the neck, where a bar is bound to be connected with the sound box and hold the tailgut for future assembly. Trim and polish the horse head and the neck, and then install the copper peg. Those two parts must be brushed with putty and primer, and then installed with the wooden peg when the primer is dry.

The production of the sound box includes the following procedures. (1) Make the front and back plates. Bind board with the same wood grain after planing the edges, and then plane both sides. (2) Make the f-holes. Mark their position first, place the mold drawing the contour shape by placing the mold on it, and then cut the holes. (3) Bind the bass sound bar. Plane the contact surface between the bar and the front plate and shape the bar into a structure with a slightly arc shape protruding in the middle. Mark its position on the inner surface of the front plate, bind, and clamp it with a fixture. Then trim the outer end of the bar into an arc shape, which is slightly higher near the upper end and slightly lower on both sides. (4) Bind four painted maple wood boards to form the sides of the morin khuur and tie them tightly with ropes. Bind angle blocks at the four corners of the plates after the glue on the sides dry thoroughly and then trim the boards to make them circular arc shaped. (5) Assemble the sound box. Draw the outer profile lines of the four sides on the back plate, bind them with the back and clamp them with an axial fixture. Bind the four sides with the front plate in the same way. Then trim the four sides of the front and back plates to make them flush with the frame of the side plates. Next, drill two coaxial holes in the middle parts of the upper and lower side plates for installing the neck. Install the saddle at the lower end of the front. (6) Draw decorative patterns on the painted maple wood boards based on the design and paste them on the four corners of the side plates. (7) Install the soundpost between the front and back plates inside the sound box with a special device, and then polish and trim the sound box and f-holes. For professional and high-grade morin khuur, it's also necessary to tune the back plate. Place the sound box on a special bracket, trim the back plate with the soundpost as the center until satisfactory sound effects are obtained. Then finely polish the surface of the back. (8) Paint the sound box by brushing the putty first, and then the primer. After they have been dried thoroughly, draw the contour lines of decorative patterns at the four corners of the front, and draw black and white patterns with black pigment and white latex. Then paint the surface of the sides and the back, and polish them with sandpaper dipped in water to make them smooth after the paint has dried.

Once the parts and accessories are ready, they can be assembled, then the strings and copper peg are installed. The alto morin khuur has two nylon strings, the bass one with 160 hairs and the treble one with 120 hairs. The higher the tone, the thinner the string. The bow stick used to be made of Brazil wood, but now it is made of wood imported from Myanmar. The bow hair is made of the horse tail.

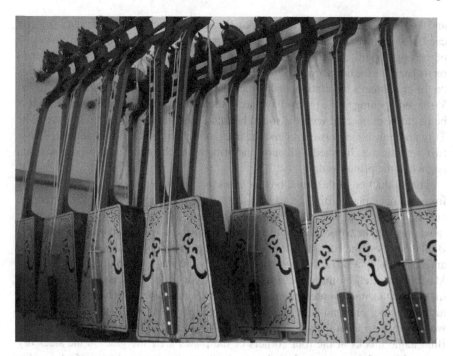

Picture 1.72 Finished morin khuur

The shape, decoration, production, and timbre of the morin khuur (as shown in Picture 1.72) have distinct and rich national characteristics, which is deeply loved by Mongolians. Its music and production skills have been listed in *National Intangible Cultural Heritage List.*

1.6 Household and Folk Utensils

1.6.1 Lock

The spring lock is the most common example of the use of a spring to lock something in ancient China. The earliest known lock was excavated from tomb No.5 of Caojiagang, Dangyang county, Hubei province, which was made before the late Spring and Autumn period. It has a long concave bolt with two 8-shaped sides, decorated with *tao* patterns (similar to zigzag patterns) and triangular thunder patterns. The bolt can twitch but cannot come out of the lock.

The structure of the primitive lock was quite simple, with only one bolt attached to the door and sliding on a wooden block. To prevent the lock from coming out, a pellet and two locking rings were added, and then locking rings were fitted onto the wall as well. Later, a hole was opened on the door for the hand to reach in for convenience. Then more improvement was made by narrowing the hole only for a small device (such as a key) to secure safety.

However, the first major breakthrough in the design and making of ancient locks was the invention of the lock peg. It means that a movable object made of wood or metal fell into the lock hole due to its own weight to make the lock closed tightly. When the key is inserted into the hole, the lock peg will be jacked up by the protrusion on the key and the lock will be opened. Such wooden locks and keys are found at the site of Bezeklik Thousand Buddha Caves in Xinjiang Uygur autonomous region.

The second big breakthrough was the application of springs. The lock peg is locked and opened with the help of the opening and closing of the springs. The locking spring is made up of several springs, one end of which is fixed on the metal bar called the lock stem, and the other end is scattered in the shape of an umbrella around the bar. When the opening end of the springs is squeezed into the lock peg, it is compressed and closed tightly, and then opens like an umbrella after entering it completely; thus, the lock is automatically locked. When the key is inserted into the keyhole, the springs are tightened and pushed to the other end, and the lock will open.

The spring lock (as shown in Picture 1.73) is popular among the people because of its simple structure and practicality and can still be seen in rural China today.

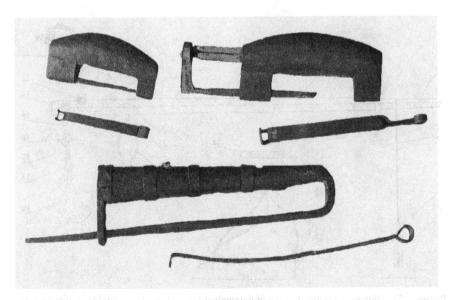

Picture 1.73 Padlocks, cited from *China at Work: An Illustrated Record of the Primitive Industries of China's Masses, Whose Life is Toil, and Thus an Account of Chinese Civilization* written by Rudolf P. Hommel and translated by Dai Wusan, et al.

1.6.2 Steelyard

Steelyard (as shown in Picture 1.74) is a commonly used weighing tool designed and made according to the law of the lever. It was invented as early as in the Eastern Han dynasty. Liu Chenggui, an official in charge of the Imperial Treasury made a small steelyard, which is commonly known as *dengzi* or *dengcheng*, during the Jingde period (1004–1007) in the Northern Song dynasty. He also clearly defined the weight of the weight and the plate, the beam length, the maximum weighing capacity, and the division value of a steelyard. The government of the Republic of China promulgated the *Weights and Measures Law of the Republic of China* in 1929 and revised the implementation rules for the manufacture and verification of steelyard in 1931. In recent years, with the popularity of electronic scales, the use of steelyard has been greatly reduced, but it is still in use in many places and the making techniques are preserved. Zhang Baichun, director of the Institute of Natural Science History, Chinese Academy of Sciences, and Jurgen Renn and Matthias Schemmel, researchers in Max Planck Institute for the History of Science, Germany, investigated the making techniques of the steelyard in Tongzhou district, Beijing, and Changsha, Hunan province in 1998.

Taking Tongzhou as an example, several rooms in the weighing instrument maintenance office in this area were originally used as workshops for making steelyard.

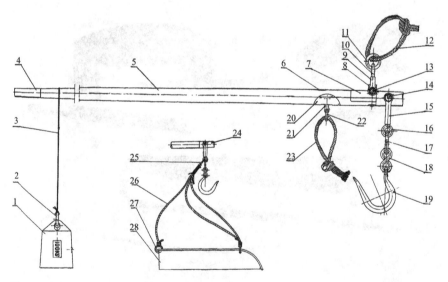

Picture 1.74 General assembly drawing of the steelyard. 1. weight, 2. weight ring, 3. weight rope, 4 (big) cap (big head band, protective cap), 5. weigh rod, 6. label, 7. cover piece, 8. knife pile, 9. fulcrum tool rest, 10. activating ring, 11. connecting ring, 12. rope button, 13. fulcrum knife, 14. key knife, 15. knife bearing, 16. activating ring, 17. connecting ring, 18. connecting ring, 19. weigh hook, 20. cover piece, 21. hanging hair (worn on hidden knife), 22. connecting ring, 23. rope button, 24 (small) cap (small head band, protective cap), 25. coil rope ring, 26. coil rope, 27. coil tie ring, 28. plate

Later only two workers continued the work due to the sharp drop in the output of the steelyard because it was replaced by electronic scales. The method of making a steelyard weighing no more than 15 kg of load is described below.

1.6.2.1 Making the Balance Beam

The balance beam is made of hardwood with good performance, such as rosewood and mahogany, or jujube wood, which is preferred in Beijing and other places north of it. In Xincheng county, Hebei province, the jujube wood should be boiled in alkaline water for one hour, and then air-dried for two months before use. The wood should be straight, without knots or cracks. When processing it, first saw the piece of wood into a rectangular section and plane it round to get the required straightness and shape. Then sand it down to make it smooth.

1.6.2.2 Installing the Beam Sleeves

Fold two pieces of galvanized sheet iron into two cylindrical sleeves, and cut the two ends of the beam into a cone to fit into the sleeves. Divide the subulate thick end into 10 equal parts and burn a square hole (A) along the radial direction with a red-hot iron rod at the seven-tenth part away from the outer end for installation of key knife. Drill two square holes opposite the square hole with a saw and a punch on the sleeve. Trim the thin end of the beam and install the sleeve likewise.

1.6.2.3 Installing the Fulcrum Knives

This process, also known as "divide or adjust the balance", means to determine the positions of the fulcrum knives, and install the buttons. The so-called "knife" is in fact a pin made of carbon steel. Install the barycenter knife in the square hole at the thick end of the wooden pole to determine the "rear fulcrum". The space between the two feet of a pair of dividers is regarded as one "step". Try to adjust the spacing until it is exactly 22 steps from the barycenter knife (A) to the middle of the sleeve at the thin end of the beam. Draw a mark on the point of the first step, which is the position of the rear fulcrum (B), that is, the fulcrum for 15 kg (as shown in Picture 1.75).

Fix the knife rest to the barycenter knife and hang the looped string of the plate on the loose-jointed ring of the U-shaped knife bearing. According to the specification, the length of plate string is two-thirds of that of the beam, and the length of the weight string is one-half.

The "front fulcrum", also known as "the second *hao*", is also determined by "dividing the balance". Hang a weight of 750 g, add weights of 3 kg, and lift the beam with a kitchen knife with the blade as the fulcrum. When the beam is in a horizontal position (the tail is slightly higher than the head), the position of the blade will be the "front fulcrum" (C) of the steelyard, that is, "the second *hao*".

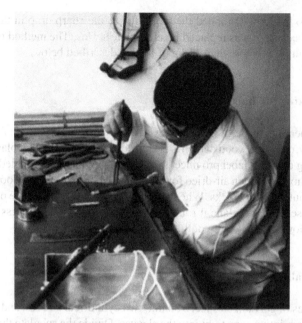

Picture 1.75 Dividing the steps and carving the graduation marks (Photo by Zhang Baichun)

1.6.2.4 Determining the Graduations

Determining the graduations is also known as "carving the graduation marks", or "determining the division value of the graduation marks", including zero correction. The wooden beam isn't as uniform in material, size, and weight as a metal one, so graduations should be determined for each one. The method is as follows.

Hang the knife button of the rear fulcrum (B), put the weights of 15 kg on the hook and the plate, and hang a weight of 750 g on the beam.

Move the weight until the beam is in the horizontal position, and the string of the weight stands exactly at the graduation mark of 15 kg. Then find the marks of 3 and 5 kg in the same way. Hang the knife button of the front fulcrum (C) on the hook or plate with no weight, the position of the weight string will be the zero point of the graduation. Hang up the weight of 3 and 1 kg respectively, determine their positions and mark them.

Draw a longitudinal line in the center of the beam, and then draw a line on the other side, and the graduation marks will be on those two lines.

Carve the graduation marks along the straight lines divided by 22 steps.

It is necessary to determine the space of the steps and draw the marks through repeated measurement.

1.6.2.5   Making Graduation Marks

Drill the star holes between 1 and 3 kg based on the marks and patterns of the stars, and then drill another hole in the middle of the adjacent holes. Thus, the minimum graduation will be 0.5 *liang*, or 25 g. Drill the star holes between 3 and 15 kg, and the minimum graduation will be 2 *liang*, or 100 g.

Inlay short copper wires in the star holes one by one and polish them with oil stone to smoothen the beam. It is also necessary to fill the gap between the copper star and the star hole with sawdust to make it firm. Wax the beam and rub it repeatedly to make it bright and beautiful. If it is white, it must be painted with pigment before waxing, and then painted with varnish after drying to prevent moisture and fading.

1.6.2.6   Assembling

Install the plate and hang the weight. After being reexamined, the work will be handed over and verified by another craftsman.

1.6.2.7   Verification

In the 1980s, "three fixes" were implemented, that is, fixed beam length and weight, fixed weight of the sliding weight, and fixed weight of the plate. The graduation marks and the zero point are determined by the sliding weight. This kind of regulation made it easy to replace the new parts.

The steelyard and *dengzi* should be verified regularly. There are ten kinds of steelyard made in Beijing, which are those weighing 3, 5, 10, 15, 30, 50, 80, 100, 150, and 200 kg, respectively. Some of the steelyards have a square plate, a round one, and a hook; however, those weighing above 30 kg have only the hook but no plate.

Changsha Hengyuan Metrology Equipment Shop makes the steelyard weighing 10 kg in the same way as the craftsmen in Tongzhou, Beijing. Wen Zhifei, a craftsman in Changsha city, has been engaged in steelyard making for 40 years. He has never made mistakes in determining the fulcrum position and graduations in the traditional method. According to the old rule, apprentices need to learn to plane the beam, fit in the sleeve, and then learn to files the knives, and then they can start work two years later. Chinese handicraft industry has always attached great importance to professional ethics, and the steelyard makers also have their own moral rules. It is customary in the industry to say that if the steelyard is short of one *liang*, the maker will live one year shorter. Therefore, the rules passed down from generation to generation cannot be violated (as shown in Picture 1.76).

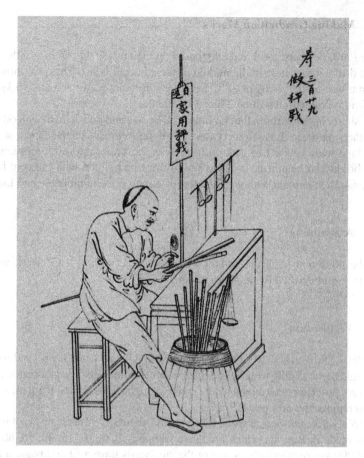

Picture 1.76 Steelyard making, cited from *Photos of Old Peking*

1.6.3 Umbrella

Legend has it that the umbrella was used as early as the Yellow Emperor (a legendary ruler in China) period. According to the records of *Selections from Lu's Commentaries of History* and *Discourses of the States*, bamboo umbrellas with handles were in use in the Spring and Autumn period, and now they are still necessary tools in our daily life.

The earliest umbrella found in China was unearthed from Wangjiang tomb No. 1 of in Jiangling county, Hubei province, which was made during the Warring States period. It was used as part of the imperial ceremony with a diameter of 3 m. The components of the umbrella for the imperial chariot were quite complete in the Han dynasty. During the Wei and Jin dynasties, a large number of hand-held umbrellas emerged, with various forms. The umbrella in the Sui and Tang dynasties had a simple cover and a shortened handle, which was more suitable for folding and unfolding.

The structure of a paper umbrella was finalized during the Northern Song dynasty. The painting *Along the River During the Qingming Festival* by Zhang Zeduan of the Northern Song depicts umbrellas being folded or unfolded in many scenes. The standard components and linkage mechanisms formulated at that time have been widely used in umbrella-making industry. Colored umbrellas with finely painted flowers appeared in the Qing dynasty.

1.6.3.1 West Lake Silk Umbrella

According to *Industry of Hangzhou Annals*, West Lake silk umbrella (as shown in Picture 1.77) was created by Du Jinsheng in 1932. From 1928 to 1929, he went to Japan several times and got the idea of using the scenery image of the West Lake as the umbrella cover inspired by the popular silk umbrellas there. In 1932, Du ordered bamboo from Jilong Mountain in Fuyang, Hangzhou, and asked skilled umbrella makers in Fuyang and Wenzhou to make umbrella ribs. Zhu Zhenfei, one of the workers, made the first Hangzhou silk umbrella with the cover decorated with patterns of the West Lake landscape. Such kind of umbrella received warm welcome and became popular for quite a while. The West Lake silk umbrella rose again from stagnation with the support of the government after 1949. In 1960, the West Lake Silk Umbrella Research Office was set in Hangzhou Institute of Arts and Crafts, and Zhu Zhenfei was appointed as the director. Together with his wife You Jingzhi, they trained 21 craftsmen, who became the technical backbone of the silk umbrella production.

The construction process of the West Lake silk umbrella involves three steps including selecting bamboo, making umbrella ribs, and installing the cover, which needs dozens of detailed processes, all of which are completed by hand.

Selecting bamboo. Zhejiang province is rich in bamboo, but only the henon bamboo is suitable for making umbrella ribs. The tubes of this kind of bamboo are slender, the nodes are flat, the grain is thin and delicate, and the color is pure and luster. All these qualities make it the best material for making the ribs of umbrellas. Every year before the White Dew (about September 7–9th), one of the 24 traditional Chinese solar terms, experienced umbrella makers were sent to places around Hangzhou such as Fuyang, Yuhang, Fenghua, Anji, Deqing, which are rich in henon bamboo. They would select plants more than three years old with uniform color, suitable thickness, and proper internodes and without any notch. Only the second to fourth internodes in the middle of the bamboo were taken, and those that were too tender, too old, too big, or too small could not be used.

Making umbrella ribs. There are more than ten steps to process the henon bamboo into umbrella ribs, such as wiping bamboo, splitting long ribs, weaving, shaping, and splitting the green bamboo veneer strips. There are 35 ribs, each with a width of 4 mm, for an umbrella. When splitting the bamboo veneer strips, the thickness of the green and yellow ones must be separated, uniform and symmetrical.

Installing the umbrella cover. The cover is made of special Hangzhou silk, which is as thin as cicada wings, finely woven, breathable, sun-resistant, and easy to fold.

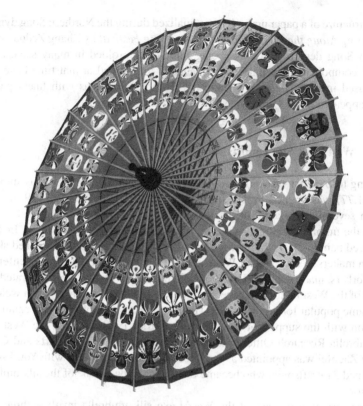

Picture 1.77 West Lake silk umbrella

It has more than 20 colors such as light green, dark green, fruit green, and orange, and dozens of patterns such as "ten scenic spots of the West Lake", "ancient ladies", "dragons and phoenixes", and "magpies on plum branches". The whole procedure includes more than ten processes, such as stitching corners, stretching the surface, fixing it on the frame, cutting the edges, threading colored thread, painting patterns, folding the umbrella, and sticking the green bamboo strips. Take stitching colored thread as an example. The maker cross-weaves reticulation between the ribs by stitching among the fine gaps as many as 296 times. It also requires skilled crafts-manship to stick the bamboo skin strips onto the umbrella surface. The finished umbrella weighs only half a *jin*, or 250 g. When folded, the strips retain the form of the round bamboo with obvious nodes, and the cover is completely covered by the strips; when unfolded, the colorful cover is magnificent to see. The requirement for a classy silk umbrella is "three neats and one round", namely, neat head, neat node, neat edge, and round body when folded. The umbrella is equipped with a wood carving head in the shape of "Three Pools Mirroring the Moon", one of the famous scenic spots of the West Lake. The handle is carved from hardwood, and the heads and handles of high-grade umbrellas are inlaid with cattle horns.

In 2008, this skill was listed in *List of the National Intangible Cultural Heritage*.

1.6.3.2   Luzhou Oil-Paper Umbrella

The traditional-making techniques of oil-paper umbrellas were originally popular in Yunnan, Guizhou, Sichuan, Guangdong provinces, and Chongqing city, but there exists only one such factory called Luzhou Fenshui Oil-paper Umbrella Factory in Sichuan, which is good at making lithographic tung oil paper umbrellas by hand. According to *Annals of Luxian County*, Fenshui oil-paper umbrella originated in the late Ming and early Qing dynasties with a history of more than 400 years. According to the genealogical book of family Xu in Fenshui county, they have made such umbrellas for more than 100 years, and the techniques have been inherited by eight generations including Xu Shaokai, Xu Tongsheng, Xu Futing, Xu Zifu, Xu Changqi, Xu Xueming, Bi Liufu, Dong Qinghua, and Hu Tianzhen. During the Republic of China, Xu Tongsheng founded the brand of "Xu Tongsheng Old Umbrella Shop", and then registered the trademark of "Meimei Brand". At first, they only made red oil-paper umbrellas, but now they make a series of umbrellas of more than 20 types such as the red ones, various hand-painted and lithographic ones, with diameters ranging from 8 *cun* (27 cm) to 12 m.

The oil-paper umbrella consists of a holder, ribs, and a handle. It is made of 36 ribs, which are made of tortoise-shell bamboo, and fibrous leather paper, tung wood, water bamboo, tung oil, and other materials. The tortoise-shell bamboo is selected from the bamboo grown in the forest belt exposed to the sun in the petrographic regions in southern Sichuan for more than 3 years; the paper is the bamboo fiber paper produced with the local method, and tung oil is also locally cooked. The production of oil-paper umbrella nowadays still follows the tradition way, which must go through more than 90 processes, such as sawing the holder, threading (as shown in Picture 1.78), trimming the edge, pasting paper, painting, applying tung oil (as shown in Picture 1.79), hooping and baking, etc. Such umbrellas are of high quality. The holder can endure more than 3,000 times of folding and unfolding, the cover doesn't take off from the ribs even if is soaked in water for 24 h, and the head will not break again force five wind. There are also Chinese poems and paintings on the umbrella cover to express people's emotions. The umbrellas can be used for shading as well as for special occasions such as weddings, funerals, and ornamental decorations. Those Chinese umbrellas are sold to France, Britain, Germany, Singapore, the United States, Japan, and other countries, as well as Hong Kong, Macao, and Taiwan and are very popular among users and tourists, especially folk collectors.

1.6.3.3   Jialu Paper Umbrellas in Wuyuan, Jiangxi

Jialu village is located in Wuyuan county, Jiangxi province. Hu Zhouxin, an old umbrella artisan in Jialu, is called "Jia San Zhen Ren", literally meaning "the real artisan of Lujia umbrellas". He became the apprentice of a certain Master Wu at the age of 14 and worked in an umbrella shop in the Laojie Street after finishing his apprenticeship. During the ten years between 1966 and 1976 of the "Cultural

Picture 1.78 Threading five-color silk thread, which is the most unique "full wearing" skill. The maker stitches more than 2,000 times with five-color silk thread, which is regarded as the nonsuch skill in umbrella making

Picture 1.79 Applying tung oil, which means the tung oil, boiled by traditional technology, is evenly applied on the umbrella cover until it is smooth and translucent

Revolution", the umbrella makers of Jialu were concentrated in Fuchun town, 15 km away, and Hu's shop was forced to close down for a period of time. The umbrella making in Jialu village was restored after the reform and opening up of China in 1978. In 1990, several artisans headed by Hu Zhouxin established Jialu Arts & Crafts Umbrella Co., Ltd. Its umbrella-making processes have been finely separated, with a number of workers specialized in making the shafts, the ribs, or the paper in the surrounding villages such as Meicun and Duiwu. There are sales outlets in Wuyuan county, Nanchang city, and Hangzhou city, thus forming a complete set production and sales system. At present, the company has an annual output of 600,000 umbrellas, which fall into 10 series and 60 models and are sold to Japan, the United States, Australia, Singapore, and Europe.

A Jialu paper umbrella consists of a shaft, ribs, stretchers, a notch, a runner, a spring clip, a cover, and a handle (as shown in Picture 1.80).

The shaft (as shown in Picture 1.81) is the pillar of the whole umbrella, and all the other parts are constructed based it, so the fine skill and patience of the craftsmen are needed. It is generally made of straight fine bamboo with thick culm which is many years old. After the bamboo is cut off, the marks of the nodes are removed and ground flat, and then it is insolated. After that, the workers will drill the spring clip hole and the pin hole, and then brush clear paint on the shaft.

The structure of umbrella ribs and stretchers is quite complex, and the combination and connection between them (as shown in Picture 1.82) are the core part of the whole retracting structure. The size of the cover is determined by the length of the ribs. All the ribs of an umbrella are taken from the same bamboo tube, so that they can be closely matched, and are suitable for integrated and standardized processing. One

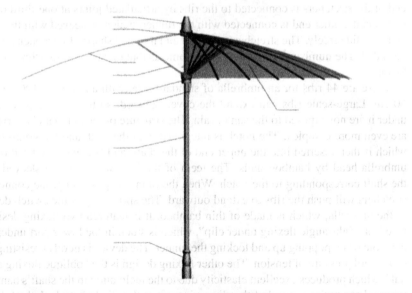

Picture 1.80 The structure of a paper umbrella, cited from *Moving and Retracting: Field Investigation and Research on the Structural Design of Chinese Paper Umbrella* by Li Lixin

Picture 1.81 Cutting off the shaft, cited from *Moving and Retracting: Field Investigation and Research on the Structural Design of Chinese Paper Umbrella* by Li Lixin

end of the stretchers is connected to the ribs by articulated joints at one-third of the ribs, and the other end is connected with the runner, which is sleeved with the shaft and can slide freely. The stretchers are short and thin and shaped like a bucket when unfolded. The number of stretchers is the same as that of ribs and fits with ribs one by one.

There are 44 ribs for an umbrella of standard size, with a radius of 1.8 *chi*, i.e., 60 cm. Large-scale ribs can extend the cover outward, so that people and things under it are not exposed to the sun or rain. The structure of the notch and the runner are even more complex. The notch is integrated with the teeth and the gourd head, which is then inserted into the upper end of the shaft, and becomes the head of the umbrella head by bamboo nails. The teeth of the hollow runner are sleeved into the shaft corresponding to the notch. When the runner is pushed up, the connected stretchers will push the ribs to extend outward. The spring clip is the switch device of the umbrella, which is made of thin bamboo. It is featured two flexing designs. One is a "right-angle flexing under clip", which is stuck in the lower part under the clip hole when popping up and locking the runner. The device is good at resisting the downward pressure of tension. The other flexing design is the "oblique flexing at its tail", which produces excellent elasticity due to the inclination in the shaft's bamboo tube and can make one end of the clip spring out of the clip hole to lock the folding and unfolding of the umbrella. In addition, bamboo pins are fixed through the shaft

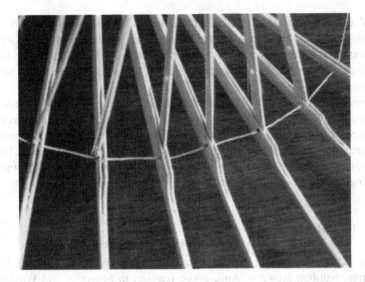

Picture 1.82 Connections between the ribs and the stretchers, cited from *Moving and Retracting: Field Investigation and Research on the Structural Design of Chinese Paper Umbrella* by Li Lixin

and outcrop 5 cm above the spring clip, to prevent the umbrella from cracking in reverse due to exceeding the critical point when pushing up the runner.

The cover is the main function of shading an umbrella. The paper cover consists of three layers of good tissue paper, which is easier to process than cloth or silk, and folds more freely. The downside is that it takes more time. However, this problem can be solved by pasting multiple layers and lacquering. Persimmon lacquer, tung oil, and clear paint make the paper cover not easy to damage, as well as waterproof, sunproof, and moth-proof.

The bottom of the handle is round, and the upper part is bell-shaped, which makes the holder feel comfortable and ensures the umbrella meets the requirements of ergonomics.

1.6.4 Fans

Fans date back to a very early time. It is recorded in *Notes on the Antiquity and Present Days* by Cui Bao of the Jin dynasty, "Emperor Shun made *wuming* fan to encourage the free airing of views and seek sages to assist himself. This was the beginning of the fan". The Shang dynasty witnessed the appearance of *sha*, a kind of feather fan used by the emperor's honor guard or as a shield again sun and dust, and later in the Western Zhou dynasty there were feather fans. During the Warring States period as well the Qin and Han dynasties, a semi-circular hand-woven paddle fan made of thin bamboo skin was invented, which had a short handle and was called "*bianmian*" or "*hushan*" (door fan) as it looked like a single door. The *hehuanshan* in the Western Han dynasty was also known as gong fans (*gongshan*), silk fans (*wanshan*), or round fans (*tuanshan*). It was symmetrical with the handle as the central axis and looked like a full moon. This type of fans has been used for quite a long time and has become a traditional type in China.

However, it was inconvenient to carry the round fans since they couldn't be opened and closed, so folding fans came into being. The production of folding fans in the Southern Song dynasty reached a considerable scale, and it was more popular in the Ming dynasty. Fans made according to local conditions in various places formed their own unique styles. The famous ones are Sichuan fans, Su fans, Yuezhou fans, and Jinling fans, with Hangzhou fans being the most famous.

1.6.4.1   Hangzhou Fan

Mulberry paper produced in counties around Hangzhou such as Fuyang, Lin'an and Yuqian, bamboo grown in Anji, goose feathers in Shaoxing and Wuxing are all indispensable raw materials for making the Hangzhou fan. In the long process of development, the Hangzhou fan experienced two heyday periods. The first was in the Southern Song dynasty. After the Jingkang Incident in 1127, skilled fan makers from the north gathered in Hangzhou and integrated their production technology with the local technology, promoting the increase of the varieties and styles of fans as well as the good and exquisite materials. Thereafter, the fan making became an important industry in Hangzhou, the new capital. The two-mile-long Fan Lane between Qingtai Street and Hefang Street was the hub of fan shops, and the name is still used today. The other heyday period was after the middle of the Kangxi period (1662–1722) in the Qing dynasty. According to *Records of Industries in China*, "There are about 50 fan workshops in Hangzhou city, with four or five thousand workers". A memorial hall, also the guild hall of the fan industry is located in Xingzhong Lane in the city. It was rebuilt in the 14th year of the Guangxu period (1888), and there are 462 artisans enshrined and worshiped. Hangzhou fans are nicknamed "elegant fans" because they are perfect in workmanship and elegant in shape and are known as the "Three Musts of Hangzhou Specialty" together with Hangzhou silk and Longjing tea. They not only sell well at home but are also popular abroad. The following is a brief description of the fan-making skills of Wang Xing Ji in Hangzhou.

Wang Xing Ji Fan Workshop was founded in the first year of the Guangxu period (1875) by Wang Xingzhai, whose grandfather and father were both experts in fan making. Wang learned the skill with his father since childhood and worked at Qian Ji Fan Workshop near the Sansheng bridge in Hangzhou after finishing his apprenticeship. Nearby was a golden paint appliqueing fan workshop set up by Chen Yizhai, a famous fan maker, whose eldest daughter Chen Ying was also a master in fan making. Later, after Wang Xingzhai married Chen Ying, their golden paint appliqueing fans were quite famous for a while. In the 27th year of the Guangxu period (1901), Wang set up another fan workshop in Yangmeizhu Xiejie Street, Beijing. In the first year of the Xuantong period (1909), his son Wang Ziqing continued his career after his death. In 1929, Wang Xing Ji Fan Workshop, with "three stars" as its trademark, was founded in Taipingfang Lane in Hangzhou and the business started booming day by day. After the outbreak of the Second Sino-Japanese War in 1937, the fan industry in Hangzhou experienced a sharp decline, with an output of 14,000 in 1944. During the early days after the founding of PRC, the production stagnated, and it was not until March 1958 that Wang Xing Ji Fan Factory was restored, with an annual output of 480,000. At present, it produces 13 main fan categories, which include over 300 kinds of different designs. The ribs of the fans are made of moso bamboo in Anji and Lin'an and lady palm in Guangxi and Guizhou; the panels are made of mulberry paper produced in Rui'an and Fuyang, which is tough and difficult to break. The panels of the fans are painted with alpine persimmon paint produced in Zhuji city, Zhejiang province. This kind of paint, which is extremely exquisite in production, must be stirred hard until it droops continuously like thread when lifted, and the color is dark and transparent before use. From material selection to finished products, there are about 80 steps to follow, such as rib making, pasting the surface, folding the panels, coloring, shaping, and finishing. Qualified black paper fans had to be durable; they should not be able to get drenched by the rain nor warped by the sun. When they are inspected after finishing, the fans should be soaked in water for four hours and still be as strong and new as if they had never been used; they should also be exposed to the sun for four hours and still be as flat as before, neither warped nor cracked. The fan cannot only cool people down, but also shelter them from the rain and sun, so there is an old saying "a fan equals to half an umbrella".

The sandalwood fan was created by Wang Xing Ji Fan Workshop in 1920. It is made of sandalwood produced in India and Southeast Asia for more than 100 years. Only the tree core is used, so the fan will still retain the fragrance even after 30–50 years. If put in the wardrobe, it can also prevent damage from insects and moths. Most of the decorations of fans are painted, carved, hollowed out and pyrographed, and figures, landscapes, flowers, birds, fish, and insects are generally engraved on two large ribs. Hollowing out means that the craftsman pulls a bamboo bow with a toothed steel wire to saw fine and delicate hollow patterns on the fan. In 1982, a senior artisan of Wang Xing Ji made a sandalwood fan named "Romance of the West Chamber", on which were nearly 20,000 holes sawed, showing exquisite, wonderful patterns of the story. Pyrograph is to iron various patterns on the fan. Special effects such as water waves and clouds can also be created by taking the advantage of the natural wood grain of fan.

The decoration on the fans made by Wang Xing Ji Fan Workshop is rich and colorful, and the artisans skillfully integrate the beauty of literature, Chinese characters, and fans themselves, making Hangzhou fans the best in China. The production technique has been included in *List of the Second Batch of National Intangible Cultural Heritage* in 2008.

1.6.4.2 Suzhou Fan

Ming and Qing dynasties were the heyday of Suzhou fan industry. According to *Annals of Gusu* by Wang Ao of the Ming dynasty, the ribs of folding fans were made in Lumu town of Suzhou, with some famous craftsmen such as Ma Xun, Ma Fu, Liu Yutai, and Jiang Sutai. In the Qing dynasty, Suzhou fans already formed their unique artistic style and became a tribute to the Imperial Palace. After the Taiping Rebellion (1851–1864), fan-making workshops shifted from Lumu town to the Suzhou city, with Shantang Street near Changmen and Hanyazhuang Street of Taohuawu district as two hubs. Folding fan industry associations were also set up and their fans were sold all over China. In the early years of the Republic of China, the workshops expanded from Taohuawu to the vicinity of West Street, and many fan workshops such as Huang Rong Ji, Sun Henghe and Xu Jingshan from Yangzhou city as well as Zhang Duo Ji from Changzhou city also moved there. After 1949, the Suzhou fan industry revived, and Suzhou Sandalwood Fan Factory and Suzhou Fan Factory were established. The products made there include folding fans, sandalwood fans, and silk gong fans, among which the annual output of sandalwood fans is as high as 300,000. There are more than 300 kinds of designs of fans, which sell well both at home and abroad.

According to *Compilation of Unofficial History of the Wanli Period* by Shen Defu of the Ming dynasty, "Those Suzhou folding fans that are made of rosewood, ivory or ebony are considered common, and only those made of lady palm or moso bamboo are deemed as elegant objects for people with refined tastes". Suzhou folding fans are exquisite in workmanship and beautiful in shape. They are firm and durable with pleasing bamboo color, straight edges, even panels and compact structure, which are easy to use.

The sandalwood fan (as shown in Picture 1.83) has a graceful shape, with elegant panels and a natural fragrance. It is famous for its decoration, such as hollowing out, pyrograph, painting, and engraving.

The history of the silk gong fan can be traced back to the Song dynasty, when it was called *gongshan*. As the panels were made of silk, satin, or brocade, it was also known as *wanshan, luoshan,* and *juanshan*. In addition to cooling, it was also used to cover faces, so it was also named "face covering" (*bianmian*).

The folding fan's production follows several steps. The ribs should be made of the moso bamboo with thick culm and good color grown in Anhui and Zhejiang provinces. Bamboo with clean color and without any spots or black lines is selected, and then boiled, split, scraped, pulled, deburred, baked, polished, carved, and lacquered. There are 58 processes to go through before a fan is completed. The number and size of fan ribs generally range from 9 to 19, with landscapes, flowers and birds, and Chinese characters engraved on them. What's more, there are various high-grade lacquered fans. The panels are made of super-grade *xuan* paper, which are mounted on the ribs with alum, and the surface is coated with golden paint made of gold foil. The skill was listed in *List of the First Batch of National Intangible Cultural Heritage* in 2006.

Picture 1.83 Sandalwood fan made by Shen Jie (Provided by Shan Cunde)

1.6.4.3 Gong Fan

Gong fan is a well-known type of fan in Sichuan, which was named after its inventor Gong Juewu, a fan-making artisan in Ziliujing (nowadays Zigong city) of Sichuan. During the Guangxu period (1871–1908), Gong made fans with thin woven bamboo strips. Those fans with patterns of auspicious meaning such as bats ("*fu*" for luck), deer ("*lu*" for fortune), peaches (for longevity), magpies ("*xi*" for happiness) were popular among customers. In 1908, the bamboo strip round fan was honored the "High-grade Handicraft" of Sichuan together with Chengdu lacquerware and Liang-ping bamboo curtain. Gong Yuzhang, the son of Gong Juewu, managed to weave famous paintings on the fan surface, and their fans reputation skyrocketed since then.

The raw material of the fan is the one-year-old local yellow bamboo grown on Yinshan Mountain. The bamboo strips processed with special tools are crystalline and a bit glossy and are as thin as a cicada' wings. Then, they are carefully threaded, hung, inlaid, and broken according to the original painting and calligraphic work. The final product not only looks as smooth as silk but also reproduces the charm of the original work. People would mistakenly assume it is made of cotton threads as it sounds like a drum when the handle is beaten. The handle is made of white cattle horn and decorated with a silk tassel, which makes it more noble in texture. This kind of fan made of natural materials with simple style and superb skills are the representative of Chinese folk crafts, which can be described as wonderful art.

Gong fan is peach-shaped, with a diameter of about 26 cm. It is all about manual work from material preparation, strips making to weaving. The diameter of bamboo strips is only 0.01–0.02 mm, and a craftsman must weave 700–2,000 strips to create a vivid calligraphic work and painting on the fan. In addition to weaving the fan surface, there are four other processes including weaving the edge, mounting the wooden frame, lockstitching the edge, and gluing, which are completed by the maker alone.

Gong Yuzhang, the second-generation descendant of Gong fan, invented a complete set of special tools such as scrapers, sharp knives, and teasing needles. He showed the sparrows at different ages by making the bamboo strips fluffy; sometimes he poked dots on the thin strips to form human eyebrows; he also used green and white bamboo strips to deal with the lighting and shade of different patterns, calligraphic works, and paintings. Natural scenery, figures, birds, animals, flowers, and calligraphic works all look vivid and life-like on the fan, which shows the exquisite skills and ingenious craftsmanship. "The Story of Diao Chan Worshipping the Moon" made by Gong Changrong, the third-generation descendant, created a scene where the light smoke of the incense could faintly be discernible by taking advantage of the front light or back light. The unique Gong fan-making skill was included in *List of National Intangible Cultural Heritage* in 2011.

1.6.5   Flying Car

Miscellaneous Thoughts, Inner Chap. 15 in *The Master Who Embraces Simplicity* by Ge Hong of the Jin dynasty, records that "Some build a flying vehicle from the pith of the jujube tree and have it drawn by a sword with a thong of buffalo hide at the end of its grip". In 1956, during the Eighth International Congress of the History of Science held in Italy, Joseph Needham (1900–1995), an English scholar, asked Zhu Kezhen and Liu Xianzhou, two Chinese scientists, about the flying vehicle. He thought the above recording was related to the working principle of helicopters. After returning home, Zhu and Liu told Wang Zhenduo about it, hoping that he would make a study on the vehicle described by Ge Hong. Three years later, Wang made a model of the flying vehicle (as shown in Picture 1.84), which managed to fly as high as the eaves of the side tower in front of the Meridian Gate of the Forbidden City in Beijing based on his own research. His paper "Restoration of the Flying Vehicle in Ge Hong's *Bao Pu Zi*" was published in the Journal of the History Museum of China (nowadays National Museum of China) in 1984.

According to Wang, the pith of the jujube tree mentioned in *The Master Who Embraces Simplicity* is hard and mostly brown–red and is suitable for making small woodwork. In Xu Shen's *Discussing Writing and Explaining Characters, radical che* explains "车" as "the general name of a vehicle's wheel (舆轮)" and explains "轮" as "a wheel with spokes". Ancient books such as *Book of Agriculture* by Wang Zhen and *Heavenly Creations* by Song Yingxing describe the machinery with shafts as "车" (vehicle). In *Biography of Yue Fei* in *History of Song*, the wheel-driven warships made by Yang Yao are also called "vehicle ships". Therefore, the "flying vehicle" (飞车) mentioned by Ge Hong is in fact a flywheel. The actual meaning of "[…] have it drawn by a sword with a thong of buffalo hide at the end of its grip" is that the rope made of the buffalo hide is tied into rings around a shaft, and the bow-shaped "sword" is used to drive the shaft-shaped screw thread mechanism to make the flying vehicle fly. This kind of mechanism, similar to the well windlass, is installed vertically on a wooden handle, with a vertical shaft in it, and bears the flying vehicle above it. When the rope

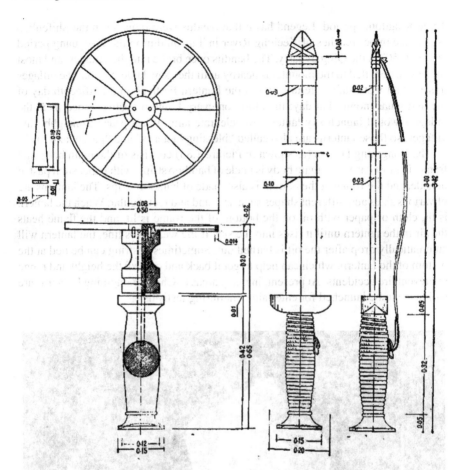

Picture 1.84 Structure diagram of the restoration model of the flying vehicle

rings pull the windlass, the rapidly rotating shaft generates force to make the vehicle fly. Bamboo dragonflies, which are often considered a toy for amusement and pleasure nowadays, are similar to the flying vehicle both in structure and function, except that their original wheels are changed into bamboo blades, which undoubtedly evolved from the flying vehicle. They were introduced to Europe as Chinese tops, which are regarded as the predecessor of helicopters.

1.6.6 Kongming Lantern

The Kongming lantern is also known as sky lantern, which is said to be invented by the military strategist Zhuge Liang, with the courtesy name of Kongming, during the

Three Kingdoms period. Legend has it that bandits were rampant in the Shifenliao area in the upper reaches of Keelung River in Taiwan during the Daoguang period (1821–1850) in the Qing dynasty. The bandits once broke into the villages, and most of the villagers fled to the mountains nearby until they saw lanterns from the villages to signal all were safe to return. It was the Lantern Festival on the fifteenth day of the first lunar month that day; therefore, on every Lantern Festival since then, the villagers would launch sky lanterns to celebrate and inform peace to neighboring villages, so those lanterns are also called "blessing lanterns" and "peace lanterns".

The Kongming lantern (as shown in Picture 1.85) consists of the main body and frame. In most cases, the main body is made of bamboo strips, with paper shell pasted outside, and the frame at the bottom is also made of bamboo strips. The lanterns are of various sizes and different shapes such as round and rectangular. When the lantern is lit, cloth or paper with oil on the bottom of the frame is lit, and the flame heats the air in the lantern until it rises into the sky. If the weather is fine, the lantern will automatically drop after the oil is burned out. Sometimes, a string can be tied at the bottom of the lantern, which can help to get it back and control the height and range to prevent fire accidents. At present, in many areas in China, Kongming Lanterns are still made to be launched for entertainment during festivals.

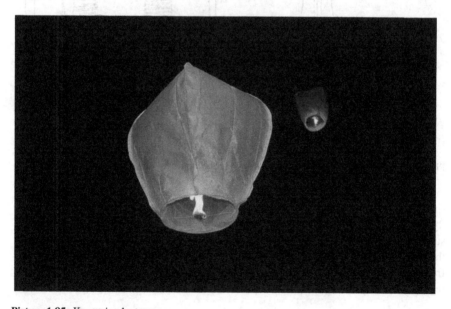

Picture 1.85 Kongming lanterns

1.6.7 Incense Burner in the Quilt

The incense burner in the quilt (as shown in Pictures 1.86 and 1.87) is used to perfume the bed with burning incense. There are two or three layers of concentric rings between the shell and the furnace body in the center, which is connected to the inner ring by an axis and can rotate freely. The inner ring is supported by the outer ring, which is supported by the outer shell. Due to gravity, no matter how the spherical shell rolls, the furnace body inside always keeps a horizontal state, and the incense ash won't fall out.

The incense burner in the quilt was first recorded in *Miscellaneous Records of the Western Capital*, which mainly writes about the stories in Chang'an, the capital of the Western Han dynasty. It says, "Ding Huan, a skillful craftsman in Chang'an … made an incense burner in the quilt. It had been invented before, but the technique was lost, and Ding recovered it. It was named so because the burner could be placed inside the quilt as the furnace body could remain horizontal by the supporting rings outside". Since the Tang and Song dynasties, the burner has often been mentioned in many works. For example, Zhang Shu (772–846) writes in his poem *Lyrics to the Melody of Sandy Creek Washers,* "The smoke of the incense burner in the quilt is lingering inside the embroidered curtains". Tian Yiheng of the Ming dynasty also records in volume 22 of *Daily Notes on Green Bamboo,* "The gold-plated incense burning ball looks like an armillary sphere. There are three layers of movable round rings, each with different weight. When it is put inside the quilt, the incense inside still burns and the smoke flows out from the hollowed-out shell in exquisite flower patterns".

Picture 1.86 Round silver incense burner in the Tang dynasty, unearthed in Shapo village, Xi'an city, Shaanxi province in 1963

Picture 1.87 Internal
structure of the incense
burner in the quilt

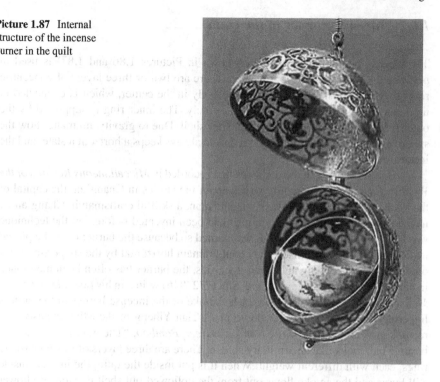

What is marvelous about this kind of incense burner is that no matter how the outer layers rotate, the furnace body in the center always remains horizontal. It operates on the same mechanical rule as the modern gimbal, with the gyroscope as an example.

Girolamo Cardano (1501–1576), a versatile Italian scholar, was the first to design the gimbal in the West, which is also called Cardan's suspension or Cardan's ring. The author of the book *Le Machine*, published in Latin in 1629, proposed to transport patients with vehicles equipped with the gimbal so as to absorb the shock of vehicles on rugged roads. It was Jean Bernard Léon Foucault (1819–1868), a French physicist, who used gimbals in modern scientific research and made important discoveries. In 1851, he proposed to show the rotation of the earth with a gyro rotating at high speed.

In the mid-nineteenth century, steel was widely used to build ships driven by steam engines. However, the magnetic compass was unreliable near steel, but the gyroscope was not affected by it, so it could replace the magnetic compass. In 1908, Hermann Anschütz-Kaempfe (1872–1931), a German inventor, designed the gyroscope that could be used for navigation, and soon the gyroscopic navigation instruments were installed on the submarines and armored ships of German Navy. In 1921, Elmer Ambrose Sperry (1860–1930), an American inventor, designed a control device that automatically controlled the direction of a ship by gyroscope, and then made a stabilizer to reduce the turbulence of the ship by using its function of orientation. During World War I, Germany and the United States applied gyroscopes on airplanes as indicators of tilt and turn. In 1929, J. H. Dolit, an American, used the gyro horizon and the directional gyro to control aircrafts. All those devices are typical examples of practical application of the gimbal.

1.6.8 Trotting Horse Lantern

The trotting horse lantern, or revolving lantern, (as shown in Picture 1.88), a Chinese folk lantern with moving objects inside, can be traced back to the shadow lantern in the Tang dynasty. Cui Ye of the Tang dynasty wrote in his poem *Second of the Six Poems on the Night of Lantern Festival*, "Shadows of a hundred wheels shift in the magic light, seven treasures of figure patterns are concealed in the lanterns; in the shadow, it's like sayings from the golden mouth of Buddha, in the air, it's like rays of jade light shining in the sky". The "hundred wheels" and "seven treasures" depicted in the poem are supposed to be the impeller device and the images of paper cuttings projected on the screen.

According to *Collected Talks of a Drunkard* by Jin Yingzhi of the Northern Song dynasty, on the Lantern Show in Kaifeng were "horse-riding lanterns", which were in fact trotting horse lanterns. *Old Things about Wulin* by Zhou Mi in the Southern Song dynasty records that on the Lantern Show of the capital Lin'an (also known as Wulin, nowadays Hangzhou) "different figures on horses were spinning at a high speed as if they were flying in the shadow lanterns". *Annual Customs and Festivals*

Picture 1.88 Trotting horse lantern

in Peking, printed in the 32nd year of the Guangxu period (1906), writes that "in the trotting horse lantern, the paper cuttings on an axis, which is in the shape of figures, horses and chariots, will spin continuously because of the air flow produced by the burning candle. It won't stop until the candle goes out".

The structure of the trotting horse lantern is generally like this. An iron wire is installed in the center as a vertical axis, with an impeller on the top and two horizontal crossbars in the middle. Paper-cuttings such as figures and horses are pasted on the outer ends of the crossbars. The blades of the impeller are inclined along one direction, and a candle is installed at the lower end of the vertical axis. When the candle is lit, the hot air causes the impeller to drive the crossbars to rotate, and the paper cuttings also rotate. Their shadow is projected on the lampshade, resulting in the image of figures and horse "rotating as if flying" and "spinning round and round".

The lantern converts the heat energy generated by burning candles into mechanical energy, which is similar to the working principle of modern gas turbines. At the end of the fifteenth century, impellers with axis and transmission devices are placed in chimneys to drive barbecue forks in Europe. Joseph Needham thinks that there is speculation that they may originate from the Chinese trotting horse lanterns.

*****

To sum up, there are many inventions and creations of ancient Chinese tools and instruments that were once the most advanced in the world. As a large country in agriculture and the birthplace of silk, China boasts its outstanding achievements in agricultural machinery and textile machinery, such as the design and production of crank-connecting rod mechanism and double-action bellows, which fully demonstrate the wisdom and creativity of Chinese ancestors.

Meanwhile, there are also some structural defects in ancient Chinese as well as Western machinery manufacturing industry. For example, China's traditional manufacturing skills relied more on casting than forging, while it was just the opposite in ancient Europe. In mechanical connection, China is known for its mortise and tenon joints while bolts and nuts are never seen. Nevertheless, the production of advanced tools and instruments has always been a powerful pillar of China's social and economic development in its long history. The prosperity of Chinese civilization is unimaginable without it. The machinery manufacturing industries in both the East and West accomplished their historical missions brilliantly in ancient times. It's just as what is said in *The Great Treaties II* of *The Commentaries on the Book of Changes*, "In the world there are many different roads, but the destination is the same. There are hundred deliberations, but the results are the same".

The intelligence and ingeniousness of the Chinese people are manifested in the flying vehicle, Kongming lantern, incense burner in the quilt, and trotting horse lantern mentioned above, which are similar to modern helicopters, hot air balloons, gyroscopes, and gas turbines in their working principles. After all, however, the Chinese utensils and toys weren't developed into its modern function as machines. Such development and transformation are only possible when the society develops to the stage of modern civilization.

Chapter 2
Agricultural and Mineral Processing

Li Jinsong

After thousands of years of continuous development, the traditional processing technology of agricultural and mineral products is rich in content, involving many manual production techniques. This chapter will be focused on traditional skills that are closely related to people's daily lives, such as the production of salt, sauce, alcohol, vinegar, tea, oil, sugar, leather, incense, and fireworks.

2.1 Production of Table Salt

The main component of table salt is sodium chloride. Its chemical composition is NaCl. The salt that we see daily is made up of white transparent cubic crystals. When mixed with insoluble impurities such as sand, the color becomes yellowish brown, or grayish brown and dark white. In the market, salt is divided into granular salt, brick salt, cylindrical salt, flower salt according to the shape of table salt. Based on its source, people divide salt into sea salt, pond salt, well salt, rock salt, and soil salt.

Table salt is one of the most important substances for human survival, as Tao Hongjing (456–536) said, "Salt is the most important of the five flavors, and it is indispensable". When humans were in the age of fishing and hunting, the salt needed for the human body could be supplemented from the flesh and blood of animals. After entering the agricultural era, grains became the staple food. At this time, the salt needed for human growth and development had to be found in nature. Because of this, early human settlements were often adjacent to salt producing areas, such as the famous Lake Turkana in northern Kenya and the Xiao Qaidam Salt Lake on Qinghai-Tibet Plateau in China (as shown in Picture 2.1). The most prestigious source of pond salt in China is Xiechi in southern Shanxi. As the legend goes, Pingyang, Puban, and

L. Jinsong (✉)
The Institute for the History of Natural Sciences, Chinese Academy of Sciences, Beijing, China
e-mail: lijs@ihns.ac.cn

© Elephant Press Co., Ltd 2022
H. Jueming et al. (eds.), *Chinese Handicrafts*,
https://doi.org/10.1007/978-981-19-5379-8_2

Picture 2.1 Xiao Qaidam Salt Lake on Qinghai-Tibet Plateau

Anyi are the homes of the Yao, Shun, and Yu tribes. It is by no means accidental that they are all close to Xiechi, and along the upper reaches of the Yellow River there is a group of salt ponds at the junction of Gansu, Ningxia, and Qinghai provinces. In the Yangtze River Basin, well salt and rock salt can be found in eastern and northern Sichuan, and sea salt in Jiangsu and Zhejiang in the lower reaches. People discovered that adding salt to food made it more palatable, and gradually learned how to add salt to their food and process it as well.

Salt is indispensable to the national economy and people's livelihoods, and huge profits can be obtained from its production, transportation, and sales, so it was the most important and the most stable special commodity in ancient times. Guan Zhong stipulated that all mines and sea salt belonged to the government and promoted trade on the basis of sea salt after he was appointed the prime minister of Qi state in the Spring and Autumn period. Since then the salt tax had been a major revenue for courts of the past dynasties. Emperor Wudi of the Han dynasty implemented the policy of government-run salt and iron. Later, all succedent governments set up salt officials to monopolize salt profits. During the reign of Emperor Daizong (762–779) of the Tang dynasty, Liu Yan enacted the salt monopoly law, which was in use for 1200 years. During this period, the salt tax was second only to land tax, as *Treaties of Foods and Commodities* in *New Book of Tang* records, "Of all the taxes, salt accounts for half, and the imperial expenses, annual salary, and official salaries are all paid with it".

Since the salt ban was strict, illegal salt production and trafficking came into being. During the Yuan, Ming and Qing dynasties, the dispute between official and illegal salt became more severe. The first reason was that exploitation made the livelihood of salt workers difficult and risks had to be taken; second, the monopoly of table salt caused serious supply and demand issues, which influenced the start of profitable trafficking; third, people who had rich trafficking experience had a much higher chance of succeeding. The darkness of officialdom and the inefficiency of the system made illegal salt more widespread. Salt merchants colluded with the government to monopolize sales channels to make huge profits by buying "*yanyin*" (salt certificate). This inevitably led to a fierce conflict with illegal salt dealers. Despite that the government strictly forbade the sale of illegal salt, and stipulated that the offenders should be exiled and even buyers should be flogged as punishment, the illegal trade never stopped. The salt industry became much more profitable than other industries, and being a salt merchant became a popular career choice. The salt industry also had a profound impact on the economy, commerce, and culture of the city where it was produced. The most prominent city was Yangzhou. The prosperity of Yangzhou salt industry began in the Tang Dynasty. According to *Comprehensive Mirror to Aid in Government*, "Yangzhou is rich in the world", known as "Yangzhou ranks first and Yizhou (now Chengdu) ranks second". "It is said Yangzhou and Shu (nowadays Sichuan province) are the first and second most prosperous places in the whole country". In the Qing dynasty, Emperor Qianlong inspected the south reaches of the Yangtze River for several times. Every time he went, the salt merchants spent money like running water to accommodate the party. Yangzhou's cultural prosperity and the emergence of talents were also related to the salt industry economy.

In the history of Chinese technology, the production process of table salt occupies a unique position. *Treaties of Foods and Commodities* in *History of Ming* describes the production methods of table salt in salt-producing regions across the country in detail: "Salt is produced in different ways: Xiezhou's salt is crystallized by wind and water; Ningxia's salt is made from scraping the soil; salt of Anhui, Jiangsu, and Zhejiang is obtained by boiling; Sichuan and Yunnan' salt is drawn from well brine; Fujian's salt is drawn from deposit brine; salt in Huainan area is decocted; salt in Huaibei area is dried in the sun; Shandong's salt is either decocted or dried".

The well salt production technology in southwest China, the sea salt production technology in coastal areas, and the pool salt production technology in northwest China had their own characteristics, reflecting the wisdom of people in different regions in taking advantage of natural resources. Since the twentieth century, under the impact of modern technology, these salt-making processes that have been passed down for thousands of years have quickly disappeared. Today, only a few places still follow the traditional salt-making process.

2.1.1 Production of Well Salt in Tibet

Yanjing town of Mangkang county is located at the southeast tip of Tibet, immediately adjacent to Deqin county in Yunnan province. It is a township of the Naxi ethnic group and is the only place that the ancient Tea-Horse Road must pass through. In history, the salt well in Mangkang county was very famous, and the place name of "Yanjing" salt well) is found in Du Changding's *Trip to Tibet* during the Kangxi period (1662–1722) in the Qing dynasty. *Yanjing Local Records* written during the Guangxu period

Picture 2.2 Salt field on the bank of the Lancang river

(1875–1908) records 52 brine wells, 689 brine ponds, and 2763 salt fields in this area. In the first year of the Xuantong period (1909), Cheng Fengxiang, an army officer of the Qing court, recorded the production process of salt wells and the construction of salt fields in *Trip to Kamu, Southwest China*.

On the third day of lunar October of the same year, Zhao Erfeng, the Minister of Frontier Affairs of Sichuan and Yunnan, recorded that he tried to crystalize salt with fire to improve the quality of the product, but he gave up because of the high cost of burning wood.

The above records show that the well salt in Yanjing town has a long history, original craftsmanship and local characteristics. What is rare is that this craft has been passed down to the present time (as shown in Picture 2.2).

The production process of Mangkang well salt is as follows.

2.1.1.1   Extracting Brine

According to historical records, a ladder is put in the well, workers go down into the well along the steep ladder, scoop brine into wooden barrels with small basins, and carry it out. Nowadays, workers also pump brine to salt fields.

2.1.1.2 Transporting Brine

After taking the brine out of the well, one way is to pour the brine into your own brine storage tank, scoop up the brine when necessary, and pour it into the salt pan in the shallow pond; another way is to pour the brine directly into the shallow pond. In the salt field of the pond, or during the drying process, the brine is introduced into the salt field from the brine pond with a thin tube.

2.1.1.3 Drying Salt

Drying salt in salt fields mainly relies on sunshine and wind, which is stated in the literature as "with sunshine and wind, the brine will become salt". Generally, it takes about 1–2 days to dry the salt in spring, and 2–5 days in other seasons.

2.1.1.4 Scraping Salt

Scrape the crystallized salt particles with an iron knife-like tool. Sweep them into a container with a broom and pour it into a bamboo basket. At this time, there is still a certain amount of water in the salt granules, which are gradually drained in the basket. The salt scraped first is whiter and is for human consumption, while the salt scraped later contains more soil for livestock consumption. Each salt field can scrape over 10 *jin* (5 kg) salt at a time.

Scraped salt particles are piled on the ground and measured with wooden buckets. Men come with mules to pack and carry the salt and sell it in town or in other places. Mangkang's well salt is similar to that of sea salt, but is different in color, has coarser particles, and contains a small amount of soil, but it has a strong taste, for which it is praised by the locals.

2.1.2 Production Process of Well Salt in Nuodeng Well

Nuodeng well is located in a Bai village of Guolang township, Yunlong county, Yunnan province, six or seven kilometers away from the county (as shown in Picture 2.3). The well has a long history. Fan Chuo's *Chorography of Yunan* in the Tang dynasty records: "There is the Nuodeng well in Jianchuan". According to *Illustrated Record of Yunnan* of the Ming dynasty, Nuodeng had a salt tax office that was in charge of the salt tax of the five wells in Yunlong county. *Salt Extracting Method in Southern Yunnan* written during the Qing's Kangxi period (1662–1722) contains special maps depicting and describing the Yunlong well area.

Before 1949, among the 350 permanent households in Nuodeng village, 230 were stove households with brine resources. They depended on the production of well salt for living and turned over salt products to the government (as shown in Picture 2.4).

Picture 2.3 Old Nuodeng village

After 1949, when People's Republic of China was founded, the government nation-alized the brine resources, abolished the stove household system that had lasted for thousands of years, and concentrated scattered family workshops into salt factories.

Picture 2.4 Salt field mantled by wooden frame

Picture 2.5 Brine well

From 1949 to 1996, Nuodeng Salt Factory produced 1.5–2.5 tons salt per day. In order to promote production, Nuodeng Salt Factory carried out many technological innovations, but they did not let go of the traditional model of well salt production technology, and have always been regarded as "salt production by indigenous methods". In 1996, the Nuodeng well was shut down and salt production ceased.

The production process of Nuodeng well salt is as follows.

2.1.2.1   Salt Well Structure and Brine Extraction Technology

The Nuodeng salt well is 22 m deep and has two sub-wells, saltwater and freshwater well (as shown in Picture 2.5). The saltwater well is a channel for drawing saltwater, and the freshwater well is a channel for drawing fresh water. The essentials of the downhole technology are to form a brine extraction system and a water extraction system to effectively separate saltwater and freshwater. There are two reservoirs respectively for storing saltwater and freshwater under the wellhead, which are separated by wooden boards and plugged with cement to prevent leakage. The source of brine is in the deepest part of the well. The saltwater channel is paved by bamboo tubes, through which the brine is drawn and transported to the saltwater reservoir with the use of height difference of the tubes. Then draw out the brine with wooden barrels and a well windlass for future use (as shown in Picture 2.6).

2.1.2.2   Boiling Salt

Nuodeng's salt stove has a large pot in the middle, with barrel-shaped pots next to it. The number of barrel pots varies from 10 to 12, and they surround the large pot

Picture 2.6 Wooden frame
at the wellhead

in a circle. Once the stove is hot enough, the boiling can start. From the beginning of boiling to the end of the frying, the heating cannot be stopped.

After reaching the boiling point, the water evaporates and the brine becomes concentrated. In order to produce salt as soon as possible, the water in the middle large pot must be scooped into the surrounding pots and more cool brine is added to the large pot. This process is called "spraying water". During this process, the brine in the barrel pots starts to thicken.

After salt in the barrel pots begin to crystallize, no water should be scooped from the large pot into the barrel pots. Thicker brine in a certain barrel pot is transferred into the other pots near it to make the barrels with concentrated brine extract salt first. This process is called "*guiguo*" (returning to pot) or "*bingguo*" (merging into pot). After salt is extracted, the female workers began to scoop it up. After it's been scooped up, the small remaining amount of thick brine is scattered into other pots. Then cool brine is added to dissolve the remainders in the pots. This process is called "de-potting". This de-potted water is also high-concentration brine and must also be transferred to other pots.

Pounding salt: Pound 2 or 2.5 kg at a time. After pounding the salt into fine flour, put it in a dustpan and let it air for 3 or 4 days.

Pinching salt: To pinch salt, a technical female worker is needed. Add an appropriate amount of "salt clarifying water" to the fine salt, and stir it evenly. Put the supporting mold on a wooden or iron plate, stuff the mixed salt into the supporting mold, pat the salt tightly with the plate, and pour a little "salt clarifying water" on it. When disassembling the mold, remove the hoop and the supporting mold, and use the cushioned board to put down the supported salt.

Firing salt: Sprinkle the charcoal flour mixed with the salt left over from the previous burning of salt on the ground, and then sprinkle a layer of chaff on top of the salt. After arranging 40–50 lumps, take the soaked and dried foam charcoal out of the fire and put it on the salt. Then, shovel the reddish charcoal on the foam charcoal with a shovel, and keep fanning the flames. After the top is hardened, pick up the salt and burn the bottom. It is very important to control the temperature and it is better to boil dry, hard but not let it burn.

After the salt has cooled, put a layer of hay and then a layer of salt in the bamboo basket, and so forth. There is a basket for every 10 lumps, and there are 24 tubes for each transportation.

The Nuodeng salt was transported out by horse or carried by hand. It was mainly sold to Baoshan city, and the area along Tengchong county, Yunnan province, China to Myanmar. Nuodeng tube salt had a mellow taste and was very popular among people of all ethnic groups in northwestern Yunnan.

Because of its excellent salt quality, the "Nuodeng ham" and "Nuodeng soy sauce" pickled and brewed with it are also famous.

2.1.3 Mining and Production of Shenhai Well Salt

Shenhai well in Zigong, Sichuan province is located next to the Changyan pond at the foot of Ruanjiaba Mountain in Da'an district, covering an area of 0.2 ha. The well was drilled in the 15th year of the Daoguang period (1835) in the Qing dynasty, using the traditional percussion drilling method (as shown in Pictures 2.7 and 2.8), including sinking, logging and correction, repairing well walls, salvaging, and repairing wood pillars. It took three years to complete drilling the well. With a depth of 1001.42 m, the well was the world's first ultra-kilometer deep well at that time. The well is a high-yield well that produces both natural gas and black brine, and created considerable economic value for more than 100 years.

In the early stage of completion of the Shenhai well, there was a spectacular blowout, which produced both brine and gas, which completely solved the problem of fuel for boiling salt. At that time, it produced 8500 m^3 of natural gas, 14 m^3 of black brine, and more than 80 salt pots were needed.

After 1875, the output of natural gas gradually decreased. There were more than 20 salt pots, and the daily output of salt was about 1500 kg. In 1944, the natural gas output of the well once increased to 3200 m^3 per day, and there were 30 salt pots. Its daily output of natural gas now is 1500 m^3 per day. The excavation of

Picture 2.7 Salt production site of the well recorded in *Heavenly Creations*

Picture 2.8 Derrick of Shenhai well

Shenhai well brought salt merchants from all over the world to dig the well and set up stoves, presenting a prosperous scene of "cranking cranes" lined with pots and stoves. The Shenhai well stove was renamed several times in 100 years. First it was called Yuanchang Stove, then Ronghua Stove; after that it was Qianyuan Stove, and it was also called Siyi Stove. Later it was renamed Yiji Dexin Stove, Xinji Tongsen Stove, Junji Tongsen Stove, Yiji Tongsen Stove, Jinhe Dexing Stove, Fuji Tongyi Stove, and Jianji Tongsen Stove.

The equipment of the Shenhai well and the process of collecting brine and boiling salt is as follows.

2.1.3.1 Equipment

The equipment includes the well, crown block, cart, stepping frame, etc. The Shenhai well is covered by a well house and is cemented with stone rings and wooden columns, with a well diameter of about 16 cm. The black brine produced in Triassic strata (about 1,000 m away from the surface) is black because it contains organic matter and sulfide. In the past, bamboo was used to get through joints and then brine was extracted, but now steel pipes are used. Originally, the bucket was made of bamboo strips, but now it is made of steel ropes.

The crown block of the Shenhai well is 18.3 m high, which is tied with hundreds of Chinese firwood poles from bottom to top with hemp rope, and is used for bailing brine.

On the right side of the wellhead is a cart for lifting brine (as shown in Picture 2.9). When lifting, the one-way piston in the cylinder automatically opens, and brine is poured into the cylinder. After lifting the bamboo tube, hook the piston, pour the brine into the bucket, carry it to the kitchen and pour it into the salt pot to make salt.

The Shenhai well was first driven by manpower to lift brine by cart, which required 8–12 people, but later it was pulled by four cattle. Electric winches are now used as power.

The manual stepping frame is an important equipment for filing wells, and a rectangular derrick is erected from logs. Below the well is the treading board, and one end of the treading board is the tilt hammer head. The workers stepped on the treads at the same time, using the principle of leverage to tilt the head and raise the filing tool. When the workers jump off the tread, the filing tool hits downwards by its own gravity, repeatedly breaking the rock because of the movement. At the same time, water has to be poured into the well and the mud has to be lifted out.

2.1.3.2 Boiling Salt

The raw materials used for salt making are yellow brine, black brine, and salt rock brine, and the fuel is natural gas, so it is commonly called a fire well. The gas transmission facility is made of bamboo knots and buried in the ground. A round pot stove is used to make salt.

Picture 2.9 Cart

The salt produced by the Shenhai well is flower salt (refined fine and white salt). Its production method is recorded in *Recording of Sichuan Salt Making* in the Qing dynasty:

(1) Concentrated brine: Pour the brine into the pot and boil it for about 8 h. There are three rounds of salt in total for 24 h.
(2) Pouring soy milk: For impurities that are invisible to the naked eye, use soy milk to purify it.
(3) Sprinkle *huashui* (high concentration brine): Shovel the salt in the pot onto the bamboo bun, and shovel and loosen the salt on the upper part of the bun. Let the brine leak in the ditch and let it drain for more than 12 h. After shoveling off the salt, put the brine in a pan again to turn it into salt, and the brine in the ditch can be recycled.
(4) Put in *muzizha*: The purpose is of this process to crystallize the brine into salt with the help of muzizha, which is the already crystallized salt that has been boiled in another pot. Finally, the salt is poured and tested.

The well can now produce more than two tons of salt per day. The raw salt produced is high in sodium chloride content, low in moisture, less impurity, white in color and dry. The salt boiled with natural gas is called flower salt. The local kimchi in Zigong must use this kind of salt. The soaked vegetables are fresh and crispy, especially tasty.

2.1.4 Production of Sea Salt in Xiangshan

The sun-dried salt in Xiangshan county, Zhejiang province is distributed in the coastal area of the county border. It starts from Qiancang in the north, from Juexi River to Shipu and Sidu in the south, with a roundabout of over two hundred *li* (100 km). It is described in history book that those places "are surrounded by stoves".

Treatise of Geography in *New Book of Tang* records the salt production in Xiangshan. In the late Northern Song dynasty, a salt field was set up under Yuquan Mountain in the northeast of the county, and a salt official was set up. From the Southern Song dynasty to the Qing dynasty, salt fields gradually increased. At the end of the Republic of China, Yuquanchang once governed the three counties of Sanmen, Ninghai and Xiangshan. After 1949, the salt area (field) was adjusted and abandoned several times. By the end of the 1970s, five major salt fields, Changguo, Hua'ao, Baiyanshan, Xinqiao, and Danmen, had been formed, mainly concentrated in coastal areas.

The sea salt production in Xiangshan has gone through several stages. Originally, it was a salt-making method called "*aobo*" by Yuan people. After the Yuan dynasty, the methods of scraping mud to drench brine and throwing ash to make brine were gradually adopted. Since Jiaqing period (1796–1820) of the Qing dynasty, the plate drying method was introduced from Zhoushan, and the jar drying method was introduced in the late Qing dynasty, which was a great change in salt production technology. After the 1960s, the flat beach sun drying method was successfully tested, and the manual operation was gradually replaced by machines, but the traditional salt drying technique was still reserved.

The sun-dried salt takes seawater as raw material, which is poured onto of white mud (salty mud) or lime soil (mud) on the beach and is evaporated by sun and wind power. The brine (fresh brine) is crystallized into salt with different thickness by boiling, drying or blowing (as shown in Pictures 2.10, 2.11 and 2.12).

2.1.4.1 Beach Yard

Build ponds near the sea to resist tide and build sluices to take in seawater and discharge freshwater; a square beach field is formed with ditches and ridges as the boundary. The ditches around the field are used to store seawater, while pools are also dug to store seawater.

Picture 2.10 Picture of spraying ash to get brine recorded in *Heavenly Creations*

Picture 2.11 Salt field by the sea

Picture 2.12 Yangpu salt field

2.1.4.2 Lime Soil Preparation

Cut loose the mud in the beach yard with sharpeners and smash it, and then use a bamboo pole to collect the mud into a fine shape, which is like ash. Bail the seawater from the pool, sprinkle it evenly with a wooden ladle onto the mud (ash), make it absorb the salt in the water; sprinkle it again at noon and dry it in the sun till sunset; gather the mud (ash) with a knife, and clip it into a dike shape with wooden boards to prevent night rain. The mud (ash) is turned over and flattened, and then grilled and loosened using the same method as before the next day if it is sunny. Generally, the mud (ash) is full of salt for 2 or 3 days in summer and 4 days in winter.

2.1.4.3 Brine Production

A soil circle like a cabinet is built at a convenient position in the center of the beach yard, which is called ash chute; dig a well or put tanks by the chute; spread short wood boards under the bottom of the chute, cover the chute with slim bamboo and then firewood ash. Fill the mud (ash) into the chute, tread it solid, cover it with straw, and pour the seawater in the pools onto the ash, so that it can slowly infiltrate into the chute, thus producing salty brine (fresh brine).

2.1.4.4   Crystallization

(1) Decocting: Set up a mud stove, put an iron dish, bamboo dish, or iron pot on it, pour brine into the cooking utensil and heat it, put into Chinese honey locust powder and half processed wheat grain, and stir it. Salt will be produced soon.
(2) Tedding: Build a trellis at a suitable place. The soil in the trellis is compacted. Put into broken tiles to separate the trellis into grids and pour the fresh brine into them. The fresh brine will crystallize into salt exposed to sunshine and wind.

In the long-term practice, the operating procedures for seawater absorption, brine production, crystallization, and piling-up have been formed (as shown in Picture 2.13). The finished product has two kinds of thickness. The salt cooked in Yuquan, Zhejiang province in the old times was white, fine-grained, and delicious, which is commonly known as fine salt. Later the sun-dried salt is white in color and coarse in grain, which is called coarse salt. The by-products include bittern and halogen ice, which are the bitter remaining mother liquor of salt making and can be used to make tofu and used as fertilizer. Due to the low production cost and high output, the proportion of sea salt has been increasing.

Picture 2.13 Stacking sea salt

2.2 Brewing Techniques

For a long time, in the daily lives of Chinese people, drinking liquor has been more than material enjoyment. It has also become a carrier of ritual communication and emotional expression. Liquor drinking could be found in sacrifices, celebrations, banquets, and gatherings of relatives and friends, forming a rich liquor culture.

Regarding the origin of liquor brewing technology in China, it was recorded in *Selections from Lu's Commentaries of History* written in the third century BC. that "Yi Di (an official in charge of liquor in the Xia dynasty) invented liquor". There was another point of view later that it was Du Kang who invented liquor. The two have been regarded as "God of Liquor" by the liquor brewing industry.

Liquor brewing technology in some influential documents includes the ancient six methods described in *Typical Weather in a Given Season* in *Book of Rites*, the method of making yeast described in *Essential Techniques for the Peasantry* in the Southern and Northern dynasties, the yeast making and liquor brewing process in *Dongpo's Experience on Liquor* of the Northern Song dynasty and *Beishan Liquor Book*.

2.2.1 Liquor Brewing

With the help of microorganisms that are ubiquitous in nature, fermentation was used in early China to turn grains and other starch and sugar-containing substances into liquor, vinegar, and sauce as beverages, condiments, or food. In addition, steamed buns, noodles, yogurt, and Chinese pickle are also made with the fermentation method.

Chinese ancestors had a major innovation in fermentation technology, which is the invention and use of yeast. Yeast is a medium for excellent molds and a key element that dominates the fermentation process. Rich in artificially cultivated molds and saccharomycetes, it can simultaneously saccharify and alcoholize starch in grains by double fermentation. Chinese people mastered this advanced technology to brew liquor, vinegar, and sauce thousands of years earlier than Europeans.

The main ingredient of liquor is ethanol. In nature, substances containing sugar (glucose, sucrose, maltose, lactose, etc.), such as fruits, animal milk, etc., will produce ethanol when subjected to the action of saccharomycetes. Moderate intake of ethanol can excite the nerve center, so there is a good story of "drinking a *dou* of liquor can make Li Bai write one hundred poems". However, excessive intake of ethanol can cause nerve center paralysis and cause harm to the human body, which is often referred to as drunkenness.

During the Yangshao culture period more than 5,000 years ago, Chinese ancestors already imitated natural fermentation and used grains to make liquor. In practice, they gradually learned about distiller's yeast and its manufacturing methods.

Distiller's yeast is a carrier for cultivating and storing microorganisms that are conducive to fermentation. The preparation and scientific use of distiller's yeast are the key to liquor-brewing technology. During the Warring States period and Qin and Han dynasties, people developed from using powder yeast to preparing a variety of brick yeast. Along with the improvement of distiller's yeast-making technology, people have explored a complete set of temperature control, humidity control, and environmental reservation technologies that will change fermentation for producing good liquor. The production process of yellow rice wine, a kind of fermented wine, was mature in the Song dynasty, but the production of distilled spirits was relatively late. The descendants of Genghis Khan introduced the grape shochu production technology from Central Asia to the Central Plains during their Western Expedition during the thirteenth to fourteenth centuries. Taking this as a lesson, the liquor makers quickly produced distilled spirits with rice wine, rice wine vinasse, and rancid rice wine, and the spirits thus made were commonly known as shochu or fire wine, now commonly known as liquor (*baijiu*). The production technology of liquor was quickly promoted in the Ming dynasty. Today's famous Chinese liquor brands such as Luzhou Laojiao, Fenjiu, Wuliangye, and Moutai all originated from that time. Liquor is developed on the basis of rice wine production technology and is unique in the world brewing industry. In addition, medicinal wine and tonic wine, which enjoy a high reputation among the Chinese, have long been popular, and they also have a place in China's medical treasure house.

Yellow rice wine, a kind of mild filtered wine, was the main drink in early China. It was made from rice, broomcorn millet, and millet, brewed with *Xiaoqu* and Chinese yeast. The brewing of liquor has inherited *Daqu* brewing and solid-state fermentation techniques while applying distillation technology. Due to the differences in raw materials, water, and microbial systems, temperature, humidity, and other ecological environments, as well as differences in cultural accumulation and technical traditions, famous liquor brands produced in various places have their own regional characteristics. *Jiaoxiang* flavor Chinese *baijiu* is aromatic and sweet; strong flavor is harmoniously aromatic; *jiang*-flavor is elegant and delicate with a long aftertaste; mild-flavor is sweet with a refreshing aftertaste; honey flavor is fragrant, elegant, and soft; rice-flavor has a pleasant aftertaste (as shown in Pictures 2.14, 2.15 and 2.16).

Typical Weather in a Given Season in Book of Rites makes a general and strict regulation for liquor brewing technology. For example, "Glutinous rice must be up to standard" is the requirement for raw materials; "a right time must be chosen for making *qu* (yeast)" is the requirement for the use of yeast; "soaking and steaming must be clean" is the requirement for raw material processing; "water must be sweet" is the requirement for water; "pottery must be good" is the requirement for liquor-brewing utensils; "fire must be under good control" is the requirement for temperature. The quality of the liquor is guaranteed by strictly following the above six requirements.

These six elements constitute the core of liquor brewing technology, and later generations consider them as the "Six Ancient Methods". They are not only technical specifications for traditional liquor brewing, but also help promote the development of the industry by more detailed requirement enriched by future generations.

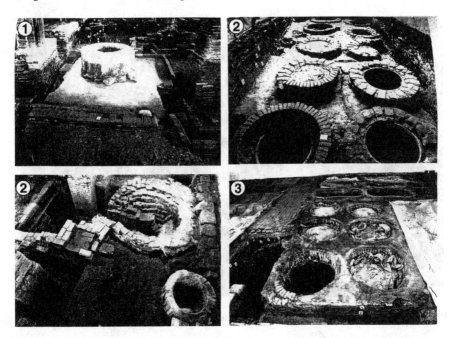

Picture 2.14 Brewing sites. 1. Well of the Ming dynasty, J1 (South → North). 2. Brewing site in Lidu, Jiangxi province. 3. Yuan dynasty wine cellars C10-C19 (Northeast → Southwest). 4. Wine cellars in the Ming dynasty C1-C7 (Southeast → Northwest)

Picture 2.15 Traditional Chinese *baijiu* brewing

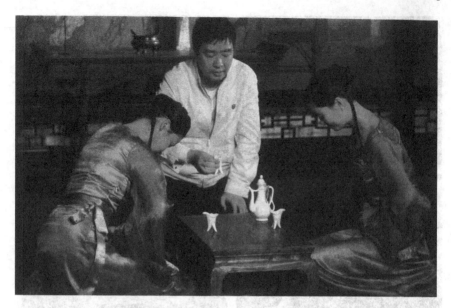

Picture 2.16 Demonstration of Chinese *baijiu* ritual

The history of yellow rice wine in China is the longest and there are many varieties. There are rice (glutinous rice, japonica rice) yellow rice wine and millet (broomcorn millet) yellow rice wine according to the materials used; there are Shaoxing yellow rice wine, Jimo yellow rice wine, and Fujian mature yellow rice wine according to the place of production; there are sweet (such as Fujian *Chengang* rice wine), semi-sweet (such as Jimo mature yellow rice wine), semi-dry (such as Shaoxing *Jiafan* rice wine) and dry (such as Shaoxing *Yuanhong* rice wine) according to the sweetness; there are also *Jiafan* rice wine and *Chengang* rice wine named according to the brewing technique.

Liquor is produced throughout urban and rural areas in China and is divided into five flavor types: (1) strong flavor, (2) *jiang*-flavor, (3) mild-flavor, (4) rice-flavor, and (5) mixed flavor, with over 20 subtypes. The following is an introduction to the representative techniques.

2.2.1.1   Brewing Techniques of Shaoxing Yellow Rice Wine

Shaoxing, east of the Hangjiahu Plain in Zhejiang province, is a well-known place in the south of the Yangtze River, with crisscrossing rivers and paths, fertile land, a mild climate and an abundance of products. During the Spring and Autumn period, with the development of agriculture, wine brewing became one of their handicrafts. According to *Selections from Lu's Commentaries of History*, "The king of Yue suffered in shame of his defeat in Kuaiji Mountain [...] If there is liquor, pour it into the river and enjoy it with the people".

The main ingredient for brewing Shaoxing wine is the high-quality refined white glutinous rice of the year it is produced in. The yeast is made of high-quality yellow-skinned wheat mixed with 10–15% barley. The original straw bale yeast must be tied with straw ropes, the volume should not be too large, and the greater viscosity will not affect yeast making and fermentation. Later, as the output increased, the brick yeast replaced the straw bale yeast. But if the bricks were too sticky, they were likely to rot. Therefore, a certain amount of rice was added when wheat grains were ground not too small, which was beneficial to make yeast. About the production technology of Shaoxing wine and its straw bale yeast, there are related records in *Southern Vegetation* written by Ji Han in the Jin dynasty (as shown in Picture 2.17).

To brew Shaoxing wine, distiller's wheat yeast and Chinese yeast must be used. Distiller's wheat yeast is made from raw wheat around the Autumnal Equinox. The essentials of making it emphasizes accuracy, evenness, and orderliness. "Accuracy" means that the making recipe should be accurate; "evenness" means that the wheat grains should be crushed and stirred evenly with water, and the size of the yeast bricks should be even; "orderliness" means the stacking of yeast bricks shall be in the correct order and the temperature of the upper and lower, the front and back of the yeast should be approximately the same during the cultivating process.

Chinese yeast (*jiuyao*), also known as *Xiaoqu*, liquor cake, or *baiyao*, is a saccha-rification and fermentation agent for preparing the distiller's yeast. It is cultivated by a wine brewing culture that is unique and excellent and contains a large amount of rhizopus that has a strong ability to convert starches into sugars, also known as saccharification. Chinese yeast is divided into white yeast and black yeast. The former has a stronger effect and is suitable for use in winter while the latter, which is made of early long-grain nonglutinous rice flour, with polygonum hydropiper powder, tangerine peel, Chinese pepper, licorice, and atractylodes, has a milder effect and is suitable for warm seasons (as shown in Picture 2.18).

The traditional brewing techniques of Shaoxing wine has the following major procedures: (1) soaking rice, (2) steaming rice (as shown in Picture 2.19), (3) putting rice into the vat (as shown in Picture 2.20), (4) fermenting (as shown in Picture 2.21), (5) mixing and cooling (*kaipa*) (as shown in Picture 2.22), (6) squeezing, and (7) decocting.

Among them, fermenting and *kaipa* (mixing and cooling) are the unique skills of Shaoxing wine brewing and are key to making good wine. According to the local customs in Shaoxing, the craftsman who masters the mixing-and-cooling technique is called the "*tounao*", the master of wine brewing. When the steamed rice, distiller's wheat yeast, and distiller's yeast are put into the vat, saccharification and alcohol fermentation take place simultaneously. During the process, heat is released and a large amount of carbon dioxide is produced. At this time, a sterilized wooden rake (*pa*) must be used to move up and down inside the vat, stirring the mixture so that the rice is sent to the bottom of the vat. This is called *kaipa*. The purpose of this process is to lower the fermentation temperature and make the upper and lower temperatures uniform, as well as remove the accumulated carbon dioxide and supply the yeast with fresh air. While maintaining the proliferation of the yeast, it destroys the anaerobic condition of the mixture and inhibits the growth of acid-forming bacteria.

Picture 2.17 Yellow rice wine brewing tools: ① rice steamer, ② wine pottery, ③ wine presser

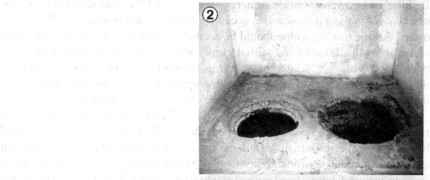

Picture 2.18 1. Polygonum hydropiper 2. White yeast

Picture 2.19 Steaming rice

Picture 2.20 Putting rice in the vat

Picture 2.21 Fermenting

Picture 2.22 *Kaipa*

The five traditional varieties of Shaoxing rice wine are *Yuanhong* wine, *Jiafan* wine, *Shanniang* wine, *Xiangxue* wine, and *Huadiao* wine. *Yuanhong* wine, also known as *Zhuangyuanhong*, is the most common Shaoxing wine.

There are two methods of Shaoxing wine brewing, namely, *Linfan* method and the *Tanfan* method.

The *Linfan* method (as shown in Picture 2.23) is to drench the steamed glutinous rice with water during brewing, and then add Chinese yeast, water, and wheat yeast into the vat for fermentation. The finished product is called *Linfan* rice wine. More often, rice was used to produce "distiller's yeast", also known as *jiuniang*. The preparation and use of the distiller's yeast not only improve the quality of the finished product, but also give it a special flavor.

The *Tanfan* method (as shown in Picture 2.24) is to spread the steamed glutinous rice on the bamboo mat to cool down, then add slurry (soaking water of rice), distiller's yeast, wheat yeast, and water, and put it in a vat to fermentation. After a certain period, the alcohol content and aroma of the fermented mixture gradually increases until finished. Then the water is pressed out, sterilized, packed in a wine jar, and sealed with lotus leaves and clay. The seal can prevent light and air from entering. The post-fermentation during the aging period makes the wine more tasty.

Picture 2.23 *Linfan*

Picture 2.24 *Tanfan*

2.2.1.2 Brewing Techniques of Chinese Liquor

Traditional Chinese brewing techniques are different from those in other areas of the world. One difference is the use of distiller's yeast to promote fermentation, and the other is that the fermentation process is mostly carried out in a state close to a thick mixture (solid or semi-solid).

Distiller's yeast is a kind of microbial product with many molds and bacteria, rich in beneficial molds and saccharomycetes that have been screened and cultivated for a long time. In the process of grain fermentation, the mold and saccharomycetes proliferated by the distiller's yeast are added to promote the simultaneous alcoholization of the grain during the saccharification process, that is, the introduced microbial strains have completed the double fermentation, and will also produce aging, so that the alcohol has a fragrance. The unique style is rich, sweet, and mellow. Distilled spirit brewing also regards yeast making as the key to a good product (as shown in Pictures 2.25 and 2.26).

At the end of the Han dynasty, Cao Cao recommended the "*Jiuyunchun* liquor brewing method" to Emperor Xiandi. During the entire fermentation process, steamed rice was added to the fermented mixture in several batches. This technique could not only make full use of the microbial strains that multiply in the fermented mixture, but it was also close to solid-state fermentation, and the alcohol content of the liquor was relatively high. *Essential Techniques for the Peasantry* by Jia Sixie of

Picture 2.25 Workers treading on the yeast at the Moutai Distillery in the 1950s

the Northern Wei dynasty records 40 liquor brewing methods, as well as many kinds of solid state or high gravity fermentation. People have already acquired a wealth of experience in solid state or high gravity fermentation techniques in traditional wine making, and have used these experiences for the development of distilling liquor.

When distilling liquor, the distiller's yeast is key to creating good spirit, since, without it, nothing can be produced. The 13 kinds of distiller's yeast introduced in the *Beishan Liquor Book* adopt three yeast-making methods: masking, aeration, and wind. These three methods are somewhat random in terms of living at the mercy of the elements, and the operation is not standardized. In contrast, the distiller's yeast-making technique of modern distilled liquor is more scientific in controlling temperature, humidity, and fermentation time, and the operation is more standardized. Additionally, to better collect beneficial microbial strains, the changes in temperature and humidity are used to promote the propagation of molds. *Daqu* (hard distiller's yeast) cultivated at high temperature, medium temperature, and low temperature, and *Daqu* of different strains are produced in this way. For example, Fenjiu uses the well-known "two airing and two heating" processes to produce three types of distiller's yeast including *Qingcha, Hongxin,* and *Houhuo*. Then they are used in the brewing process according to a specified ratio. Moutai uses high-temperature distiller's yeast. Yeast bricks need to be cultivated for 40 days with a product temperature of 60 °C or higher. This kind of high-temperature distiller's yeast naturally formed yellow, white, and black yeast due to the difference in temperature of the glutinous rice.

Picture 2.26   Yeast bricks used in brewing Moutai *baijiu*

Yellow yeast is the most fragrant and has the best quality with the highest proportion being over 80% in the mixture. It can be seen that the yeast-making technology of distilled liquor not only incorporates the traditional yeast-making process, but also has developed, thus reaching a new level in production methods.

The second key technique for distilled liquor inherited from traditional brewing techniques is fermentation (as shown in Pictures 2.27, 2.28 and 2.29). The fermented natural liquor is separated from the fermented grain by squeezing or filtering, since the fermentation is mostly carried out in the liquid or thick mash. However, distilled liquor adopts distillation techniques that allow water vapor to extract the liquor from the mash. The fermented grain does not need to be in a liquid state, and a loose solid state is preferred. For this reason, adding chaff and rice bran to the fermented grain can not only prevent it from hardening, but also plays a role in loosening and storing nutrients, which is conducive to fermentation and distillation. Since solid state fermentation is used, the fermentation equipment is not limited to pots, wooden cellars, stone cellars, and mud cellars can also be used. In particular, the old mud in the mud cellar can continuously domesticate and enrich the microbial population and improve the quality of the brewed liquor (as shown in Pictures 2.30, 2.31, 2.32 and 2.33).

The technique for producing yellow rice wine that makes full use of mother liquor is still widely used in today's distilled liquor production. Many distilleries use

Picture 2.27 The iron liquor distilling utensil of the Fenjiu Distillery during the Republic of China

the original cellar layered grain stacking method, the reserved empty cellar layered distillation method (the production process of Sichuan strong-flavor *Daqu* liquor), or the co-firing old five steamer method (the production process of *Daqu* liquor from Jiangsu, Anhui, Shandong and Henan provinces), which all adopt replenishing grains and ingredients, and co-steaming and co-firing is the development of this technology. During the production of Moutai, the grain is put in use in two batches and the liquor is taken in 7 batches. The same batch of raw materials has to go through 9 rounds of steaming, 8 times of yeast adding, stacking for fermentation, and fermentation in the cellar. After 7 times of distilling, the distillers' grain is discarded. The whole process could be found in the traditional liquor brewing process. Based on solid-state fermentation, another two fermentation methods, which are combined with airing house stacking fermentation or closed-cellar and in-cellar fermentation, have been developed. This fermentation method, being obviously different from other flavored liquor, is an important innovation of Moutai. With this process, the fermented grains are in full contact with the microbial strains in the environment when they are piled up, which is equivalent to adding a new strain with certain saccharification and

Picture 2.28 The fermentation workshop with underground vats of Fenjiu

Picture 2.29 Putting the fermented grain into the distilling utensil is one of the key manual techniques in Fenjiu brewing

Picture 2.30 Luzhou Daqu Old Cellar, the brewery of more than 400 years

Picture 2.31 The tin *tianguo* (pot for liquor distilling) at Luzhou Laojiao

曲药制作、鉴评技艺流程图

1-小麦　　　　　　　　　2-润麦　　　　　　　　　3-拌料

4-踩坯　　　　　　　　　5-晾曲　　　　　　　　　6-安曲

7-培菌　　　　　　　　　8-翻曲　　　　　　　　　9-生香

10-储存　　　　　　　　　11-鉴评

Picture 2.32   Yeast-making process of Luzhou Laojiao *baijiu* 1. Wheat 2. Wetting wheat 3. Mixing materials. 4. Stamping base 5. Airing yeast 6. Placing yeast bricks. 7. Cultivating bacteria 8. Turning over yeast 9. Producing fragrance. 10. Storage 11. Evaluation

泸州老窖酒传统酿造技艺流程图

1-泸高粱　　2-挖槽　　3-下粮
4-拌粮　　5-上甑（蒸酒蒸粮）　　6-看花摘酒
7-出甑　　8-打量水　　9-摊凉
10-下曲（用脚踢手摸测试温度）　　11-入窖　　12-封窖
13-滴窖　　14-起槽　　15-堆糟
16-洞藏　　17-尝评勾兑　　18-包装成品

Picture 2.33 The brewing process of Luzhou Laojiao *baijiu* 1. Lu sorghum 2. Ditching 3. Laying grain. 4. Mixing grains 5. Steaming (steaming alcohol and steaming grains) 6. Observing hop and gathering distillate. 7. Taking out of the steamer 8. Adding water 9. Spreading to cool down. 10. Putting in yeast (use feet and hands to test the temperature) 11. Laying it in the cellar 12. Sealing the cellar. 13. Yellow water absorption 14. Taking out distiller's grain 15. Stacking distiller's grain. 16. Cave storing 17. Tasting and blending 18. Packaging

Picture 2.34 Adding cold water to the *tianguo* (pot for liquor distilling) (the brewing process of Beijing Niulanshan Erguotou Liquor)

alcoholic power to accelerate the fermentation and liquor brewing process. As there are many enzymes in the high-temperature yeast that decompose protein and fat, the huge microbial system fluctuates and grows, and produces a variety of metabolites during the entire fermentation process, resulting in an outstanding *jiang* flavor, with a pure and mild body, and long aftertaste (as shown in Pictures 2.34, 2.35, 2.36, 2.37 and 2.38).

2.2.2   Vinegar Brewing

The main component of edible vinegar is acetic acid, which is formed by further oxidation of ethanol. People have long discovered that liquor with very low alcohol content that is stored in an environment above 30 °C will become sour quickly, so after having discovered how to make liquor, people learned how to produce vinegar.

During the Shang and Zhou dynasties, sour food, including vinegar, was collectively referred to as *xi* (醯), and the words *cu* (酢), *cu* (醋), and *kujiu* (苦酒) were later derived from this. At that time, grains were used to make vinegar.

The seasoning vinegar used in the West is mainly fruit vinegar, while the Chinese ancestors learned from the experience of liquor making and took the lead in using

Picture 2.35 Observing hops (the brewing process of Beijing Erguotou Liquor)

grains as raw materials, and used the solid-state fermentation process to produce a variety of brands of vinegar with excellent color, fragrance, taste, and body, such as Chinkiang vinegar, Shanxi mature vinegar, Fujian red yeast rice vinegar, and Sichuan Baoning vinegar. They not only have a strong mellow and vinegary aroma, soft acidity, a mellow aftertaste, a thick and clear body, but they also contain a variety of amino acids, which can buffer and harmonize the sharp stimulation of acetic acid. In short, good vinegar should not be sour. Chinese vinegar is not only an important condiment, but also has the functions of increasing appetite, helping digestion, eliminating fatigue, stabilizing blood pressure, and preventing arteriosclerosis.

There are many types of vinegar in China, which can be divided into rice vinegar, wine vinegar, sweet vinegar, and acetic vinegar, according to the raw materials used in the production process. The famous vinegars in history are rice vinegar, made by the traditional solid-state fermentation process. The temperature of the solid-state fermentation product is controlled at about 40 °C, which is conducive to the progress of biochemical reactions such as low-temperature saccharification, medium-temperature alcoholization, and moderate-temperature acetification. In this process, various beneficial bacteria and their secretion enzymes coexist, so that saccharification, alcoholization, and acetification can take place simultaneously, thereby alleviating the interference of starch and ethanol on saccharomyces and acetic acid bacteria, promoting the growth of beneficial microorganisms, and making the fermentation process more complete. The coexistence of various microorganisms and the

Picture 2.36 The well cellar of Shandong Zibo Bandaojing Distillery

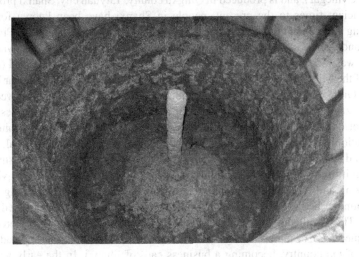

Picture 2.37 Inside of the well cellar (the cellar core is erected in the center)

Picture 2.38 The pit mud of Xuanjiu's small cellar accumulates layer by layer, up to seven layers

co-fermentation of multiple enzymes is an important feature of traditional Chinese solid state fermentation of vinegar, which precisely guarantees the excellent quality and unique flavor of vinegar.

2.2.2.1 Shanxi Aged Vinegar Brewing Skills

Shanxi aged vinegar ranks first among China's four famous vinegars (namely, Shanxi mature vinegar, Chinkiang Hengshun vinegar, Sichuan Baoning vinegar, Fujian red yeast rice vinegar), and is produced in Qingxu county, Taiyuan city, Shanxi province. From ancient times to the present, people in Shanxi have been known for both producing and consuming good vinegar. Qingxu and Jinzhong's Fenyang, Pingyao, Jiexiu, and Xiaoyi are even more famous. Before the Ming and Qing dynasties, Jinyang was famous for its aged white vinegar. During the Hongwu period (1368–1398) of the Ming dynasty, Wang Laifu, the master of the "Meiheju" Vinegar Workshop in Qingyuan, Taiyuan prefecture, added a fumigation of vinegar between the "vinegarization" and "drizzling vinegar" of the original white vinegar process. The effect not only enhanced the color, but also promoted esterification, inhibited the proliferation of some bacteria, and made the finished vinegar more flavorful. It had a mild and sour taste. After that, he placed the vinegar jar in the yard in the summer, exposing it to the weather. In cold days in winter, he was not afraid of the vinegar jar freezing and took off the ice on the surface. This method of "exposing to the sun in summer and scooping up ice in winter" is a continuously concentrated aging process. This is called mature vinegar. Since then, with the rise of Shanxi merchants, the unique flavor of mature vinegar has come to the forefront and has spread to all corners of the country, becoming a business card of Shanxi. In the early years of the Republic of China, Wang Liangcai and others took over "Meiheju". In 1924, the Shanxi mature vinegar sent by Meiheju won the first prize for high-quality goods at the International Exposition and has since become famous both at home and abroad (as shown in Picture 2.39).

Picture 2.39 Meiheju, the birthplace of Shanxi mature vinegar

The climatic characteristic of the mature vinegar brewing environment is that the four seasons are distinct: cold and dry in winter, windy and dry in spring, hot and rainy in summer, and cool in autumn with good air fluidity, low humidity, sufficient sunshine, and heavy rainfall.

The raw materials are sorghum, barley, peas, caving, and bran. The brewing process of mature vinegar is as follows. Grind the sorghum consequently into four, six, and eight pieces with a stone mill, use liquid alcohol fermentation, solid acetic acid fermentation, and then smoke and leach the vinegar grain. The aging period should be at least one year (as shown in Picture 2.40). The mixed fermentation of various grains, beans, and bran, complementing each other's biochemical and physical and chemical reactions, and various nutritional supplements of microbial strains enhance reproduction and transformation, making the quality of aged vinegar significantly different from other vinegars, which could be summarized in five words: "mild, sour, fragrant, sweet, and fresh".

2.2.2.2 Chinkiang Vinegar Brewing Skills

Chinkiang vinegar uses high-quality glutinous rice as the main raw material and uses *Daqu* as the saccharification starter. It is brewed through three major processes (brewing, fermentation, and leaching) with more than 40 procedures in more than 60 days.

The main components are glutinous rice, bran, and crude bran. Alcohol fermentation is divided into pre-fermentation period, main fermentation period, and post-fermentation period. When the distiller's yeast and mixture enter the fermentation tank, the pre-fermentation period begins. During this period, the mixture is rich in nutrients and has a small amount of dissolved oxygen, so the yeast can multiply rapidly. After 8–10 h, the yeast is produced in large quantities and reaches a certain concentration, entering the main fermentation period dominated by alcohol fermentation. In the post-fermentation period, most of the sugar has been consumed, and the fermentation effect slows down.

Under the action of acetic acid bacteria, alcohol is first oxidized to acetaldehyde, and then to acetic acid. This process is collectively called acetic acid fermentation. It is the most important part in vinegar production. Generally, the operation is undertaken in an open environment, which promotes multi-strain mixed fermentation and is quite complicated. Hengshun vinegar is produced according to the solid state layered fermentation method. This is a unique process. As with alcohol fermentation, accurate temperature control is key.

After precipitation, the leached raw vinegar is concentrated at a high temperature. This process has the function of sterilization, during which protein can be denatured, coagulated and precipitated, and then removed. The concentrated vinegar is cooled to 60 °C and aged in a storage tank for 1–6 months. During the aging process, esterification significantly improves the flavor of vinegar. The longer the storage time, the more obvious the effect.

2.2.3 Paste, Soy Sauce and Chinese Douchi

Paste and soy sauce are also fermented products of grains. *Origins of Articles* says "Zhou Gong (an outstanding thinker in the Zhou dynasty) makes paste", *The Analects of Confucius* contains such words as "no paste, no eating", and *Rites of Zhou* records that there are 120 urns of paste served for the king, indicating that the Chinese have been making paste for a long time. In the Han dynasty, paste became a staple in their food culture. As the old saying goes, "Such basic necessities as oil, salt, paste, vinegar, and tea are indispensable in the daily lives of Chinese people". It seems

Picture 2.40 The brewing
process of mature vinegar:
steaming, fermentation,
smoking, leaching and aging

Picture 2.40 (continued)

clear that paste is a must-have for every family. First, meat can be turned into meat paste, fish can be made into fish paste, and vegetables can be made into vegetable paste. Making paste became one of the methods of processing food. Later, people discovered that using flour, soybeans, and broad beans as raw material, the flour paste, salted and fermented paste, and bean paste made through the process of moisturizing → steaming → yeast making → fermentation → making unstrained paste, were not only good dishes for easy consumption, storage and carrying, but could also be used as condiments and for pickling vegetables. Due to the various uses of these kinds of paste, their production skills have continuously been inherited and developed, and even become an essential food for every household. Eating and making paste spread throughout East Asia. In Japan and Korea, sauce soup was served in almost every meal. Therefore, eating and making paste is a feature of East Asian food culture, and its birthplace is China.

Soy sauce was originally the juice skimmed from the paste for flavoring. Later, it was discovered that it had some functions that sauce could not replace, so it gradually became independent and formed the soy sauce drying process.

Douchi is a semi-fermented product processed from soybeans and other crops. Because it is semi-fermented, the starch, fat, and protein contained in it are partially decomposed because of the mold and its enzyme system to form rich sugars, amino acids, peptides, and other nutrients, which have a special flavor when eaten, and it became the special ingredient of certain dishes and got very popular.

2.2.3.1   Bean Paste-Making Skills

Soybean paste uses soybeans as the main ingredient, sometimes supplemented by flour, barley, or rice. Soybean paste that uses soybeans alone as the main component is rare in other places except for the rural areas in northeast China. Soybean paste with flour as the auxiliary ingredient is called flour and bean paste, which is currently a more popular type of paste.

Soybean paste is a fully fermented product. With its beneficial microorganisms, the starch in the beans and their adjuvant decomposes to produce sugars and dextrin, while the protein also decomposes to produce a variety of amino acids, altogether resulting in something flavorful (as shown in Pictures 2.41 and 2.42).

Picture 2.41 Yeast-making bed with soybean paste

Picture 2.42 Soybean paste becoming yeast

2.2.3.2 Yellow Paste-Making Skills

Yellow paste is a combination of flour and beans and is made from soybeans as the main ingredient and fermented with flour. The process flow is as follows.

Soybean → measuring → impurity removal → rinsing → soaking → steaming material → cooling → mixed inoculation → ventilating yeast → first turning → second turning → forming yeast → entering vat → overturning → thinning → mixing → maturing and grinding → flour and soybean paste.

There are two ways to process the ingredients. One is the steaming method: Put the soaked soybeans into the pot and steam the ingredients, first steam over a big fire for an hour, then simmer over a slow fire for two hours. The second is the cooking method: The soaked soybeans are directly cooked in the pot. Generally, small-scale production often adopts the cooking method (as shown in Pictures 2.43 and 2.44).

The steamed soybeans should be cooled quickly to reduce the contamination of bacteria. Mix the pre-prepared seed yeast with the flour, and mix well with beans to complete the inoculation process.

Traditional paste-making technology has used two methods to make yeast: One is to use mixed fermentation with yeast like liquor brewing, which was more common in ancient times. The other is fermenting in a vat. After the yeast is put in the vat, salt water is added, and the mixture is mixed and fermented. Stirring is commonly known as "*dapa*", and it happens during the first of the three ten-day periods of the dog days (mid-July).

Picture 2.43 Paste-making low-temperature fermentation tank

Picture 2.44 Drying yellow paste in the fermentation tank

2.2.3.3 Flour Paste-Making Skills

The sweet flour paste in Beijing has the best quality made by the natural sweet flour paste method. The main reason is the unique technology: Making yeast in spring and making paste in summer. The paste is made through the fermentation, offset by exposure to the respective elements during the day and night. Because the enzyme system of yeast making is more complex, there are more microorganisms involved in fermentation, and the metabolic products are also abundant. Additionally, the fermentation temperature is low and takes a long time, which not only can produce a large amount of flavor substances, but also produces more non-enzymatic reaction flavor substances, so the flavor of the finished product is good.

First, blend the dough into noodles according to the raw material ratio of flour: yeast: water = 100: 30: 30, knead the dough and cut it into pieces like making steamed buns, and put them in a basket for steaming.

The steamed buns are cooled to 40 °C and placed on the reed foil of the yeast shelf. After stacking, the yeast shelf must be tightly wrapped with two layers of reed mats, and then sprayed with water to maintain the humidity of the yeat room, which is conducive to mold germination and reproduction. Aspergillus oryzae is hidden in the long-used yeast shelf, straw and reed mat, which is the equivalent to natural inoculation.

The next process is the mixture's fermentation. First brush the leavened dough to remove the pili (facilitating the entry of saline into the yeast), smash it, weigh it, and put the leavened dough into a tank; add 150 kg of yeast per tank, then add 150 kg of saline with a concentration of 14%, stir well, and then expose it day and night, lightly stir once every 1–2 days, replenish 3 kgs of saline every 4 days, and replenish the yeast 4 times. It matures in 2 months in summer.

The saline used for the fermentation process has a Baumé (salinity) of 14, which is lower than that of soybean paste, because high salinity inhibits enzyme activity. Therefore, the flour paste is fermented in a low-salt solid state. After about 7–10 days, add a small amount of saline and stir the yeast to form a thick mixture. The temperature of the product is controlled at about 30 °C, and the product is matured at a low temperature. After that, while controlling the temperature, stir once a day to keep the temperature consistent. Proper oxidation can promote the production of pigment and aromas.

The mature flour paste will inevitably have lumps due to incomplete stirring, so it needs to be finely ground and filtered to ensure its quality. It also needs to be sterilized and an appropriate amount of potassium sorbate solution needs to be added to prevent deterioration.

The finished flour paste should be bright yellowish or reddish brown in color, shiny, aromatic, tasting moderately salty and sweet, fresh and mellow, with no acid, bitter, astringent, burnt and other peculiar smells, and moderate viscosity.

2.2.3.4   *Douban* **Paste-Making Skills**

Douban paste, or "thick spicy broad bean paste", is made with broad beans as the main ingredient, so it is also called broad bean paste for short. It is mainly produced in Sichuan. *Records from the Kitchen* written by Zeng Yi during the Guangxu period of the Qing dynasty describe how spicy bean paste in the author's hometown was produced: "Soak the big broad beans and take them out once they are ready. Then, grind them and remove the shells. Pour boiling water over the beans and then gather them with a dustpan. Mix it to make dough and stretch it until it is thin and even, and place it in a dark room when it is slightly cool. Cover it with a straw or reed mat and wait seven to eight days. When the yellow mildew (aspergillus flavus) develops, expose it to the elements throughout the day and night. From the end of July, put the broad beans into a saltwater tank and air it until the red pepper is ripe. Chop the red pepper and mix with it the broad bean. After being exposed for another two or three days, store the mixture in a jar. Add a little sweet liquor and the broad bean paste can be kept for years". This record shows that Sichuan spicy *douban* paste is made by mixing a small amount of flour with raw broad beans, first making yeast and then fermenting. Today, the production of *douban* paste in Pixian county, Sichuan province follows this process (as shown in Pictures 2.45 and 2.46).

According to whether the broad beans are steamed or cooked or not before making yeast, the craftsmanship of *douban* paste is divided into two types: raw and cooked.

The former is mainly for cooking, and almost all of them are spicy. The latter is meant to be eaten together with rice or bread.

Picture 2.45 Broad bean yeast under cultivation

Picture 2.46 Broad bean sauce made with mashed salt water

The spicy *douban* paste produced in Pixian county is a famous historical brand. It was founded in 1873 and is characterized by a mellow scent, no spices, crispy texture, strong spicy flavor, and a bright red color. The traditional spicy *douban* paste is fermented naturally. One method is sealed fermentation, and the other is exposing it to the elements. Pixian adopted the latter. Put the yeast into a jar with salt water, and mix a certain proportion of sliced fresh red peppers. After exposure to the elements day and night and regular tedding, the material put into the jar in summer will have to ferment from the hot summer to late autumn, and they will mature in about 6 months (as shown in Pictures 2.47 and 2.48).

2.2.3.5 Chinese *Douchi*-Making Skills

There are many varieties of Chinese *douchi* (fermented black beans). Two varieties will be introduced here.

(1) Liuyang *douchi*

Hunan's *douchi* can be represented by Liuyang, which belongs to the dried *douchi*, plain *douchi* and light *douchi* of the aspergillus oryzae type.

It uses natural aspergillus oryzae to fermentation, and the specific operations are as follows.

Sift the soybeans and steam them in the steamer for 20 min until they are half-cooked. After washing away the mud and sand impurities, soak in water for 20–40 min

Picture 2.47 Spicy *douban* paste exposed to the elements day and night

and re-steam for 1 h. After steaming, spread them out on the reed mat to cool down to disperse the moisture. When there is residual heat, move the beans into the yeast basket of the yeast chamber, and close the door tightly to get them moldy. When a layer of yellow mold grows on the surface of the beans, ventilate them and cool them down. A few days later, the prepared bean yeast needs to be washed to remove the mold spores on the surface, and poured into bamboo strips to allow it to ferment naturally. After 2 days and 2 nights, the bean yeast is then stripped, rubbed and scattered, and then fermented in a wooden barrel for a certain period of time. Choose a sunny day, spread the matured bean yeast in the drying yard, dry out the surface moisture, and then put it back into the storage room for further ripening and natural fermentation. It can be sold at the market as a finished product within 1–3 months (as shown in Pictures 2.49 and 2.50).

(2) Tongchuan *douchi*

Tongchuan *douchi* is the representative of mucor-type *douchi*. It is produced in Santai county, Sichuan province. The douchi made there was thus named since Santai was called Tongchuan in ancient times.

It's said that the production of Tongchuan *douchi* began in the early Qing dynasty and has a history of more than 300 years. Later, Chongqing, Chengdu, Deyang,

Picture 2.48 Naturally fermented spicy *douban* paste

Shehong, Jiangyou, Wenjiang, and Pixian all produced this type of *douchi*. Tongchuan *douchi* is black, with a granular texture, incompact, shiny, fresh, and strongly fragrant and sweet. It easily dissolves in soup and is suitable for cooking. Tongchuan *douchi* is best produced in winter, because the cold temperature is suitable for the reproduction of mucor.

There are two main ingredients of Tongchuan *douchi*: One is locally produced black beans and the product is called big *douchi*; the other is soybeans, and the product is known as small *douchi*. The production techniques are basically the same, but big *douchi* is superior. The production process is as follows.

Soak the soybeans for about 10 h, and drain them. Steam on the steamer for about 2 h until they are half-cooked, then turn the steamer upside down to ensure the beans are evenly cooked. Stop heating the beans when the beans can be crumbled. The steamed beans are cooled on a bamboo mat to room temperature, and then transferred to the bamboo board of the yeast chamber to leave them to naturally mold.

The prepared bean yeast is poured on a bamboo mat, the agglomerated is opened, and then put in a dustpan and mixed with salt. For every 50 kg of bean yeast, 4.5 kg of salt is added and 0.5 kg of alcohol, and mixed well. If the bean yeast is too dry,

Picture 2.49 Chinese *douchi* yeast after *yeast* is produced

Picture 2.50 Fermentation tank of *douchi*

water needs to be added. The bean yeast mixed with the salt and liquor is put in a jar for fermentation, and the upper layer of the bean yeast is sprinkled with a layer of salt and the jar is sealed with oil paper. The bean yeast is generally placed in the open air, and the fermentation is promoted by sunshine. In the third month of the lunar calendar, the jar containing the beans is moved indoors to avoid high temperature, rain and dew, so as not to ruin the fermentation process. The entire aging period takes about half a year; it is not matured enough until the fifth month of the lunar calendar.

2.2.3.6 Soy Sauce-Making Skills

The production of soy sauce can be regarded as the extension and development of the bean paste technology. The main difference between them is that one is a liquid and the other is a thick paste. Their different physical states are suitable for different occasions. This can be confirmed by relevant historical facts.

Soy paste appeared first in the Han dynasty and was called "*qingjiang*" (clear sauce). In the Southern and Northern dynasties, it was also called "*jiangqing*" (paste sauce) and "*chizhi*" (*douchi* sauce). It was called "*jiangzhi*" in the Tang dynasty. It was not called soy sauce until the Song dynasty.

After the Ming dynasty, there were more and more records of soy sauce-making skills. Daixi's *Monthly Ordinances for Superabundance* in the Ming dynasty detailed the Nanjing soy sauce recipe: "Each *dou* (about 6 kg) of soybeans use 10 kg of flour. The beans must be cooked first with water about 20 cm above the bean surface. After cooking, spread the beans cool them down afterward. The cooking water must not be discarded. Put the beans with the flour in a large basin and mix them thoroughly. Add the cooking water if it is dry, and mix the beans, flour, and cooking water into granules. Spread the granules on the door plank with a reed mat underneath and another at the top of the granules, and then use the quilt to cover them; remove the quilt after they become hot. Three days later, remove the mat covering them, and take the granules out after 7 days; then spread them on a bed sheet and air them for another 7 days; leave them with dust and mildew, and do not wash them away. When putting them in the container, mix the black soybean granules with salt and cold well water at a ratio of 1:1:6, stir well, expose them to the air until they are sunburnt and ready for use. Then sieve the mixture to get the sauce, and the sauce may be edible after the precipitation becomes clear. The sediment and fermentation residue is still added with salt and water according to the previous ratio, and then air it again, and take the soy sauce. The bottom residue is very salty. They could be mixed evenly with chopped vegetables or radishes, and aired to make dry vegetable. It can be used as *douchi*, but it has impurities". This record indicates that the soy sauce brewing technique had taken shape and had already become mature by that time. The following is a brief introduction to the production techniques of Hunan Xiangtan Dragon soy sauce (as shown in Picture 2.51).

Dragon brand soy sauce uses soybeans and flour as main ingredients, and the ratio of soybeans to flour is 2:1. Soybeans selected must be soaked for 3 h in summer and 5 h in winter until the beans swell without wrinkles on its skin. Drain and cook under high pressure for 30 min, cool to 40 °C, mix well with flour, and move them to a ventilated yeast-making tank to make yeast. When the surface of the bean yeast is covered with hyphae and green spores are formed, the yeast has aged sufficiently.

Picture 2.51 Natural sun-dried soy sauce

The fermented beans are then transferred into a large tank, mixed with salt water, and placed in the sauce field to air them. A canopy is placed over them to prevent them from rain. The mash is generally made in early summer. After the three periods of the hot summer (dog days), under high temperature, the mash is fully fermented, and meanwhile the moisture volatilizes. The sauce gradually thickens, and finally becomes a reddish brown thick mash. During this period, you need to stir it one to two times a day, until the sauce turns into an oily black red mixture with a fragrant smell, the so-called "drying sauce in the three ten days period of the hot summer, and it becomes soy sauce in autumn". In order to obtain high-quality soy sauce, it needs to be exposed to the sun for a long time and then fermented to improve its flavor.

2.3 Tea-Making Technique

2.3.1 Chinese Tea

China is the hometown of tea. Both making and drinking tea have a history of thousands of years. The custom of tea-drinking has penetrated into many areas of people's lives, giving birth to a rich and colorful tea culture. *Shennong's Herbal Classic* states: "Shennong collected a hundred types of herbal medicine, found 72 poisons in one day, and got tea to help detoxify the body". The book was a work in the

Eastern Han dynasty. The original book was long lost, but people still believe it. For example, Lu Yu in the Tang dynasty says in *The Classic of Tea* that "tea originated from Shennong".

Lu Yu (733–804), with the courtesy name if Hongjian and literary name Dong-gangzi, was born in Jingling (now Tianmen city, Hubei), Fuzhou prefecture, in the Tang dynasty. He was addicted to tea all his life and was good at tea ceremonies. He is famous for the world's first tea monograph *The Classic of Tea*, and made outstanding contributions to the development of the tea industry, and is regarded as the "tea sage" and worshipped as the "tea god".

Due to climate, environment, and other factors, Southwest China, especially Bashu (area around Sichuan Basin), was an early tea producing area, tea making, and tea drinking center. In the Qin dynasty, tea affairs centered on tea drinking began to spread to the east. After the mid-Tang dynasty, the tea-drinking style was popularized in northern areas, and the tea-drinking style prevailed in the court and the commonalty. "Tea is food, nothing different from rice and salt", tea became one of the main daily necessities: firewood, rice, oil, salt, soy, vinegar, and tea. "Present the visitor with a cup of hot tea as soon as he steps into the house" has become a part of etiquette and social customs. As a result, people pay more attention to the quality of tea, the way of drinking tea, and the utensils, so there is a practice of tasting tea.

With the addition of cultural elements to the custom of tea drinking, its function is not limited to quenching thirst and health care, but has also become a cultural display and spiritual enjoyment. When tea drinking focuses on drinking skills, "tea art" is developed; when tea drinking focuses on etiquette, "tea ritual" is developed; when morality is emphasized, "tea ceremony" is developed. This requires not only the fine processing of tea, with attention to its color, flavor and shape, but also the quality and drinking skills of tea. This developed from a simple "sampling of tea" to the social and commercial custom of "*doucha*", which refers to "tasting tea to assess its quality". The tradition of *doucha* came into its own in the Song dynasty, which centered around the appraisal of tea and the display of tea art. Tea tasting and *doucha* are deliberately pursued as artistic activities that show elegance and self-expression. Tea drinking is integrated into cultural activities and constitutes a tea culture. In turn, it promotes the popularization of tea drinking, and promotes tea-making skills and the prosperity of the tea industry (as shown in Picture 2.52).

Through Sino-foreign trade and exchanges, China's tea-picking, tea-making techniques, and tea-drinking customs have spread to all over the world. Today, 2 billion people in more than 150 countries around the world have forged an indissoluble bond with tea. In Japan, a unique Japanese tea ceremony evolved. In South Korea, offerings in honor of the tea culture ancestors are held every year, and May 25th is designated as tea day every year. After the introduction of tea to the United Kingdom, it changed the lives and daily habits of the British population. By the eighteenth century, tea had become one of the most fashionable beverages in Britain. In other European countries and North America, tea has become an indispensable beverage like coffee and hot chocolate. India and Sri Lanka have become major tea-producing countries after China due to the European demand for tea.

Picture 2.52 Painting of Zhu Tong making tea

Due to the differences in geographical and natural conditions, there are many types of tea in China, which are only classified according to the color of the tea as shown in Fig. 2.1: green tea, black tea, oolong tea, white tea, yellow tea, and dark tea. According to the manufacturing process, there is a distinction between crude tea and refined tea. According to the degree of manufacturing fermentation, as shown in Fig. 2.2, there are three types: non-fermented, semi-fermented, and full-fermented (as shown in Picture 2.53).

2.3.2 Traditional Tea-Making Technology

The traditional tea-making techniques are all derived from green tea making, and the tea-roasting tools are also similar (as shown in Picture 2.54).

2.3.2.1 Green Tea

Regarding the preparation method of roasted green tea, most tea people in various places adapt to local conditions, so they have many unique skills to produce varieties with their own characteristics in appearance and qualities. The production processes of green tea are similar, mainly consisting of high-temperature blanching, twisting, re-roasting, and baking.

The craftsmanship of roasting green tea was quite perfect in the Ming dynasty. According to *Stir-Roasting Tea* in *Tea Commentary* by Xu Cishu's in the Ming dynasty, "When the raw tea is picked at the beginning, the aroma is not clear, and fire must be used to bring out its fragrance. However, it is not durable and should not be roasted for a long time. If you add too much to the pan, it will roast unevenly because the pan is too densely packed. Leave it in too long, and it will become over-ripe and the fragrance is dispersed. It might even burn and be unsuitable for brewing. Using a new iron pot is a taboo for roasting tea. Once the iron rust of the pot enters the tea, the tea will no longer be fragrant. Even worse is the grease, which would do more

Green tea	Roasted green tea	Mee tea (Roasted Green, Special Mee, Chun Mee, Fengmei, Gongxi)
		Pearl tea (Zhucha, Rain Tea, Xiumei)
		Delicately roasted green tea (Longjing, Dafang, Biluochun, Yuhua Tea, Pine Needle)
	Oven-dried (baked) green tea	Ordinary oven-dried green tea (Fujian, Zhejiang, Anhui, Suzhou oven-dried green)
		Tenderly oven-dried green tea (Huangshan Maofeng, Taiping Houkui, Huading Yunwu, Gaoqiao Yinfeng)
	Sun-dried green tea (Dianqing, Chuanqing) Steamed green tea (Sencha, Yulu)	
Black	Souchong black tea (Lapsang Souchong, Yan Souchong)	

Fig. 2.1 Classification by tea color

tea	Gongfu black tea (Yunan, Qimen, Sichuan, Fujian black tea) Broken black tea (leaf tea, broken tea, sliced tea, powdered tea)
Oolong tea	Oolong in northern Fujian (Wuyi Rock Tea, Narcissus, Dahongpao, Cinnamon) Oolong in southern Fujian (Tieguanyin, Qilan, Narcissus, Golden Ssmanthus) Guangdong Oolong (Phoenix Dancong, Phoenix Narcissus, Lingtou Dancong) Taiwan Oolong (Dongding Oolong, Baozhong Oolong.)
White Tea	White bud tea (Silver Needle) White leaf tea (White Peony, Yemei)
Yellow tea	Yellow sprout tea (Junshan Silver Needle, Mengding Yellow Bud) Yellow small tea (Bei Maojian, Weishan Maojian, Wenzhou Huangtang) Yellow big tea (Huoshan Yellow Big Tea, Guangdong Dayeqing)
Dark tea	Hunan dark tea (Anhua Dark Tea) Hubei old green tea (Puqi Old Green Tea) Sichuan brick tea (South Road

Fig. 2.1 (continued)

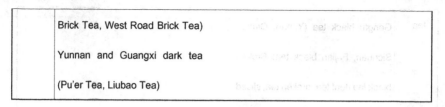

| Brick Tea, West Road Brick Tea) |
| Yunnan and Guangxi dark tea |
| (Pu'er Tea, Liubao Tea) |

Fig. 2.1 (continued)

Non-fermented tea	Semi-fermented tea					Fully-fermented tea
Green tea	Oolong tea					Black tea
0%	15%	20%	30%	40%	70%	100%
Longjing, Biluochun	Light Oolong	Jasmine Tea	Dongding Tea	Tieguanyin	Pekoe Oolong	Black tea

Fig. 2.2 Classification by fermentation degree

harm to the tea fragrance than the iron rust. A pan must be reserved specially for roasting tea and shall not be used for other purposes. For roasting tea, only branches should be used as firewood instead of dried leaves. Because the dryness of dried leaves will lead to a fierce fire, while regular leaves would easily burst into flame and extinguish. The pan must be clean; after the tea leaves are picked from the tea trees, quickly roast the leaves. Only 4 *liang* (200 g) of tea can be roasted in a pan at one time. First use a gentle fire to bake it, then use the strong fire to expedite it. Wear finger cots to quickly stir and roast the tea until it is medium-baked and becomes fragrant, which is the right time to stop heating. Then, immediately take out the tea and place it in a bamboo cage, put large pieces of cotton paper underneath to wait for the tea to dry and wait for it to cool, then store it in a bottle. If there is enough manpower, more pans and cages could be used to make tea at the same time; but if there is not enough manpower, only one pan or two pans are needed, while still needing four or five bamboo cages. Because it takes a short time to stir-fry the tea but it takes a long time to bake the tea. The dry and wet tea cannot be mixed together, and the mixing of them will greatly reduce the fragrance. If one leaf is slightly burnt, the whole pan of tea is ruined. Fierce fire is a taboo especially when the pan is cool and the tea leaves baked will not be soft. It is the most difficult to achieve the ideal effect".

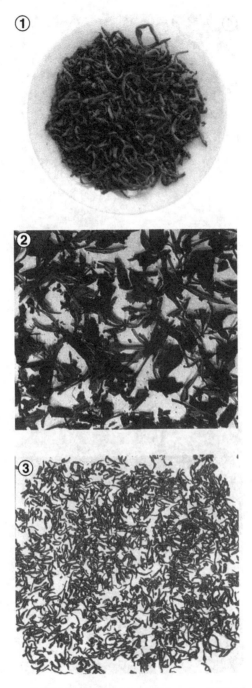

Picture 2.53   1. Green tea 2. White tea. 3. Black tea 4. Oolong tea. 5. Dark tea brick 6. Scented tea

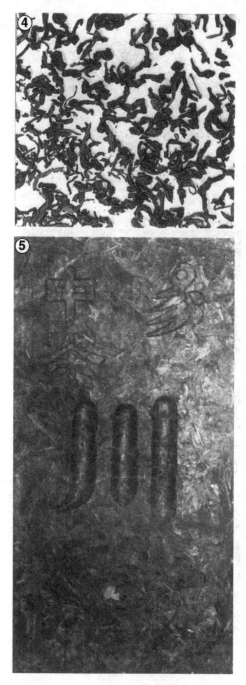

Picture 2.53 (continued)

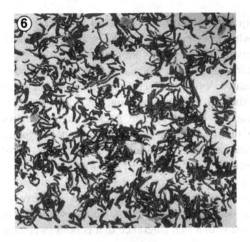

Picture 2.53 (continued)

Picture 2.54 Manual tea-roasting tools

West Lake Longjing is the most famous green tea. The fresh leaves are required to be picked early, when both the leaves and the buds are still tender. The fresh leaves are thinly spread, giving off the grassy air, reducing the bitterness and astringency, which increases the fragrance of tea. Then, according to the size and tenderness of the leaves, and the formation degree of the green tea in the pot, they are baked in a special smooth iron pot by shaking, slipping, banding, pinching, tossing, grasping, pushing, buckling, pressing, and rubbing. It roughly goes through three procedures: The first one is the *qingguo*, that is, keeping the temperature of the furnace at 80–100 °C, first grasping and shaking the fresh tea leaves, then slipping, pressing, shaking, and throwing to kill green by hand; roast until it is medium well to well-dried, take it out of the pot and spread it thinly for the tea to regain moisture and cool; the third procedure is re-roasting, that is, sieving the tea leaves and roast them in the pan again for drying and taking shape. If the roasted tea is stored properly, it becomes high-end Longjing tea with a rich fragrance that is both sweet and fresh (as shown in Picture 2.55).

2.3.2.2　Yellow Tea

Yellow tea production process: Picking fresh leaves → de-enzyming → first-step roasting → packed on a small pile for oxidization, fermentation and yellowing → sorting → drying.

While roasting green tea, when the temperature is too low, the time is too long, or when it is not spread out in time after the drying is finished, it is not dried in time after being spread out, or the accumulation time is too long, then the color of the tea will turn yellow, and the tea infusion will appear yellow during brewing and so will the tea leaves. This type of tea is called yellow tea. Yellow tea has a special, good flavor. Therefore, the tea workers consciously highlight the above-mentioned improprieties in the tea-making process, especially the addition of braising methods, forming a unique production technique for yellow tea. Xu Cishu's *Tea Commentary* records the history of this evolution: "Gu Bi is not good at tea making in the mountains. He roasts and bakes tea in a pan with a big fire. Before the tea is taken out of the pan, it has already been burnt and is not worthy of use. He then made a giant bamboo trap and stored tea when it was hot. Although the green branches and purple bamboo shoots of the bamboo trap will turn yellow, such tea is inferior and is not suitable for tasting or competition". It can be seen that the green tea cannot be made into a good one once it's been roasted too long; if so, it might just as well to store the tea when it is hot and have it turn into yellow tea this way.

2.3.2.3　Black Tea

The term "black tea", which literally translates to "red tea" in English, first appeared in the book *Various Abilities in Rustic Matters*, which was written in the sixteenth century. The earliest production of black tea was in the Wuyishan area of Fujian

Picture 2.55 West Lake Longjing tea and its processing techniques. 1. Longjing Tea Garden. 2. Spreading new tea. 3. Manual roasting

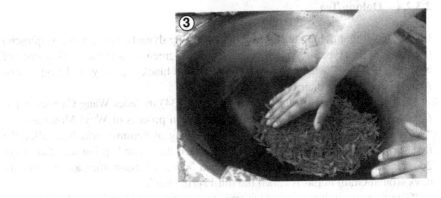

province. According to the legend, during the turmoil in the late Ming dynasty, an army entered Fujian from Jiangxi, passed through Renyi township, Chong'an county, and occupied the tea farm. The tea leaves could not be dried in time, and turned red. In order to recover the loss, the tea growers use combustible pine wood to heat and dry and obtain a kind of tea with a strong mellow pine aroma and excellent taste with longan flavor. This kind of tea could be put on the market after a little sieving. This is Lapsang Souchong black tea. In the seventeenth century, this kind of tea was transported by Shanxi merchants to Wuhan, Tianjin, and Datong, and then exported to Europe by the Dutch and East India Company. It soon gained popularity among the upper ten and royal families and replaced wine as a favorite drink. At the beginning of the eighteenth century, a group of missionaries came to China. They traced this black tea to the Wuyi Mountains, and then brought tea tree seeds and tea-making techniques to India, the British colony for planting, producing Indian Assam black tea and Darjeeling black tea, and other varieties. At the same time, Fujian 's black tea production techniques gradually extended to 12 other provinces. In the first year of the Guangxu period (1875) in the Qing dynasty, Yu Ganchen, who had served as the magistrate of Chong'an county, resigned and returned to his hometown in Anhui province. He successively set up tea houses in Dongzhi and Qimen county to produce black tea according to Minhong craftsmanship, turning Qimen, which originally only produced green tea, into a black tea production base (as shown in Picture 2.56). The export volume of black tea accounts for the largest amount of Chinese tea exports. Lapsang Souchong black tea is very popular in some European countries and is called "BOHEA", which is a homophone of "Wuyi".

Black tea production process: Fresh leaves → withering → kneading → fermentation → drying.

Souchong black tea production process: Fresh leaves → withering → kneading → fermentation → pan-roasting → re-kneading → baking → sieving and picking → re-baking and piling.

2.3.2.4   Oolong Tea

Legend has it that oolong tea appeared in the Song dynasty, also known as *qingcha* (green tea). This type of tea has the fragrance of green tea and the mellow taste of black tea. It is somewhere between green tea and black tea. Wuyi Dahongpao and Anxi Tieguanyin are their representatives.

Lu Tingcan's *Sequel to the Classic of Tea* (1734) includes Wang Caotang's *Tea Commentary* (1717), which details the production process of Wuyi Mountain tea: "Wuyi tea is picked from Grain Rain to Beginning of Summer, which is called the first spring tea, and the re-harvested tea is called the second spring tea after about 20 days around Lesser Fullness; and the tea harvested again after another twenty days around Grain in Ear is called the third spring tea".

"The first spring leaves are thick and strong, the second spring and third spring leaves become thinner, and the taste becomes less pronounced, and bitter. It is

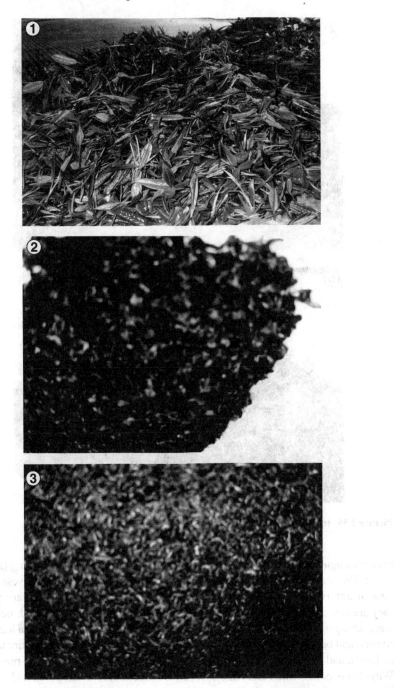

Picture 2.56 Qimen black tea. 1. Newly-picked fresh leaves. 2. Leaf color at the beginning of fermentation. 3. Leaf color at the end of fermentation. 4. Baking tea. 5. Tea after baking at high temperature

Picture 2.56 (continued)

harvested again in the late summer and early autumn. It is called *Qiushuang* (autumn frost). The fragrance is stronger and the taste is also good, but for the harvest of next year, it unfortunately cannot be picked too much. After the tea leaves are picked, they are spread evenly in bamboo baskets and placed in the sun and wind, hence the name *shaiqing* (sun-drying). The color is gradually faded and then the tea leaves are roasted and baked. Yixing tea is only steamed, not roasted, but baked on fire to make it. Usnea and Longjing are both roasted and not baked, so their color is pure. Only Wuyi tea is both roasted and baked. When it is cooked, it is half green and half red. The green color is the result of the roasting, and the red of baking. The tea is picked and then spread and swirled. While it is being roasted it will emit its aroma, but it should not be roasted for too long. Once roasted and baked (oven-dried), the old leaves and branches are picked out to make the color unified. It is vividly described by Shi Chaoquan in his poem as 'The fragrance is like the aroma of plum and orchid, and the craftsmanship is exquisite'".

The process is as follows: Picking tea leaves → withering (green drying: two processes of airing in the wind and two rounds of drying in the sun → drying green) → fine manipulation of green tea leaves (repeated shaking and airing in the wind) → baking (double baking and double kneading) → kneading → initial baking (i.e. first firing, commonly known as fire baking) → winnowing → airing in the wind (spreading out) → sorting → re-baking (big fire) → rolling into a ball → complement of firing → packing crude tea. These 10 procedures are closely linked together and cannot be changed or left out. In particular, "double withering", "watching out for fine manipulation of green tea leaves, checking out weather for fine manipulation of green tea leaves", "firing to dry", "double baking and double kneading", and "long-time roasting at low temperature" are all unique essences of Wuyi Dahongpao.

The production process of Anxi Tieguanyin: Fresh leaves → sun-drying → airing → shaking → airing → roasting → kneading → initial baking → wrapping and wrapping-kneading → re-baking → re-wrapping-kneading → drying → crude tea.

2.3.2.5 White Tea

According to *On Daguan Tea* written by Emperor Huizong of the Northern Song dynasty, "White tea is a kind of tea on its own; it is different from regular tea, and the leaves are thin. It grows naturally between the cliffs and in forests, so it is not planted by people. There are only four or five households that have such tea trees, and each household only has one or two such tea trees, so the tea cakes it could produce are only two or three *kua* (unit for calculating tea). There are not many tea buds, and it is especially difficult to steam and roast them. If the heating isn't properly controlled, the tea roasted would become a regular product. White tea must be roasted carefully and meticulously and proper production methods are needed. In this way, the tea cakes produced are completely clear from the outside to the inside, as if a beautiful jade is hidden in the rough stone, which is unparalleled by other varieties. There is also light roasted tea, but the taste is far inferior". Thus it can be seen that the yield of high-quality white tea is quite low. Being rare makes it rank top among the tribute tea.

Tian Yiheng's *Boiled Spring Essay* (1554) in the Ming dynasty described the method of making white tea: "The bud tea made using fire is inferior. Sun-dried tea is the best since it is closer to nature and also doesn't come with the smell of fire and smoke. Irregular heat can damage the tea's fragrance. A bowl of sun-dried tea contains unfolded quality tea leaves that are clear, green, and especially lovely". From this historical data, we know that the early high-quality white tea was not roasted or kneaded but sun-dried.

The production process of white tea: Fresh leaves → withering → baking (or drying in the shade) → picking → re-baking (drying).

2.3.2.6 Dark Tea

The term "dark tea", which literally translates to "black tea" in English, appeared no later than the mid-Ming dynasty. In the third year of the Jiajing period (1524) in the Ming dynasty, Chen Jiang says in his memorial to the throne that "Commercial tea is of low quality, with most being dark tea. This kind of tea is of the second grade of the upper and middle class branded on the parcels. It is tasted and inspected by the businessmen. Every 5 kg of dark tea is packed in a parcel and sent to the official tea department in exchange of horses. The officials and businessmen take the profit fifty-fifty".

Dark tea may originally be caused by excessive leaf volume or low fire temperature and long time when green tea is baked; it may also be caused by the accumulation of crude tea, which produces high temperature and humidity, causing fermentation and making the tea black. The dark tea has lost the original color and fragrance of green tea, but it has gained another rich taste. So people created a technique to make dark tea: After removing water by steaming and then kneading, an extra process of piling fermentation is added. The so-called piling fermentation is to let the first kneaded tea leaves piled up for about 10 h in a clean room with no direct sunshine, with a room temperature of 25 °C or more and a relative humidity of 85%. Piling fermentation is key to secure quality dark tea. Its taste is close to that of Tuanbing tea in the Tang and Song dynasties, and it is loved by many ethnic minorities, who have not yet been accustomed to green tea. Since the dark tea produced in the mainland is transported to the boardlands, it is packed into square, brick, and round cake-shaped compacted parcels for easy transportation and storage. Therefore, most of the dark tea adds a compaction process. Most areas that produce dark tea also produce green tea. Tea farmers use the buds of early spring to make green tea, and then pick the thicker leaves and branches, which go through the processes of fixation, kneading, piling fermentation, re-kneading, and drying. The tea is made into crude tea, which is then made into compacted tea after the processes of weighing, steaming, pre-pressing, compacting, shaping, and baking.

2.3.2.7 Scented Tea

Scented tea is a reprocessed tea that has both a tea flavor and a floral fragrance. The Tang people often added ginger, green onions, salt, orange peel, dates, and other ingredients to tea. Spices or flowers were added to tea in the Song dynasty to increase the aroma and it was only from that time onwards it was officially named "scented tea". There was considerable development in scented tea in the Ming dynasty, and a lot of experience was accumulated in the selection and matching of fragrant flowers. In the Qing dynasty, scented tea cellar manufacturing centers were formed in Fuzhou and Suzhou. The so-called "cellaring" refers to scenting. The flowers are mixed with tea leaves, so that the dry and loose tea leaves slowly absorb the fragrance of the flowers, and then the flowers are removed and the tea leaves are baked. There is more than one way for tea to absorb floral fragrance, so scented tea is also called mixed

Picture 2.57 Jasmine tea processing. 1. *Duiyin* (Mixing and piling up the tea and flowers). 2. Spreading to cool down

scented tea, scenting tea, and fragrant tea. Common scented teas include jasmine tea, Zhulan Dafang, osmanthus green tea, honeysuckle tea, white orchid tea, and rose black tea (as shown in Picture 2.57).

2.4 Oil Extraction, Cooking, and Grinding

Fat is a collection of a wide variety of substances. It mainly refers to fats obtained from living organisms. From the perspective of chemical composition, it belongs to fatty acid glyceride. In our daily life, it is often called oil which is liquid at room temperature, and grease which is semi-solid or solid, but there is no strict boundary between the two. Edible fats are divided into two categories: vegetable fats and animal fats. Different fats contain different fatty acids. Peanut oil, soybean oil, and sesame oil contain much more unsaturated fatty acids than lard and tallow.

2.4.1 Ancient Oil-Making Technology

In ancient times, people obtained fat from animals, but it is unknown when oil started to be pressed from plant seeds, and impossible to find out. *Intriguing Things* of the Ming dynasty records that "Shennong made oil", but there is no physical evidence. *Book of Fan Shengzhi* at the end of the Han dynasty writes that "beans have cream", which seems to indicate that oil can be pressed from beans. In *Records of Carriage and Costumes* in *Book of Later Han*, there is a record saying that "highest-ranking imperial concubines, princess consorts, princesses, and the supreme ruler rode in oil-painted vehicles". The oil used for painting decorations may be made of vegetable oil, such as tung oil.

The document that clearly records the pressing of vegetable oil is the *Essential Farm Activities in All Four Seasons* of Han E in the Tang dynasty. The third volume says: "April is the time for oil pressing. Press oil from turnip seeds as the annual household oil". The literature that clearly describes the oil pressing technology is Song Yingxing's *Heavenly Creations* in the Ming dynasty.

At the beginning of the book, he clearly points out that the oil contained in the fruits of plants and trees will not flow out on their own and can only be obtained with the help of water, fire, wood, and stone tools.

Grease can make wheels turn smoothly, putty can fill the gaps in the hull, and cooking cannot be done without grease. In the section of "types of oil", Song gives a detailed introduction to the types of oil and their use, pros, and cons, and also lists the oil content of various types based on his own tests.

According to him, the edible oil before the Ming dynasty included sesame oil, rape seed oil, and soybean oil. Peanuts and sunflowers were only introduced to China after Song Yingxing's book was published, so there is no description in the book of peanut and sunflower oil. The olive oil that prevails in the west and the palm oil that is abundant in subtropical regions were not recorded in ancient China (as shown in Pictures 2.58 and 2.59).

In ancient times, the main methods of preparing vegetable oil were pressing, grinding, and boiling. Castor oil and perilla oil are boiled. In Beijing, sesame oil is made with the grinding method, while North Korea used the pounding method. All other types of oil are pressed from seeds.

The pressing tools used should be made of thick wood the size of the span of two arms. The best choice of wood is camphor wood, followed by sandalwood and berry wood. The grains of these three types of wood are twisted, and when the pointed wood wedges penetrate into them and beat them as hard as possible, the two ends will not crack. In China's Central Plains and areas north of the Yangtze River, there

Picture 2.58 Press for making vegetable oil in South China, cited from *Oil and Fat Making,* Chap.5 in *Heavenly Creations*

are not many thick pieces of wood that span more than two arms. Therefore, people put together four pieces of wood, tightened them with iron hoops, and then strung them with horizontal bolts, hollowed out the middle, and placed the seeds inside to be pressed. Under normal circumstances, a large one could hold one *dan* (measuring unit equal to 10 *dou* or 100 L) of oil, and a small one 5 *dou*. In the hollowed out part, a flat groove is also cut, and then a small hole is cut along the bottom of the groove, and a small groove is cut so that the pressed oil can flow into the oil container through the flat groove and the small hole. This flat groove is about 3 or 4 inches long. The size depends on the presser, and there is no certain specification requirement. Sandalwood or camphor wood should also be used for the sharp wedges and sticks that are inserted into the grooves for pressing (as shown in Pictures 2.60, 2.61, 2.62, 2.63 and 2.64).

The pre-processing process of the oil: Gentle frying → crushing → steaming → wrapping into cakes → oil pressing → dry cakes → re-crushing → steaming → wrapping into cakes.

Picture 2.59 Roasting and steaming oil seeds, cited from *Oil and Fat Making,* Chap. 5 in *Heavenly Creations*

After the sundries and dust are removed, the castor seeds or rapeseeds are put into the pot and fried slowly. The seeds of woody cypresses and tung trees are only crushed and steamed instead of roasting. When the seeds give out fragrance, they are gathered, crushed, and steamed in a pot (as shown in Pictures 2.65 and 2.66).

The crushing of seeds is generally done by manual work, or by cattle power. Some seeds only need to be crushed instead of grinding, such as cottonseeds.

The crushed seeds, with husks removed, are sieved, steamed in a pot, and then poured into straw bales or wheat-straw bales fixed with iron frames or bamboo baskets, and pressed into a cake shape. The size of the cake should match the gap in the presser.

Oil is extracted by the high temperature of steam. If the cake (wrapping) process is too slow when it is out of the pot, the oil yield will be reduced. Skilled craftsmen can do fast rewinding, wrapping, and framing. The trick to high oil yield is shown in Pictures 2.67 and 2.68.

The wrapped seed-oil cakes are put into the presser. Once a wedge is used to press the oil cake inside the presser, the oil will flow out. The pressed oil cake is called a "dry cake". The dry cake of sesame and rapeseeds should be crushed, then steamed,

Picture 2.60   Oil beam (wood press) in a castor oil mill in Shenchi county, Shanxi province

Picture 2.61   Wood presser in the tea oil mill in Kaihua county, Zhejiang province

Picture 2.62 Wood presser in the tea oil mill in Changshan county, Zhejiang province

wrapped, and pressed again. The first pressed oil accounts for two-thirds, and the second pressed oil accounts for one-third. For oilseeds such as cypress seeds and tung seeds, all the oil flows out at the first time of pressing, so there is no need to press it twice (as shown in Pictures 2.69, 2.70, 2.71 and 2.72).

The above oil pressing process is the most common method.

Boiling method: Prepare two iron pots. First crush the castor or perilla seeds, put them in a pot and boil, skim the floating foam with a spoon, pour it into another iron pot, simmer the water over low heat to extract oil. The oil yield of this method is lower than that of the pressing method.

Grinding method: Sesame oil is made in Beijing and other places by putting ground sesame seeds in a cloth bag and twisting the oil out.

2.4.2 Modern Oil Pressing Technology

From the late Ming and early Qing to the 1980s, the traditional oil pressing technology described by Song Yingxing is still used in most urban and rural areas (as shown in Picture 2.73). Vegetable oil is not only a must-have for household life, but also plays an important role in foreign trade. Soybean oil, peanut oil, rapeseed oil, sesame oil, and cottonseed oil are all bulk export products.

Northeast China has high yields of soybeans. At the end of the Qing dynasty, a large number of immigrants from the inland rushed to the northeast. At a result,

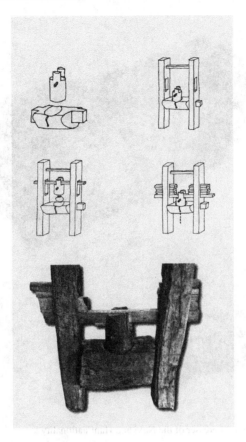

Picture 2.63 Vertical oil presser of the Yi nationality

the black land there was developed, and soybean production surged. Soybean oil is used for food, and the dry cakes (bean cakes) after pressing the oil can be used as concentrated feed, and they are also the best fertilizer. According to incomplete statistics, there were more than 300 oil mills in northeast China at that time.

The land of Qilu has always been famous for its well-developed agriculture. In cities such as Yantai, there were only a few oil pressing mills at the end of the nineteenth century. However, the number soared to 40 in 1900. Most of these oil pressing mills used animal power. The larger ones had six millstones and the smaller ones had two. There were 112 millstones in total. Each mill could press 1,150 L of soybean oil a day. At that time, there were about 1,000 people in Yantai who were engaged in the oil pressing industry, with 600 or 700 mules and horses. There were more than 30 oil pressing mills in Weixian, and the oil pressing industry in Anqiu and Qingdao were also very prosperous. When local soybeans were in short supply, they were all transported from Henan. Shandong was also the first place to grow peanuts and the main production area of peanut oil in the first half of the twentieth century.

Picture 2.64 Horizontal oil presser of the Naxi and Hani nationality

There were many oil pressing mills in urban and rural areas in Jiangsu, of which the Wujin and Xinghua oil mills were particularly famous. Cattle were used to press oil, using a wooden wedge machine. One of these machines could press 300 L of soybeans, resulting in 16.5 kg of oil and 90 kg of bean cakes. According to statistics from the 20th year of the Guangxu period (1894) to the first year of the Xuantong period (1909), there were more than 20 oil mills in towns and more than 40 scattered across the countryside. When there were not enough soybeans to be harvested locally, they were purchased in Henan and Hubei.

During the Westernization Movement in the late Qing dynasty, the technology of crushing soybeans with steam as the power and then pressing the oil with a screw-type iron press was also introduced along with other modern technology. The first machine oil pressing mill in China was established in Yingkou, Liaoning province by the British merchant Taikoo in 1868. However, due to the joint opposition of handicraftsmen, who blocked its purchase of raw materials, it had to close a few years later. It was not until after the Sino-Japanese War of 1894–1895 that foreigners were

Picture 2.65 Steaming seeds

Picture 2.66 Crushing seeds

Picture 2.67 Wrapping cakes

Picture 2.68 Oil cakes

Picture 2.69 Starting pressing

Picture 2.70 Pressing oil

Picture 2.71 The pressed oil flowing into the vat

granted the privilege of setting up mining and factories in China. Sheng Xuanhuai also invested 210,000 yuan in 1896 to set up Dade Machin Oil Pressing Mill in Shanghai.

Although the machine's oil-production rate was high, its output large and the production cost low, the necessary investment was so large that manual oil pressing mills remained an important sideline in the vast rural areas. By the 1970s, manual oil pressing mills were still common in many towns and villages, and it was only in the 1980s that they were declining.

Picture 2.72 Filling oil into bottles

Picture 2.73 The oil mill was so hot that workers were naked. In the old days, women were not allowed to enter the oil mill, and men simply worked naked.

Comparison of machine oil pressing and manual oil pressing

	Machine oil pressing	Manual oil pressing
Oilseed (soybeans)	80 *dou* (120 kg)	80 *dou* (120 kg)
Oil yield (weight)	11 kg	10 kg
Quality of bean cakes	Clean, light	Crisp, moist, and gray
Production cost	0.25 taels of silver	3 taels of silver

2.5 Sugar Production

Sugar is an important type of organic compound, also known as carbohydrate. It exists in the roots, stems, leaves, and fruits of plants in various forms such as glucose, fructose, starch, and cellulose. After people eat it, it is decomposed and digested by enzymes into glucose, which is absorbed by human body and provides energy for physiological activities.

There are more than one hundred carbohydrate substances that have been recognized by people at present, which are divided into three categories: monosaccharides, disaccharides, and polysaccharides. Monosaccharide is sugar that can no longer be hydrolyzed and is also the simplest form of carbohydrate. Common monosaccharides in nature are pentose and hexose. Ribose and deoxyribose in cells are pentoses, and each molecule contains 5 carbon atoms. Common glucose and fructose belong to hexose, with each molecule containing 6 carbon atoms. Both disaccharide and polysaccharide can be hydrolyzed into monosaccharide. Glucose is produced by hydrolysis of disaccharide or polysaccharide in many environments. Common disaccharides are sucrose, maltose, and lactose. Sucrose can be broken down into 1 molecule of glucose and 1 molecule of fructose under the action of enzymes in human digestive organs. Polysaccharide is formed by the condensation of many monosaccharides, such as starch and cellulose. Plants can convert carbon dioxide and water into monosaccharide (glucose) through photosynthesis, which is then converted into polysaccharide for storage.

2.5.1 Early Sugar Stuff

After a baby is born, it must first drink breast milk. Breast milk contains lactose, which is broken down into galactose and glucose by enzymes in the digestive tract, which becomes a source of nutrition for babies. It can be considered the first type of sugar that people consume. Dairy is the first type of sugary food enjoyed by humans.

The second type of sugary food provided by nature is various fruits. They more or less contain fructose and glucose, and two-thirds of ancient food came from collected plants and their fruits. It was only after people were aware of the sweetness and nutrition of the fruits that they gradually improved the cultivation of wild fruit trees and made them an integral part of the agricultural economy.

The discovery and collection of honey made mankind take a big step in the process of obtaining sweet food. Honey was mainly collected by burning and smoking or destroying wild beehives even during the Warring States period in China. According to *Treatise of Geography* in *Book of Tang*, there were 19 prefectures and counties paying honey as tribute honey at that time, most of which was wild honey.

The record of beekeeping began in the Eastern Han dynasty. *Biography of Talented People* by Huangfu Mi in the Western Jin dynasty describes Jiang Qi, a beekeeper in the Eastern Han dynasty. Jiang was a native of Hanyang (now Tianshui, Gansu province) and was famous for beekeeping during the Yanxi period (158–167) of the Eastern Han dynasty. He imparted this technique to many people and benefited thousands of families.

In the Yuan dynasty at the latest, beekeeping and honey processing technology had been fully developed. It is described in *Heavenly Creations* in the Ming dynasty that "80% of honey is made by bees that are found in rock and soil caves, and 20% by those kept by households". It can be seen that honey still mainly came from wild bees at that time and mostly came from the northwestern regions (as shown in Pictures 2.74, 2.75 and 2.76).

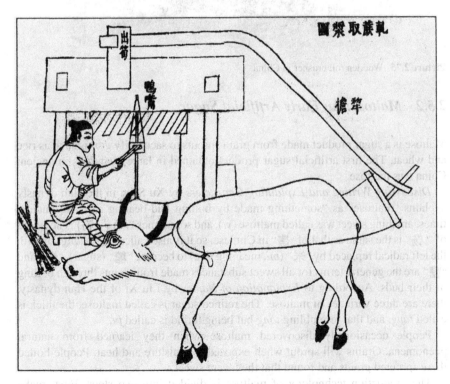

Picture 2.74 Crushing cane to make syrup, cited from *Heavenly Creations*

Picture 2.75 Wooden roll crusher in China

2.5.2 *Maltose: An Early Artificial Sugar*

Maltose is a sugar product made from grain sprouts to saccharify grains such as rice and wheat. The first artificial sugar product obtained in large quantities in ancient China was maltose.

Discussing Writing and Explaining Characters by Xu Shen in the Han dynasty explains "maltose" as "something made by boiling bud-bearing rice". In ancient times, anything sweet was called maltose (*yi*), and sometimes "*tang*" (饧). The sound of "饧" is the same as that of "唐" in Chinese, so it is also called "餹" (*tang*) or with the left radical replaced by "米" (*mi*, meaning rice) to become "糖" (sugar). "饧" and "糖" are the general terms for all sweet substances made from grains through boiling of their buds. According to *Explanation of Names* by Liu Xi of the Han dynasty, there are three varieties of maltose. The refined sugar is called maltose, the thick is called *tang*, and that resembling *tang* but being turbid is called *fu*.

People occasionally discovered maltose when they learned from natural phenomena. Grains will sprout when exposed to moisture and heat. People boiled these sprouted grains and found that they were sweet.

The production technology of maltose is divided into two steps. First, make fermenting grain, and then boil sugar. *Book of Cuisine* by Xie Feng in the Eastern Jin dynasty describes a method of making maltose. In mid-August, wheat is soaked in

Picture 2.76 Wooden crusher kept in Dehong, Yunnan province

water first for some time, and exposed to the sun after the water is poured away. The process is repeated every day until the wheat sprouts young roots. It is then spread on a straw mat, about two inches thick, and poured with water every day until the roots are long enough to be collected and dried. This bud-bearing wheat can be used to boil white sugar, which is light white maltose. If you want to get dark brown maltose, collect it when the malt is green and tangled into a cake shape, cut it with a knife, and then dry it. If you want to get amber maltose, use barley as the raw material.

2.5.3 *Sucrose*

Botanically, sugar cane belongs to the genus saccharum in gramineae, and is divided into 30 kinds (as shown in Picture 2.77). In the beginning, people collected sugar cane to sip it or squeeze it, turning it into a drink once it was squeezed. Later, it was realized that the squeezed juice could not only be drunk, but also boiled. After the Tang and Song dynasties, sucrose became the most important sugar variety.

In the early days, the gelatinous syrup was obtained by exposure to the sun, fire, or a combination of the two, to kill most of the microorganisms so it could be stored for a long time. This kind of sugar syrup is called sucrose, and this technique appeared in the Han dynasty at the latest. With the improvements in boiling techniques, people

Picture 2.77 Sugar cane production area in Dehong, Yunnan province

were able to fully evaporate the water in sugarcane juice without it resulting in coking. When the moisture drops below 10%, the sucrose will solidify into lumps after cooling, which is called refined sugar. However, it is note crystalline sugar but raw brown sugar.

Sugar-making technology developed considerably in the Tang dynasty. In addition to "refined sugar", there was also desugared granulated sugar, which was directly related to the absorption of the advanced sugar-making methods of India at that time. *Laoxue'an Notes* by Lu You in the southern Song dynasty quotes a sugar bureau official as saying, "granulated sugar was not available in China before. Emperor Taizong of Tang once asked the messenger of foreign tribute about it, and was told that it was made by boiling sugar cane juice. Later, people used the method and made sugar the same as that of foreign tribute. Since then, there was granulated sugar in China".

The "icing sugar" of the Song dynasty is known as rock sugar today, also known as rock candy. Icing sugar was first mentioned in literature by two great writers in the northern Song dynasty. When he bid farewell to Yuan Bao, a monk in Suining, Sichuan province, at Jinshan Temple in Runzhou (now Zhenjiang, Jiangsu province), Su Dongpo wrote a poem, which says, "Fujiang River and Zhongling River share the same water. Celadon plate presents amber, how beautiful it is, like icing sugar". Another writer is Huang Tingjian. During the Song's Yuanfu period (), he, in Rongzhou (now Yibin, Sichuan), received icing sugar from Monk Yongxi in Zizhou (now Santai, Sichuan), and wrote a poem for acknowledgement. In the early southern Song dynasty, Wang Zhuo of Suining prefecture, Sichuan province wrote the famous *Icing Sugar Manuals*, which comprehensively describes the history of cane sugar in China, the icing sugar in particular. Li Zhihuan, a contemporary sugar expert summarizes the method of making rock sugar in *Icing Sugar Manuals* as follows.

From October to November, peel the sugar canes, cut them into short sections the size of a copper cash string (about 10 cm), and then crush them; if there is no crushing tool, a pestle can be used instead. Put the sugar juice in a jar coated with paint on the inside and outside, and boil it in a pan. Bagasse primary crushed and pestled is called '*bo*', which is then steamed and pressed for the remaining juice. The sugar juice is boiled in a pan until it is medium cooked, which is the equivalent of 66–68% sugar content. When the temperature is about 105 °C, skim off the floating impurities. Leave it to cool and settle for 3 days. Then scoop the clarified sugar cane juice out of the pot and leave the dregs. When the sugar cane juice is boiled until 90% cooked, which is the equivalent of 85–88% sugar content, and the temperature is between 114°C to 123°C (depending on the purity of the sugarcane juice), it turns into syrup. The temperature must not be too high, as it will be too thick and will crystallize into broken rock sugar. Arrange a number of thin bamboo shoots and insert them into the urn painted on the surface and the inside, and pour the syrup in. Then cover the urn with a staw mat. After two days, rub the syrup with two fingers and observe. If it is like the shape of micro-sucrose, it can be crystallized into good rock sugar (as shown in Pictures 2.78, 2.79, 2.80, 2.81 and 2.82).

The earliest attempt to decolorize sucrose in China was to add a whipped duck egg white to the sugar cane juice and then heat it. The impurities would solidify with the egg white and float to the surface. Thanks to this method, the sugar cane juice became clear. The earliest record can be found in *Annals of Xinghua County*, compiled by Zhou Ying in the 16th year of Hongzhi (1503) in the Ming dynasty.

Picture 2.78 Wooden presser of the Dai nationality in Ruili, Yunnan

Picture 2.79 Yi people in Mangshi, Yunnan province boiling sucrose with four pots, all used to concentrate sugar cane juice

The most successful and influential ancient sugar decolorization technique is the one using yellow mud. The development of this technique can be divided into two stages, from accidental discovery to conscious use and improvement.

The first stage is the mud-covering method: The sugar cane juice is heated and evaporated, concentrated to a viscous state, and poured into an earthen bowl. This kind of bowl is wide at the upper end, narrow at the lower end, and opens downwards. Clog the bottom mouth with straw or other plugs in advance. Two to three days later, the lower part of the bowl is blocked by the crystallized sugar. Then pull out the staw, place the earthen bowl on the urn, kettle or pot, press yellow mud on it or seal the upper mouth with yellow mud. At this time, the yellow mud gradually penetrates into the syrup, absorbs impurities and sinks to the bottom of the bowl. It then gathers in the urn drop by drop with the molasses (the uncrystallized mother liquid remaining after the syrup is separated from the sugar). After a long period of time, the decolorization is complete. The upper layer of the bowl is filled with fine white sugar, and the dark brown sugar is separated from the white at the bottom.

This mud-covering method has been continuously improved and evolved into the method of adding yellow mud, which not only improves the decolorization effect, but also greatly improves the efficiency. This is the second stage of the yellow mud method. It is recorded in *Sugar Making* in *Heavenly Creations*,

Picture 2.80 Yi people in Mangshi, Yunnan province boiling syrup

Every *dan* of sugar cane juice should be mixed with half a *sheng* of lime to coagulate undesirable impurities. In boiling the juice for sugar, put three cooking pots to form a triangle and use them simultaneously. Gather the thick syrup which has been obtained after boiling the thin juice in the three pots and transfer it into one pot. [...] The heat control for refining the juice of sugar cane can be adjusted by watching the bubbles when it is boiled. When small bubbles like the bubbles of boiled meat broth appear and the juice sticks to hands while rubbing them, the boiling is sufficient. At this time the color of the boiled syrup is dark yellowish. Next put the syrup in a barrel and it will crystallize into dark granules. Set an earthenware funnel on the jar. The top of the funnel is wide while the bottom is narrow; there is a small hole at the bottom, which is clogged with straw. Pour the dark granules into the funnel. When it is congealed, remove the straw in the hole and pour yellow-earth water into the funnel. The black residue is drenched into the jar so the white sugar is left in the funnel.

The granulated sugar and rock sugar technology introduced above shows that sugar-making technology made great progress during the Tang and Song dynasties. In modern times, after the introduction of western mechanized sugar technology, large-scale sugar cane pressers replaced stone rollers and wooden rollers, resulting in cleaner cane juice. During clarification, not only a sufficient amount of lime is added, but also carbon dioxide gas is introduced to neutralize the excess lime. Boiling cane juice needs ten pots lined up. For the decolorization and bleaching of sugar solution, sulfur fumigation is used, especially the high-efficiency centrifuge, to remove the molasses in order to obtain white granulated sugar.

Picture 2.81 Yi people in Mangshi, Yunnan province boiling sucrose and precipitating crystals

2.5.4 Sugar Beet

Sugar beet was not available in China until modern times. In Persia, it was planted more than two thousand years ago, but only the leaves were foraged or used as fodder. It was the German mineral chemist S.A. Marggraf (1709–1782) who first discovered that sugar beet roots contain sugar. In 1747, he found the sugar crystals in the beet roots under a microscope, and reported his findings to the Prussian Academy of Sciences. Afterward, people consciously cultivated beet roots for high sugar, so that the sugar content of the roots gradually increased. At the beginning of the twentieth century, the sugar content of beet roots increased from 6 to 18%. At present, the sugar content of fine varieties of beet has reached 24%, and it has become the main raw material for sugar production in some countries.

The composition of beet root juice is different depending on the variety, land, and climate. The process of making sugar from beets in modern sugar factories is roughly washing, slicing, pressing, and filtering, adding quicklime or lime milk for treatment and then introducing carbon dioxide gas to precipitate the excess lime, heating and concentrating, separating honey, and crystallization.

China introduced sugar beet as a raw material for sugar production in 1906, and Russian Poles first tried planting them in the Heilongjiang River Basin. In 1905, China

Picture 2.82 Yi people in Mangshi, Yunnan province boiling sucrose and putting it into bowls

and Russia cooperated to build a beet sugar factory in Acheng county, Heilongjiang province, which was put into operation in 1908, with a daily output of 21 tons of sugar.

2.6   Leather Processing Technology

Humans in prehistoric times used animal skin to keep warm or to wrap their feet in to facilitate walking. During the Shang and Zhou dynasties, leather processing became more advanced. Both ancient oracle bone inscriptions and bronze inscriptions have the characters "裘"(fur) and "革"(leather) on them. The bronze inscriptions of the Western Zhou dynasty include records of leather shawls, leather aprons, leather rope, shoe tube leather, dyed leather and raw leather.

Because leather has a variety of properties, in addition to being a useful material to make clothes, it can also be used to sew tents, blankets, and mattresses. In the military, it can be used as parts of armor, shields, bows, scabbards, harnesses, war drums, etc., and leather boots are used in ceremonial occasions. In response to the above-mentioned needs, each dynasty set up full-time officials and organizations to manage the leather and fur industry in a unified manner.

   Leather-tanning is the process of tanning raw hides into leather through physical, chemical, and mechanical methods. The raw materials used are mainly livestock skins, such as cowhide, sheepskin, pig skin, donkey skin, horse skin, mule skin, and camel skin.

   As early as the Paleolithic age, humans have used stone tools to strip animal skins and make them useful. However, wet skin is prone to decay and becomes very hard after drying. To this end, people used the brain marrow, bone marrow, oil, and lipids of wild animals to apply to the surface of the rawhide and make it soft by rubbing it down. After the oil has oxidized, it produces a tanning effect on the rawhide, so that it remains soft after drying. This is actually the predecessor of the oil tanning method (as shown in Picture 2.83).

   Smoked tanning method: Smoking rawhides with the smoke of burning wood can repel insects as well as disinfect the hides. Later, an ancient smoked tanning method was developed, which was actually the aldehyde tanning method.

   Vegetable tanning method: The wet rawhides on the branches or wood will change color after a long time. Human beings are inspired by it, knowing that the bark contains an ingredient that can be extracted with hot water. The animal skin is soaked in this extract, and it will not shrink or rot after drying and can be made into a soft and tough piece of leather for long-term preservation.

   Fermentation hair removal method: Put the wet hides in a warm and humid place, and the hair will automatically fall off after a few days. The principle of this method is the same as that of the lime-liquid depilatory method. Both microbial enzymes are at work, but the effect of the latter is much better. After continuous improvement, an ash-alkaline dehairing method with soda sulfide and lime liquid was developed, which is still in use today.

Picture 2.83 Several
tanning tools. Hedele. Hadi,
Haronk

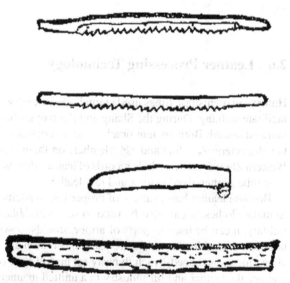

Manure softening method: The manure of poultry and livestock can be fermented with warm water to soften the hides. At first, people did not know that this was the function of microbial enzymes and regarded it as a secret, which was only passed on between masters and apprentices, and later became the key technique for making soft leather.

Bran softening method: The wheat or rice bran is fermented in warm water for one to two days, and then the organic acid produced by fermentation is used to treat the bare skin, remove lime and create favorable conditions for softening the rawhide.

Nitrate surface tanning method: The surface is covered with aluminum salt, egg yolk, and flour to keep the raw skin soft and to not let it rot after it dries. Aluminum salt plays a major role in this process, since it is the original aluminum tanning method, and has been used in Egypt, Rome, Greece, as well as in China.

People also invented leather finishing and molding techniques. Egyptian mummies dating back to the nineth century BC had embossed gold-plated leather belts, and the leather armor of the Qin terracotta warriors of the third century BC unearthed in Lintong, Shaanxi province, dyed in different colors, prove that they had already mastered leather dyeing techniques, and were at least proficient in the processing of leather and fur.

There are four main traditional leather processing techniques (commonly known as tanning) circulating in Inner Mongolia today.

One is the aforementioned tanning process in which animal fat and brain marrow are smeared on the rawhide and then softened by rubbing.

Rawhides hooking method: This is an ancient tanning method used by Inner Mongolian cobblers. The tool used is a wooden fork hook with a piece of iron in the middle that is used as a knife. After soaking the rawhides in water and re-wetting, the meat residue is scraped off with a hook, then covered with yellow rice so that the hides lose moisture during processing, after which the hooks are used to soften the raw hides (as shown in Pictures 2.84 and 2.85).

Yogurt method: The milk tanning method was invented in the pastoral area a long time ago. Add salt or mirabilite to the milk and put it in a vat. After the milk turns sour, the animal hides can be soaked. After a few weeks to a month, or even longer, the skin is taken out and dried, soaked again to regain moisture, and scraped with a scythe blade to get rid of the meat residue and make tanned leather. This tanning method is part of the acid oil tanning method. In the tanning process, it is actually lactic acid that makes the skin soft. Because the materials are easily available and the method is easy, it is widely used in pastoral areas. The tanned leather has good results, but the sour smell is not pleasant (as shown in Picture 2.86).

Flour tanning method: It is also known as the rice flour or saltpeter flour tanning method, which is actually a fermentation softening method. The principle is to use yellow rice flour containing starch and protein, then add a small amount of millet gruel and salt or mirabilite and let it ferment to produce organic acid, and then use it to tan the leather. The leather thus made has good plasticity, but it regresses when exposed to water. It is easy to be eaten by insects and has a sour odor (as shown in Picture 2.87).

Picture 2.84 Hook leather

By the 1990s, the above-mentioned methods of tanning were still in use in Inner Mongolia. However, under the impact of modern leather processing techniques, the traditional techniques disappeared quickly and became less and less visible.

2.7 Incense-Making Technology

The Chinese incense culture has a long history, and incense was used as early as six thousand years ago as a way of making offerings to deities. Incense for sacrificial rites and for daily life developed and improved along with the advancement of civilization. It has experienced several stages of development. The pre-Qin period was the beginning period, the Western and Eastern Han dynasties were the formation

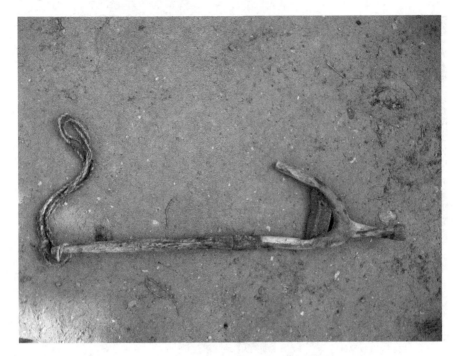

Picture 2.85   Hook

period, the Sui and Tang dynasties were the growth and perfection period, and it reached its pinnacle in the Song and Yuan dynasties, and is popularized in the Ming and Qing dynasties, forming a complete incense culture including rituals, daily life, and therapy.

The traditional Chinese incense art has a profound cultural connotation. In ancient times, the making of incense formed a whole set of theories consistent with traditional Chinese medicine theory and Taoist alchemy theory. It has a complete system of processing and is an inseparable part of traditional Chinese culture.

Traditional incense is not only very sophisticated in the compatibility with spices, but also has strict requirements for the process of spices. "When the process is not sufficient, the fragrance is barely there; when it is too much, the fragrance goes off". Therefore, whether spices are processed properly or not has a direct impact on the quality of the incense.

The specific processing method includes Modification, steaming, boiling, frying, broiling, quick-frying, baking, and water-grinding. Modification: To purify incense material as well as to cut and crush.

Steaming: Use water vapor or water to heat the material.

Boiling: Put the material into water or other ingredients to soak and boil it, with purpose of adjusting its property and remove its peculiar smell.

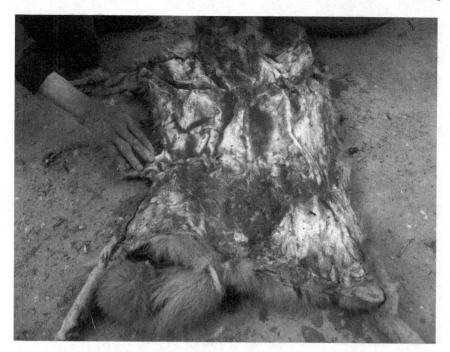

Picture 2.86 Soak the leather

Frying: Plain fry or sauté the ingredients as needed.

Broiling: Stir the material in a pan with desired liquids to make the auxiliary materials penetrate into the incense material and change its medicinal properties.

Quick-frying: Fry the materials in a pan under high heat or stir-fry it with sand and yellow cattail powder.

Baking: Put the incense material in a container and heat it to dry completely.

Water-grinding: Grind the crushed material with water, so the powder will "fly" into the water. Then the slurry is allowed to settle and the sediment is dried and ground for later use (as shown in Picture 2.88).

2.7.1 Making Skills of Beijing's Traditional Medicinal Incense

The records of Beijing's traditional medicinal incense can be traced back to the Ming dynasty, when many well-known incense makers appeared, and incense products, such as "Longlou Incense", "Furong Incense", "Wanchun Incense", and "Black Incense Cake", were made by local workshops as well as by the Imperial Court.

Picture 2.87 Folding the leather in half

At present, the medicinal incense developed by Beijing Yuliang Handmade Incense Research Laboratory is made according to the family's incense recipe and traditional craftsmanship. Its well-documented first-generation descendants are Xing Junchen and his wife. From 1880 to 1955, the two continued the Zhonghua Medicine Store opened by their ancestors, practiced medicine, and made their own medicinal incense. Later, the family members continued to produce many famous products, such as Miao'an Incense, Five Elements Incense, Yixinxiangti Incense Pills, and Baihua Incense (as shown in Picture 2.89).

From left to right, they are respectively the fifth-generation heir Li Shiliang, the third-generation heir Ma Guizhen, and the fourth-generation heir Shi Yali.

At its opening, Zhonghua Medicine Store began its cooperation with Yue Congsheng, the first-generation master of Tongrentang (a time-honored brand of medicine). Tongrentang purchased medicines including medicinal incense from Zhonghua Medicine Store. The five medicinal incenses in the Tongrentang Exhibition Hall are the products of the Zhonghua Medicine Store during the Guangxu period, and the seals showing that the incenses had been sent to other places.

The traditional medicinal incense-making process includes selection of materials, rough processing, weighing, the five methods of detoxification, grinding, processing, checking compatibility, mixing materials, tamping with tools, manual kneading, natural airdrying, degenitalizing, storing, and packaging into boxes (as shown in Pictures 2.90, 2.91 and 2.92).

Picture 2.88 Incense-making process. 1. Dipping the sticks into the incense material. 2. Swinging it like a paper fan. 3. Drying the incense

Picture 2.89 The three generations of hand-made incense discussing the process of making medicine: pastille

2.7.2 The Skill of "Rolling Incense" in Dongguang Village, long'an, Guizhou Province

The ancient water mill incense-making technique in Dongguang village is called "*guyang*" in the Buyi language, meaning "rolling incense" or "making incense". It is the wisdom of the Buyi people. What they produce is "bamboo stick incense". It can be divided into long incense and short incense according to their length, green incense, and red incense according to their color, "*ganxiang*" and "*tixiang*" according to their production process. The production of "bamboo stick incense" is very sophisticated in the selection, compatibility, and production process of the ingredients, and there is a set of rigorous and effective methods and specifications.

2.7.2.1 Raw Material Processing

The main raw materials of bamboo stick incense are camphor tree, sweetgum tree, wax gourd tree, arborvitae leaves, horse mulberry leaves, bay leaves, mugwort leaves, spatholobi, and debregeasia orientalis.

The "incense cake" (as shown in Picture 2.93) is the main ingredient for making incense, which is made of incense powder. The production method is as follows.

Picture 2.90 Tools for making medicinal incense. 1. Iron pharmaceutical roller, 2. Hay cutter, 3. Copper basin and copper spoon, 4. Porcelain mortar, 5. Copper mortar, 6. Iron mortar, 7. Wine jar, 8. Scoop, 9, Stone mortar, 10. Large molding wood mold, 11. Small molding wood mold

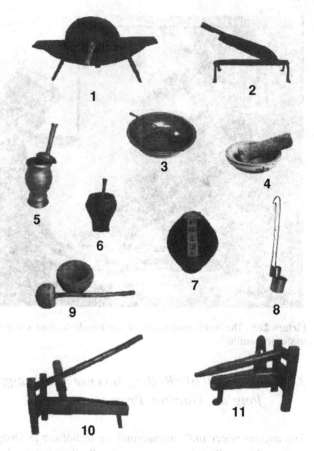

Step 1: Selecting materials, offering sacrifices to the gods, cutting trees

The Buyi people are a polytheistic people, and believe that there are gods in everything. To choose a tree to make incense, you have to respect the tree god before you can cut down the tree. Use a saw to cut the tree into small sections. Each section should be about 30–40 cm long. Holes are punched in the middle of each section to fix it on the incense cart.

Step 2: Defibrination (as shown in Picture 2.94) or using a water mill for grinding up wood pulp

The natural water flow of the Xiangche River is used as the driving power to produce wood pulp by having the incense wood on the incense mill rub against the grinding stone. After precipitation, pack the very fine wood pulp in a cloth bag, knead it into a dough, and let it air-dry to form a wood powder dough.

Picture 2.91 Spices for making medicinal incense

Step 3: Preparation of incense powder (as shown in Picture 2.95)

The bay leaves, mulberry leaves, cypress leaves, and mugwort leaves are dried in the sun, mixed in a ratio of 1:1:1:1, and then pounded into incense powder.

Step 4: Preparing the "incense cakes"

The incense powder has a cohesive effect after being mixed with water. The dry wood powder, incense powder, and water are prepared in a ratio of 3:1:5. It is kneaded into a paste, and fermented for 1 h to become a "incense cake".

Step 5: Prepare incense flour

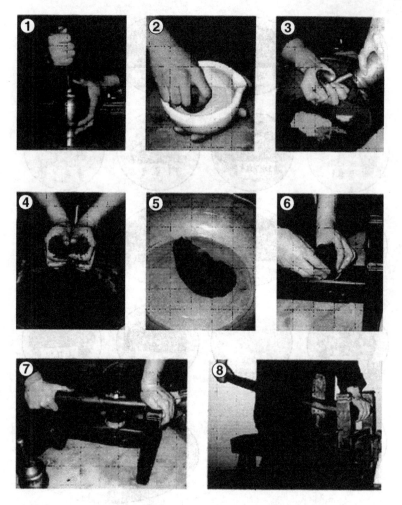

Picture 2.92 Process of making medicinal incense. 1. Pound spices. 2. Grinding. 3. Ingredients. 4. Mixing ingredients. 5. Ingredients. 6. Molding. 7. Molding in small wooden mold. 8. Molding in large wooden mold

Incense flour, also known as "*weimian*", is divided into two types according to the color of the incense. The first kind is green incense flour, which can be directly used as incense powder; the second type is red incense flour, which can be prepared by grinding an ore called "red earth stone" into powder and mixing it with dry wood powder at the ratio of 1:1.

Picture 2.93   Incense cake

Picture 2.94   Defibrination

Picture 2.95 Preparation of incense powder

2.7.2.2   The Production Process of Making Rolled Incense

To make rolled incense ("*ganxiang*"), you usually sit on a small stool and operate on a special incense table (as shown in Picture 2.96).

In the first step, use an incense knife to scrape the incense cake on the table to the bar-type incense container, leaving only a chopstick-sized strip of incense cake. Then, scrape the excess incense cake away.

In the second step, hold one end of the incense stick in the left hand and put the other end on the incense cake in the right side of the incense container. The right hand holds the incense knife to press the incense stick and the incense cake. Turn the incense stick clockwise with the left hand to let the bar-shaped incense cake coat the incense stick.

In the third step, roll the incense stick in the incense flour to coat the incense cake on the incense stick with a layer of incense flour.

The fourth step is to hold the incense stick in the left hand and place it flat on the incense table. Hold the incense plank in the right hand and gently roll the incense stick to make the incense coating even, smooth and artistic. In the fifth step, the rolled incense is sun-dried or air-dried.

Picture 2.96 Incense

2.7.2.3 The Production Process of Dipped Incense ("*Tixiang*")

First go to the hillside to collect the spatholobi root and debregeasia orientalis, put them in a pottery jar, and soak them in water at a ratio of 1:3. After half a month, the impurities are removed, and the remaining juice is called "incense juice".

Dip the incense stick into the incense juice, take it out and quickly put it in the charcoal powder and roll it so that the charcoal powder adheres to the incense stick; then dip it into the incense juice, lift it up and roll it in the incense powder to be coated with the incense powder; roll it evenly on the incense table, and then sun-dry or air-dry it to finish the dipped incense.

2.7.3 Tibetan Incense-Making Skills

Tibetan incense is an indispensable and common item in Tibetan life. People use it during their pilgrimages, to worship Buddha, to avoid ghosts, and to exorcize evil spirits. Lighting Tibetan incense can clean the air and refresh any mood.

Nimu Tibetan incense is known as "the first incense of Tibet".

The township of Tunba, Nimu county, is the hometown of thumisambhota, the founder of the Tibetan language. The Tibetan incense production techniques in this township were taught by him more than 1300 years ago. Nimu Tibetan incense is one of the most famous Tibetan incense and is favored by Tibetans since it uses a unique raw material formula and no living creatures are harmed during the making process. The production process of Nimu Tibetan incense is as follows.

2.7.3.1 Raw Material Processing

Saw the cypress trunk into small sections, peel off the bark, punch holes in the middle, and then hang the wood sections on the rocker arm of the waterwheel. Driven by the waterwheel, these pieces of wood are rubbed in the slab-paved trough day and night until they are all ground into wood paste. During this process, water is added to the trough, to prevent the cypress powder from being blown away by the wind. If there is not enough moisture, the wind will blow away the powder; if there is too much moisture, it will take a long time to dry afterward, and the aroma of the cypress will be lost if the airing takes too long, so someone must be put in charge of controlling the moisture.

2.7.3.2 Ingredients

Knead the ground cypress and spices together. Spices are made by combining dozens of Tibetan medicines. Tibetan incense is very special. The incorporation of dozens of special Tibetan medicines makes the incense have a good healing function. For example, if you burn a few in the room every day, it will heal people's spirit.

2.7.3.3 Forming

Put the wood paste mixed with various spices into the horns and squeeze it out. This is key in making Tibetan incense. The shaped Tibetan incense is required to be in straight lines. The straightness tests the perseverance and durability of the producer.

2.7.3.4 Drying

The drying process is very important. The incense should not be exposed to the sun for a long time. It can only be put in a place with sufficient sunshine that is not too warm. Due to using less water and it needing less time drying, the fragrance becomes purer. After two to three days of drying, the Tibetan incense can be packaged and sold.

2.8   Production of Gunpowder, Firearms, and Fireworks

Black gunpowder with niter, sulfur, and charcoal at a ratio of 1:2:3 as the main ingredient is an unexpected harvest of Chinese alchemists in experiments, and one of the four ancient Chinese inventions that have a great influence on the progress of world civilization.

During the Qin and Han dynasties, people started dabbling in alchemy. At the instigation of the royal family and nobles, the alchemists tried to burn minerals to obtain the magic pill that immortalized people. They combined several or even dozens of minerals through smelting, which revealed some strange phenomena, especially when combining saltpeter with the "three yellows" (sulfur, realgar, and orpiment), since that often exploded. In order to prevent accidentally being blown up, the alchemists designed a variety of safeguarding methods to modify the flammable nitrate and sulfur, while some did the opposite, either to show off, or to create a novel atmosphere, using flammable formulas to implement spells. Over time, it resulted in the invention of black gunpowder.

In the late Tang dynasty at the latest, gunpowder weapons had already proven their military power. *General Principle to the Art of War* compiled by officials in the Northern Song dynasty describes three types of gunpowder formulas and multiple types of gunpowder weapons (as shown in Pictures 2.97 and 2.98). At the same time, the appearance of a variety of fireworks indicated that the use of gunpowder had already evolved from the magic of alchemists to the technology of commodity manufacturing to entertain commoners.

In the 17th year during the Shaoxing period (1147) in the Southern Song dynasty, *Dreams of Splendor of Kaifeng* written by Meng Yuanlao records the scenes of soldiers performing a variety of shows, letting off firecrackers and fireworks in front of the imperial court. Famous fireworks artists were included in the list of well-known artists. *Old Things about Wulin* written by Zhou Mi in the Southern Song dynasty (1270) tells the grand demonstrations of firecrackers, fireworks, rolling fireworks, linear fireworks, meteors, and ground mice, designed and made by Lin'an's pyrotechnic masters during the Lantern Festival. *Rustic Talks from the East of Qi* states: In the first year of the Baoqing period (1225), "On Lantern Festival, the Qingyan palace was arranged, the Queen Mother was invited to see the fireworks show in court. The so-called ground mice firework went to the Queen Mother's holy seat, which terrified her to walk away".

Fireworks entered its heyday in the Ming dynasty, when the formula and combinations of fireworks were greatly improved. During this period, large fireworks were created, such as new types of fireworks with poles, with tubes and handles, and with handgrips, as well as rolling fireworks and rack fireworks in which gunpowder wire was cleverly connected, so that when the lead of the fire wire was lit, the fireworks would be set off automatically in a fantastic and exceptional sequence. The 42nd chapter of *Notes and Comments on The Golden Lotus*, published during the Wanli period of the Ming dynasty, contains the following description.

Picture 2.97 Fireball in the shape of tribulus terrestris

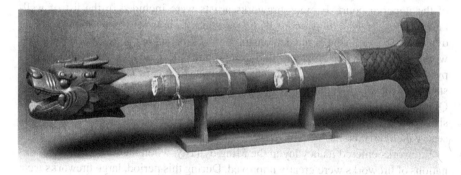

Picture 2.98 Fiery dragon spitting water

Picture 2.99 Fire dragon dancing

A one-point-five-*zhang*-high flower pile (one *zhang* equals to 3.3 m), surrounded by a lively mountain shed, and on the highest point is a crane with an imperial edict written in red in its mouth. It is in fact a fire breaking out to the thick mountain forest with a cold light penetrating halfway to the distance. Then, in the middle, a watermelon firework bursts open, and people and things in the surrounding area were everywhere with light and rumble. Fireworks in the shape of lotus boasts are as bright as the moonlight, set off one by one, like a golden lantern rushing to scatter the blue stars. There are thousands of fireworks in the shape of purple grapes, like pearl beads hanging upside down on a crystal curtain. The overlord firecrackers echo loudly everywhere; the ground rat firecrackers strung around people's clothes. Fireworks in the shape of jade table and cups rotate attractively; there are fireworks in the shape of silver moths and gold bullets, displayed ingeniously to make it hard for people to move away. The fireworks *Baxianpengshou* (Eight Immortals holding fast a peach symbolizing longevity), are well known; the firework *Qishengxiangyao* (seven saints subduing the demon) is with its whole body on fire. Yellow smoke and green smoke enshroud the glow. Lotus spits out tightly, slowly, and brilliantly, striving to open and scatter ten pieces of brocade. The sunflower is opposite the orchid, and the fire pear trees blossom altogether to compete with the peaches for its spring scenery. The towers and halls are nowhere to be seen for naught but a moment. The drums of the village seem to be drowned out. The salesman bears the burden, and the flames are bright up and down; the pinwheel smashed from top to end. The firework *Wuguinaopan* (five ghosts been sentenced) is burned appearing with ferocious features. Laying in ambush on all sides, there is no victory nor defeat. After all, it took all my heart and soul, but the fire died out and the smoke disappeared into simmering ashes (as shown in Picture 2.99).

The above description includes more than 20 kinds of fireworks. At the end of the Yuan dynasty and the beginning of the Ming dynasty, Tao Zongyi wrote *Record of Mo'e*, the part of "fireworks" in Volume 6 lists 22 kinds of fireworks, such as the silver table and golden cups, *jinsiliu*, *saiyueming*, and purple grapes, all of which have unique formulas. They are prepared with different proportions of nitrate, sulfur, and carbon, each with different ignitability and burning speed. Besides, craftsmen have made colorful fireworks by virtue of their experience with metal flame colors.

For example, the formula of *dajinxian* (yellow): *Pingman* (50 g of nitrate, 5 g of loess, 2.5 g of charcoal) and iron filings; the formula of *jinsiliu* (green): *Pingman* and verde antico; the formula of purple grapes (purple): 50 g of nitrate, 7.5 g of charcoal and 10 g of talc (magnesium metasilicate).

In the middle of the Ming dynasty, pyrotechnic screens, lined pyrotechnics, and various types of fireworks were combined to form a pyrotechnic showing with a storyline. In the early Qing dynasty, Zhao Xuemin made a special introduction to it in *A Brief Introduction to Fireworks*. The book discusses three theories on the raw materials of gunpowder, including "extracting nitrate", "making loess", and "using charcoal"; three methods of "marking", "paper dyeing" and "tube making" for the treatment of pyrotechnic materials; six theories of "backfitting", "roasting gunpowder", and "mixing gunpowder" combining with drugs; five methods such as "variety show", "software", and "multi-support" concerning parts making; and there are also four theories of component combination, namely "transformer", "connector", "flying device", and "water device", as well as various manufacturing skills and precautions.

Setting off fireworks and firecrackers has become a universal cultural phenomenon in China, which has become part of the customs of all ethnic groups since long ago. This unique culture has symbolic significance of promoting righteousness and exorcism, celebrating joy, getting rid of the old and welcoming the new, and pursuing artistic beauty (as shown in Pictures 2.100 and 2.101). The workshops that produce fireworks and firecrackers are spread all over the country, even becoming important industries in some areas. Famous cities include Liuyang and Liling in Hunan, Pingxiang and Wanzai in Jiangxi, Dongguan in Guangdong, Hepu and Beihai in Guangxi and Shulu in Hebei (now Xinji in Hebei). The following is a brief introduction to Liuyang fireworks, which is listed in the *First Batch of National Intangible Cultural Heritage List*.

Records of Industries in China, published in 1935, records: "The manufacturing of firecrackers in Hunan province began in the Tang dynasty, developed in the Song dynasty, and originated in Liuyang". "During the Qianlong period, Liuyang firecrackers were produced on a considerable scale, and they were exported to Hong Kong, Macau, and Southeast Asia during the Guangxu period".

According to legend, Li Tian is the ancestor of the fireworks industry. The 18th day of the fourth lunar month is Li's birthday, when a ceremony is held to commemorate him every year.

According to historical records, from the 10th to the 15th year of the Republic of China (1921–1926), there were more than 200 firecracker shops and 500 workshops in Liuyang and Liling counties. It had about 400,000 employees and an annual output of 750,000 containers, which were exported to Hong Kong, Australia, Singapore, Philippines, India, North Korea, Britain, the United States, France, Germany, and Japan.

The production of Liuyang fireworks is still mostly done by hand and has 12 processes, totaling 72 procedures. Among them, tube pulling, paper pasting, tube washing, bottoming mud, nitrate filling, neck sealing, drilling, inserting, whip forming and paper cutting, waist tubing, powder making, coal mixing, and sulfur and

Picture 2.100 Setting off firecrackers on New Year's Eve

earth mixing, are all done by hand. These processes are delicate and cumbersome. Although new processes, materials, machinery, products, and packaging have been gradually implemented, traditional production processes still maintain their unique advantages in terms of artistry and degree of expressiveness. In 2004, Liuyang's fireworks industry achieved an output value of 3.2 billion yuan, generated a tax revenue of 600 million yuan, and 90% of its employees came from rural areas. The external sales of firecrackers not only caused substantial economic benefits, but also promoted friendly exchanges between Chinese and foreign cultures and became an important window through which traditional Chinese culture could be displayed.

Picture 2.101   Taishun fireworks in south Zhejiang

Chapter 3
Construction

Luo Xingbo and An Peijun

Construction in China started in the Neolithic Age: It initially sprouted in the Xia, Shang, and Zhou dynasties, budded in the Qin, Han, and Southern and Northern dynasties, blossomed in the Tang and Song dynasties, ripened in the Yuan, Ming, and Qing dynasties, and gradually stagnated in the following era. Long-term practices, like the beam-column timber frame, and the unique ways cities are laid out and gardens are designed, exemplify the distinctive features and superb construction prowess of traditional Chinese architecture. It is on such a level of magnitude, that in the history of world architecture, it simply cannot be overlooked.

3.1 Characteristics of Chinese Construction

3.1.1 Grid Patterns in Urban Planning

More than 4,000 years ago, ancient Chinese people were already ramming the earth in order to build cities. It was not until the Western Zhou dynasty and the Warring States period that large-scale urban construction was carried out. At this time, the city had the function of managing residents, and each block in a residential area was seen as a single unit. The regulations for each city depended on the rank of the city manager. *Book of Diverse Crafts* in *Rites of Zhou* has different regulations on the

L. Xingbo
School of Humanities, University of Chinese Academy of Sciences, Beijing, China
e-mail: luoxb@ihns.ac.cn

A. Peijun (✉)
Department of Architecture, North China University of Technology, Beijing, China
e-mail: anpeijun@yahoo.com

© Elephant Press Co., Ltd 2022
H. Jueming et al. (eds.), *Chinese Handicrafts*,
https://doi.org/10.1007/978-981-19-5379-8_3

size, wall height, and road width of the cities under the jurisdiction of kings and princes. A city was composed of several inner city units or blocks, and these units were arranged into grid-like streets. When laid out, the central axis focused on the royal palace is very prominent.

3.1.2 Courtyards

Traditional Chinese architecture is good at grouping structures together, especially making good use of combination techniques that fit different buildings and elements together while fulfilling both the aesthetic and the practical requirements (as shown in Picture 3.1). The so-called "courtyard" not only refers to the open space enclosed by buildings but also, and more importantly, the communication between buildings and courtyard makes it the core of the whole group of buildings. Many activities are carried out in the courtyard, ranging from ceremonies honoring the ascension of a new emperor to the daily life of ordinary people. Each building complex is composed of several courtyards, with each courtyard surrounded by separate buildings. The size of each separate building and their position around the courtyard are used to distinguish the noble from the humble. Every difference and change that comes along, depending on each courtyard's size, creates a different atmosphere. This courtyard-style of grouping buildings together is a prominent feature of ancient Chinese architecture.

Picture 3.1 Beijing quadruple courtyard (*siheyuan*)

Picture 3.2 Guanyin Pavilion in Dule Temple

3.1.3 Architecture Featured in Timber Frame

In traditional Chinese architecture, timber frames are used for the roof as well as the walls of a house (as shown in Picture 3.2). The traditional architectural plane composition is generally expressed by the arrangement of column grid or roof structure, and the shape and size of the plane can be shown by the number of "rooms" or "frames", namely, the "standardization" and "modularity" of the building. When building a house, as long as the building shape and scale are determined, the construction can be organized according to the specified "materials" and "points". "Modular system" is spread among craftsmen through mnemonic chants, and houses and prefabricated components can be designed without drawing, which is an important reason why Chinese timber architecture is suitable for mass and rapid organization design and construction. The main types of timber frames are the (1) column-beam system, (2) the column-and-tie system, and (3) dense beams and flat top system.

3.1.4 Bucket Arches

The bucket arch is an important component of traditional Chinese architecture with multiple functions: structurally, architecturally, and decoratively. In fact, these three functions are typical for timber frame architecture in general (as shown in Picture 3.3).

It is formed by overlapping and assembling a variety of wood blocks with different shapes, among which the arched short wood picked out layer by layer is called "arch", so that the eaves can extend out of the house body, and a small wooden block with a bucket shape is placed between the two layers of arches, which is a component composed of multi-layer arches and buckets. The bucket arch is located between the top of each column, beam, and roof. Its use not only extends the length of the eaves but also shortens the span of the beam and evenly divides the weight onto every conjunction to ensure the stability of the column grid. This makes it an indispensable part of large and important buildings. The bucket arch is the soul of ancient Chinese architecture. Its size and complexity determine a building's specifications as well as its use. Even buildings built with brick and stone are often stacked and shaped in imitation of the timber frame and the bucket arch. The bucket arch is not only a unique component of Chinese architecture, but it's also a system, and its evolution reflects the evolution of architecture itself.

Picture 3.3 Bucket arch

3.1.5 Architecture Reflecting Hierarchy

Due to the influence of Confucianism in construction, traditional Chinese architecture follows a strict hierarchy. Confucianism emphasizes the important role of "propriety". People's status is marked by external objects such as dresses, costumes, implements, carriages, and horses. Architecture also belongs to that list, showing hierarchy through types of roofs, and the scale and color of buildings or building components.

Roof: The highest echelons of society used the double-eave hip roof (as shown in Picture 3.4), followed by the double-eave gable and hip roof, the regular hip roof (also known as a *Si'e* hip roof), the gable and hip roof (also known as a nine-ridge roof), the overhanging gable roof, and the flush gable roof, in addition to the pyramidal roof and apex.

Scale: From large cities to small building components, scale reflects hierarchical differences. For example, *Rites of Zhou* stipulates: "The surrounding area covering 9 sq. *li* (one *li* is 0.5 km) of king's palace shall not be trespassed, 5 sq. *li* for seigneur's and 3 for viscount's and baron's". In the 18th year of the Shunzhi period (1661) in the Qing dynasty, it is stipulated in *Records of Laws and Systems of the Qing Dynasty* that the base height of the house of level three officials, such as dukes or marquis, should be 2 *chi* (about 67 cm), and that of level four officials or lower, such as civil servants, should be 1 *chi* (about 33 cm).

Picture 3.4 Taihe Temple with double-eave hip roof

Color: As per the Zhou color system, cinnabar red was for the emperor, black for feudal lords, and dark green for scholars. While, in the Qing dynasty, yellow was seen as the noblest, followed by red, green, cyan, blue, black, and then gray. The highest esteemed buildings use yellow-glazed tile roofs, the lesser ones use green-glazed tile roofs, followed by black-glazed tile roofs, and all the others can only use pottery tiles.

3.1.6 Construction Planning System

In the second year of the Chongning period (1103) of the Northern Song dynasty, the national *Construction Rules* was published, which contains strict quantity limits to curb corruption. The book extensively describes the limits concerning available working times and materials depending on the average amount of daylight during each season, such as short working days (winter), regular working days (spring and autumn), and long working days (summer). The average output estimates were calculated based on regular working days; the long and short working days' output estimates were increased and decreased by 10% respectively. Military workers and hired workers also had different quotas. For each type of work, the calculation method for output estimates was stipulated according to the minimum requirements for grade, size, and quality. There were also detailed quotas for the consumption and use of various materials in the *materials* section. These regulations set strict standards for budgeting and construction and effectively controlled construction quality and costs.

3.1.7 Fengshui

People all over the country are familiar with *fengshui* (geomantic omen) and have also been using it for a long time (as shown in Picture 3.5). It comes from the ancient Chinese world outlook and holds that everything in the world is composed of metal (*jin*), wood (*mu*), water (*shui*), fire (*huo*), and earth (*tu*), and is gathered together by "Energy" (*Qi*) in different ways. To maintain or increase wealth, health, and good fortune, we must accumulate "Qi". "Qi" has the characteristics of "stopping in case of water and dispersing in case of wind". Geomancer's main duty is to select and arrange topography to create a good "Q field", so that "Qi" can gather without dispersing. The site selection, planning, and construction of architecture, as well as decoration and arrangement of furniture should follow the guidance of the geomancer, as to maximize the benefits and minimize the disadvantages. A geomancer's guidance is especially important and listened to when graves are erected. People believe that the *fengshui* of graves is related to the well-being of future generations. It is in these ways that *fengshui* still has a strong influence on many aspects of daily life in China.

Picture 3.5 Langzhong Ancient Town built according to *fengshui* guidelines

3.2 Construction and Traditional Chinese Architecture

3.2.1 Urban Planning and Construction

Whether a city is built as a military defense system or an economic hub, it cannot exist without people inhabiting it. Most of China's history was under centralized rule. For the purpose of controlling people, rulers of all ages made careful plans for cities.

According to the *Book of Diverse Crafts* in *Rites of Zhou*, a city is designed into a square with a side length of nine *li* and three gates on each side. There are nine roads running north to south and nine roads running east to west. The width of the north–south roads is nine *gui* (about 1.80 m). The ancestral temple of the emperor is in the east, the State Altar in the west, the palace in front, and the market in the back. The palace and the market cover an area of a hundred *mu*, that is, 100 steps × 100 steps (step is the ancient length measurement unit, referring to the distance of two steps forward).

Two great examples of how capital cities were constructed in ancient China are (1) Chang'an City of the Tang dynasty and (2) the Great Capital, or Dadu, of the Yuan dynasty.

Picture 3.6 Restoration of Hanyuan Temple in Chang'an city of the Tang dynasty

3.2.1.1 Daxing of Sui (Chang'an of Tang)

Daxing city in the Sui dynasty, designed and built by Yuwen Kai, was a new city built in the southeast of Chang'an city of the Han dynasty, with an area of 84 sq.km, a square outline, and three longitudinal and lateral main roads, which were called "Six Streets". The imperial palace was at the northern end of the central axis with the imperial city laid out in front of it. There were 109 blocks and 2 markets in the city. Each block was surrounded by walls, with a door on each side, which was opened and closed at a certain time. People were not allowed to enter or leave at will after closing time. The low-lying area, where Qujiang River is located, in the southeast corner of the city, was turned into a garden for residents to enjoy, which, in the history of urban planning, is something to be commended. Daxing city had a neat layout, straight streets, clear functional zoning, and showed well-organized planning and design. In the Tang dynasty, it was renamed Chang'an city and continued to be perfected as it was used as the capital city. Later, as the palace was not large enough and situated too far down, the Daming Palace was constructed outside the northeast corner of Chang'an city. The site of the Daming Palace has been excavated, and it is found that the main buildings including Hanyuan Hall (as shown in Picture 3.6) and Linde Hall have been restored and rebuilt. The size of the Daming Palace is much larger than that of the Forbidden City in Beijing during the Ming and Qing Dynasties. Even the Linde Hall in the Daming Palace, which is not the main hall, is three times as large as the Taihe Hall, the largest hall in the Forbidden City.

3.2.1.2 Dadu of Yuan

Dadu of the Yuan dynasty was a completely newly-built capital city, which basically conformed to the "system of King's City" in the *Book of Diverse Crafts* in *Rites of Zhou* (as shown in Picture 3.7). It was located in the northeast of the capital city of the Jin dynasty, at the center of present-day Beijing, and its outline was a nearly perfect

Picture 3.7 Yongcheng Site at the Yihe Gate in Dadu of Yuan

square. Besides the two gates in the north, it had nine other gates, three on each side. The Imperial palace is to the south, and to the north of the palace is the commercial area at the end of the tribute grain transportation. Imperial Ancestral Temple is to the east, and the State Altar is to the west. The layout basically conforms to the rules of "a square with a side length of nine *li* and three gates on each side…with the market in the north, ancestral temple in the left, and state altar to the right". The streets of Dadu were laid out in a grid, like a chessboard, with all grid patterns mirroring each other along the north–south axis.

The alley, which was called "*hutong*", is the passage between rows of quadruple dwellings.

The designers of Dadu of Yuan include Liu Bingzhong and Ikhtiyar al-Dīn, an Arab. It is well-planned, with straight streets, fully functional infrastructure, and the problem of tribute grain transportation was solved by drawing the water supply from the Western Hill and Changping area by Guo Shoujing.

Dongjing city of the Northern Song dynasty, present-day Kaifeng in Henan province, is also a famous historical capital. During the reign of Shenzong of Song, the city, using the original Bianzhou city as its base, was expanded by rebuilding the outer city, and outer walls and watchtowers were added. In the sixth year during Emperor Huizong's reign (1116 AD), the outer city was extended to the south by several *li* to build more administrative centers and military camps. Zhang Zeduan's masterpiece *Along the River During the Qingming Festival* reflects Dongjing's prosperous streets.

Beijing city was first built by Emperor Chengzu of Ming and expanded during the Ming and Qing dynasties, using Dadu of Yuan as its base. The whole city had a central axis with a total length of about 7.5 km running from north to south. The axis used the Yongding gate, which was the south gate of the outer city, as a starting point and first passed through Zhengyang gate, which was the south gate of the inner city, then Tiananmen gate and Duanmen gate in the imperial city, and the Wumen gate of the Forbidden City. It then passed through six doors and seven halls, then Shenwumen gate, over the central peak of Jingshan Mountain and Di'anmen gate, to finally end at the Drum and Bell Tower at the northern end. On both sides of the axis, there were buildings such as the Temple of Heaven, Temple of Agriculture, Imperial Ancestral Temple, and the State Altar, which were magnificent in size and bright in color, and stood out in sharp contrast with the surrounding blue-gray tile-roofed houses of ordinary citizens, thus using urban planning and architectural design to further highlight the supremacy of the emperor. The Forbidden City is the best example of courtyard-style buildings that use strict planning and gorgeous shapes and yet function perfectly. The Ming dynasty was responsible for building many of the altars and temples in Beijing, such as the Imperial Ancestral Temple and the State Altar, as well as altars worshipping the heaven, earth, sun, moon, agriculture, and silkworm. They also built many government offices, warehouses, temples, and residences. All important buildings were built with camphor wood.

3.2.2 Residential Buildings

Residential buildings refer to the buildings where ordinary people live, which directly reflect the living conditions of civilians. China has a vast territory, and the climate, environment, and cultures can vary greatly. The differences can range from slightly to vastly different depending on the region, thus resulting in different kinds of civil structures. These buildings clearly reflect the different regions, different climates, and different cultural backgrounds. We can see them as the true expression of different times and environments.

There are many different kinds and forms of traditional Chinese dwellings, among which the typical ones are courtyard-style dwellings, Jiangnan dwellings, Huizhou dwellings, log dwellings, cave dwellings, yurts or gers (tent), stilt style houses, Dong villages, Tibetan watchtowers, Hakka *tulou*, Kaiping watchtowers in Guangdong, and Shanghai stone gate houses.

3.2.2.1 Courtyard-Style Dwellings

Courtyard-style dwellings are the most common form of traditional Chinese civil architecture, which can be divided into *sanheyuan* (U-shaped courtyards) and *siheyuan* (quadruple courtyard) depending on the number of enclosed buildings. There are different courtyard-style dwellings in different regions, such as the Beijing *siheyuan*, Jinzhong residences, Southeast Shanxi residences, Shaanxi Guanzhong residences, Gansu Linxia Hui residences, Jilin Manchu residences, Bai residences, and Lijiang residences.

The Beijing *siheyuan* is a typical courtyard-style residence in northern China. Its layout is characterized by courtyards. Depending on the status of the owner and area of the base (the gap between the two *hutong*), there are two, three, four, or five courtyards. In addition to adding many courtyards in a longitudinal direction, courtyards and even gardens can be added in a lateral direction as well.

The most common Beijing *siheyuan* are those with three courtyards. The outer courtyard is relatively small with a house facing the north (*daozuo*), which is usually used as a gatehouse, guest room, and parlor. The front gate is in the eastern part, the southeast corner of the *daozuo* (in the *xun* location according to *fengshui* guidelines). The room next to the front gate is mostly used as the gatehouse or as a servant's room. The small courtyard just east of the front gate is used as filler, while the small courtyard west of the front gate contains a lavatory. The outer courtyard is the reception area; thus people are not allowed to enter beyond that point unless they are invited.

The inner courtyard is the main living area of the family. The outer courtyard and the inner courtyard are separated by a floral-pendant gate on the central axis (as shown in Picture 3.8); in the north of the inner courtyard is the main house, which represents the highest status and the largest scale of the whole dwelling, and is the living place for the elders; both sides of the inner courtyard are the east and west side houses, which are the living places for the younger generation; the lower houses on both sides of the main house are wing houses, and the narrow space composed of wing houses, side house gables, and yard walls is called "open ground", which is often used as a sundries courtyard, and also rockeries and flowers might be arranged here; connecting and bypassing the floral-pendant door, the side and the main houses are the veranda, which is convenient for walking in rainy and snowy days. The inner courtyard is larger than the outer yard. You can plant flowers and trees, keep fish in tanks, place bonsai, and either work or relax with your family in a quiet and comfortable living environment. The back house is located in the back courtyard, the northernmost part of the complex, which serves as the kitchen, storage room, and servant quarters. If there is a back door, it is located in the northwestern corner of the back courtyard. There is also a well.

The Beijing *siheyuan* lies at the heart of the urban feel of Beijing during the Ming and Qing dynasties. All *siheyuan* embody axial symmetry, distinct hierarchy, and perfect order. They have become one of the most unique and historical parts of Beijing and thus, naturally, have become one of the characteristic images of Beijing and even China.

Although the courtyard of Beijing *siheyuan* has no roof, walls, doors, or windows, it is the most important space in the whole group of buildings and is an important activity place. The courtyard has become a space similar to the living room in today's residence. Only such a space can be called "courtyard", and only buildings with "courtyard" can be called "courtyard-style dwellings".

3.2.2.2 Jiangnan Dwellings

Jiangnan dwelling can be regarded as a special courtyard-style dwelling. Due to the difference in climate, the architectural forms of residential buildings south of

Picture 3.8 Floral-pendant gate of a Beijing *siheyuan*

the Yangtze River are significantly different from those north of it. The climate in the south is hot and humid, and thus the main problems in construction concern waterproofing and sunscreening. The climate in the north, on the other hand, is dry and windy, and thus the architectural form should be leeward and facing the sun, so as to stop the cold wind from the northwest and enjoy the warmth of the sun as much as possible. The differences between Jiangnan dwellings and northern courtyard-style dwellings are that the former's courtyard is quite small and that the surrounding roofs are often connected, turning the courtyard into a patio. Therefore, this kind of residence or dwelling is also called a closed-hall dwelling.

There are several types of Jiangnan dwellings. Urban official residences often had several courtyards laid out longitudinally and two or three lateral axes in parallel. From the front gate, the hall, bridge hall, gatehouse, main hall, and main house are arranged on longitudinal axes, with small yards between the buildings. The parlor, study room, bedroom, small garden, and stage are arranged along the lateral axes. Rural residences are the typical type of Jiangnan dwelling. They are also laid out symmetrically, but they only have one axis on which they build the gatehouse, bridge hall, gate, main hall, houses, and sometimes also warehouses. When they are built beside a river, they also contain a boat hall adjacent the bridge hall. Most of the smaller houses of ordinary people are built on an irregular plane surfaces and the main buildings are enclosed in a patio style. The main entrances are aligned with the street, but the houses are sloped, leaning sideways, and use the irregular surrounding terrain as gardens. Some buildings are constructed near the water, and therefore have doorways opening up to the street on one side and the water on the other (as shown in Picture 3.9).

Picture 3.9 Houses along the river in Tongli town, Wujiang district, Suzhou, Jiangsu province

3.2.2.3 Huizhou Dwellings

Huizhou is located in the upper reaches of the Xin'an River. In the third year of the Xuanhe period under Emperor Huizong's reign (1121), Shezhou was renamed Huizhou. It has existed throughout three dynasties, namely the Yuan, Ming, and Qing. Huizhou encompasses Huizhou prefecture as well as six counties: Shexian, Xiuning, Wuyuan, Qimen, Yixian and Jixi counties. In the old days, Huizhou's urban and rural dwelling were mostly brick-and-wood buildings (as shown in Picture 3.10). The Ming dynasty's dwelling were characterized by their spacious upper floors, while in the Qing dynasty, they were mostly three-room houses, which had one main room with one bedroom on the left side and one on the right side, and four-bedroom quadrangle houses, which had one main room. A typical dwelling contains several courtyards, and the gate is decorated with stone carvings of figures and landscapes. Each gateway arch has double eves, and there is a patio in each of the courtyards. Being well-ventilated, the patio brings in light easily. On rainy days, the roof's bamboo gutters divert the rain into the sewer, which is commonly referred to as "returning waters from all sides to the hall", symbolizing that "fortune does not flow away from this house". There is a middle partition between the courtyards, with high firewalls (horsehead walls) built around them, which makes the dwelling look like a castle from afar (as shown in Picture 3.11). Generally, when the central gate is closed, the dwelling is divided so that each entrance houses one single family in one single yard. When the central gates are opened, each family can go in and out through any of the entrance gates to the ancestral hall to pay homage to their ancestors. The climate in the Huizhou mountainous area is humid; therefore, the downstairs areas are where the residents spend their day for daily activities, while the upstairs rooms are more spacious, with halls, bedrooms, and side rooms.

Picture 3.10 Huizhou dwellings

Picture 3.11 Horsehead
wall of Huizhou dwellings

During the Ming and Qing dynasties, most Huizhou dwellings faced north, and tens of thousands of these dwellings have been preserved to this day. The residence is deep, with a vestibule at the entrance, a patio in the middle, and a house in the back for living, which is separated from the back house by a door in the middle. The back house contains one hall and two bedrooms. Once past the back house, there is a firewall, from which there is a patio and side rooms on both sides. This is the first yard. The structure of the second yard is similarly structured, with one courtyard in front and one in the back, and a partition wall in the middle, thus dividing it into four bedrooms and two halls.

3.2.2.4 Log Dwellings

In dense forest areas in Northeast and Southwest China, there are houses built with logs, called "log dwellings". These "log dwellings" require the same technique of wall building as that of ancient well walls, that is, logs are slightly chopped on the top and bottom in order to stack them on top of each other to form walls. In order to prevent the wall from letting wind through, the intersections at the ends of the logs need to be crafted into a circular groove. This kind of dwelling is made of local materials, is constructed very quickly, and is insulated well. However, due to the limitations of the natural length of the wood, the width and depth of these houses are limited (as shown in Picture 3.12). In addition, since thick logs are used for the walls, the tensile strength of the wood is not used to its fullest.

Picture 3.12 In Muli, Sichuan, the Lijiaze matriarchal families live in quadruple wooden dwellings, built with round or square timber, covered with Chinese fir boards. They are divided into four parts: (1) the main house, (2) the sutra house, (3) the boudoir, and (4) the livestock stable

Picture 3.13 Cave dwelling in northern Shaanxi

3.2.2.5 Cave Dwellings

Cave dwellings are common throughout Shaanxi, Gansu, and Ningxia in the upper and middle reaches of the Yellow River, used by over 40 million people. The cave dwelling is dug into cylindrical arches taking advantage of the soil mechanics property. The areas with thick loess layers in North and Northwest China are the most suitable for the construction of these dwellings (as shown in Picture 3.13). There are cliff cave dwellings, sunken cave dwellings, and masonry cave dwellings. There are many cave dwellings built on the edge of hillsides and soil plateaus. These cliff cave dwellings are often distributed as steps up the cliff or hillside. The lower cave roof serves as the upper vestibule's floor and offers a wide view. For sunken cave dwellings, a square pit is dug in the ground and then the actual dwellings are dug inside the inner walls of that pit, thus transforming it into an underground courtyard (as shown in Picture 3.14). Masonry cave dwellings, on the other hand, are cave dwellings built on ground level. They use adobe bricks to form an arched structure. It is one of the few buildings in China that uses this type of structure.

The characteristics of cave dwellings are that they are simple to build, easy to repair, don't require a lot of materials, are sturdy, durable, warm in winter, and cool in summer. However, during construction, special attention must be paid to proper ventilation, lighting, and preventing moisture.

3.2.2.6 Ger (Yurt)

A ger is a mobile dwelling, widely used by herders in Inner Mongolia, Xinjiang, and Qinghai. It is closely related to the nomad hunter-gatherer lifestyle (as shown in Picture 3.15).

Picture 3.14 Sunken cave dwelling in Gongyi city, Henan province

Picture 3.15 Ger (yurt)

The ger is easy to build and has a simple structure. Before setting it up, the ground is leveled with a shovel. According to the size of the ger, a shallow groove line is dug in the ground. Next, branches are tied with leather straps to form the skeleton of the walls. Then an umbrella-like arched net frame is placed on top of it. The junction between the node and the vertical skeleton is tightly tied with leather straps.

Finally, sheepskin or felt is tied to the outside of the frame with ropes, thus covering the entire frame and completing the ger. In order to make the floor inside the ger moisture-proof, a *cun* (about 3 cm) of sand or a layer of dry sheep dung is spread on the floor and then covered with leather and felt. In the center of the ger's roof, there is a circular piece of felt which can be removed to let light in during the day. The entrance of a ger is small, so people need to bend over to enter. When migrating, the ger can be easily disassembled into foldable wooden frames and felt blankets, and transported in carts.

3.2.2.7 Stilt Style Houses

The main feature of stilt-style houses are their bamboo and wooden beams and columns. The houses are elevated off the ground to protect them from moisture, insects, snakes, and wild animals. It is mainly used by ethnic minorities in Guangxi, Hainan, Guizhou, and Sichuan, such as the Zhuang, Dai, Dong, Miao, Li, Jingpo, De'ang, and Buyi people.

The dwellings of the Dai people are also known as bamboo houses, because their structure consists mostly of bamboo with thatched roofs (as shown in Picture 3.16). The "ground floor" of the bamboo house is elevated and covered in bamboo mats in order to be able to sit on the ground. It has a large front porch and an open-air terrace. It is mainly characterized by its low-hanging eaves and steep gable and hip roof. The rooms of the stilt houses of the Dong people are partly open, with more exposure. They like to use overhanging corridors in their wooden or bamboo houses supported by wooden pillars. There are drum towers with high multi-eaves in the villages of the Dong people, which serve as a place for various activities for the whole village. They are an important part of their architectural heritage. The Miao people often live in half-stilt houses, that is, the stilt of the house is half elevated and half buried underground according to the terrain. The Li people live around Wuzhi Mountain, Hainan, in a humid climate with heavy winds and heavy rain fall. Their residential buildings are built with a low roof and a slight elevation from the ground with stilts, covered with thatched semi-circular boat-shaped canopies, with no walls or windows. There are front and rear doors, with a deck outside them. From afar it looks like a long and overturned boat, so it is also called a "boat-shaped house" or "boat house" (as shown in Picture 3.17). The roofs of the Jingpo and De'ang people's stilt houses all have their own unique forms. The traditional houses of the Buyi people were originally stilt houses, but the Buyi people living in Zhenning, Anshun, and Liupanshui, Guizhou province, now build houses with stone due to the lack of other materials. However, their prototypes are still stilt houses.

3.2.2.8 Dong Villages

Dong villages are architecturally most famous for their wind and rain bridges, and their drum towers. There are corridors and pavilions on the bridge, which are used by

Picture 3.16 Bamboo house of the Dai people

Picture 3.17 Boat-shaped house of the Li people

people to pass through and take shelter from wind and rain. It has a rigorous structure, a unique shape, and clear ethnic characteristics. The whole building does not require rivets or iron components but is connected with corrosion-resistant fir mortises and tenons. Pagoda-shaped and palace-shaped pavilions are built on top of large heavy stones on the bridge, in an interlocking fashion, which makes the structure stable and firm. The drum tower is shaped like a polyhedral pagoda (as shown in Picture 3.18).

Picture 3.18 Drum tower of Dong village in Xiaohuang, Congjiang county, Guizhou province

It is generally more than 20 m high, counts 11 stories, and is supported by 16 fir columns. The center of the building is wide and flat, measuring about 10 sq.m. It has a large stone fire pit in the middle with wooden railings around it and long wooden benches for resting. The spire of the tower has either a treasure gourd or a crane mounted on top, which symbolizes and heralds the auspiciousness and safety of the village. The corners of the eaves are protruding and raised, giving people a sense of exquisite elegance.

There are many kinds of drum towers, such as hall types, stilt types, and dense-eave types. No matter what kind of drum tower it is, it consists of three parts: (1) upper, (2) middle, and (3) lower. The upper part is the steeple top, which stands in the center of the roof with a wooden column or an iron column of about 3 m long which is covered with 5–7 ceramic beads arranged from large to small in the shape of a gourd, just like a spire standing in the air. The roof is a colorful one, mostly umbrella-shaped, with four, six, or eight corners. The herringbone bucket arch on the lower slope of the roof is like a honeycomb window. The wood carvings around it look like mud spots of bird's nests. It is exquisitely crafted and the shape is unique. The middle part is built layer upon layer, like a pagoda. Generally speaking, the eaves are hexagonal, but there are also shapes like simple quadrangles or complex octagons. Each eave has a warped angle and is stacked layer upon layer with double eaves. Each layer is stacked on top of each other from biggest to smallest. The building has four thick, straight, long fir columns, which impressively lead from the ground all the way to the roof. Inside the building are sculptures and paintings of lifelike fish, insects, birds, and animals.

3.2.2.9 Tibetan Watchtowers

Most areas in Tibet are high in altitude, cold, and dry. The main materials people there use to build their houses are the stones in the wasteland. Tibetans build three or four-story houses with these stone blocks, which are named watchtowers or *diaolou*, because they resemble actual watchtowers (as shown in Picture 3.19). Watchtowers initially were built as defensive military buildings. The ground floor generally contains a pen for livestock with movable Tibetan stairs, which are retracted at night to prevent livestock from climbing up. The protruding balcony behind the watchtower, which serves as a place to keep watch, is used as a toilet as well. The excrement falls from the balcony into the livestock pen on the ground floor. Threshing ground is built on the roof, because it prevents livestock from being able to steal food. It is also a place to do housework or bask in the sun.

Picture 3.19 Earth-rock watchtower on Buwa Hill in Aba Tibetan and Qiang autonomous prefecture, Sichuan province

The outer wall of a Tibetan watchtower is made of solid stone, which is very defensible. The livestock pen at its bottom is only provided with small windows or ventilation holes. The neatly arranged windows in the rooms are not large so that they are conducive to resisting the cold wind. For safety reasons, there is usually only one gate. The main room is the most important space and covers a large area. It is used as the kitchen and dining room with a stove or fireplace. It's the place where daily chores such as extracting yak butter and grinding highland barley are done. It also functions as the living room. The most important sutra hall in Tibetan religious activities is located in the relatively closed space on the top floor.

Watchtowers take advantage of the properties inherent to mountainous areas. Being mountainous, the rocks are of slate or gneiss structure which are easy to cut off and process, thus making materials easily accessible. The external wall of the tower is thick and tall, a construction made of stone. The inside is built with dense beams and wooden floors. The floors are made by densely lining wooden beams with battens, followed by a layer of twigs with about 20 cm of solid soil layer underneath. The roof is also covered with a solid soil layer of about 30 cm thick. The use of a rammed earth surface on the roof of a building is locally called "da'aga" and is a traditional Tibetan construction method. The workers of "da'aga" are divided into two groups, men and women, each holding a special tool that consists of a long rod inserted into a flat stone. Men and women take turns rhythmically beating the soil layer, while moving their feet along with the rhythm, just like they're dancing. In this fashion, it takes about 40 days to tamp a roof.

3.2.2.10   Hakka *Tulou*

Hakka *Tulou* or Hakka earth buildings are found in Fujian and Guangdong provinces, among which the Longyan and Zhangzhou earth buildings in Fujian are the most famous and categorized under "courtyard houses". Hakka people mainly live in groups on the first floor of a tall building with thick walls. The building, named *tulou*, is made with packed earth and is therefore called "earthen buildings" (as shown in Picture 3.20).

The forms and building practices of earthen buildings differ from region to region, but their shapes have some things in common: First, an earthen building is centered around the ancestral hall, which is used as a communal meeting place for Hakkas and contains a table holding memorial tablets of ancestors in the middle; second, whether it is a round, square, or curved building, it is always axisymmetric and keeps the traditional pattern of using quadrangles in northern China; third, every earthen building is constructed to contain a multitude of unit residences.

The construction methods of the Hakka *tulou* were created by northerners after having moved south and needing to deal with the different local climates and situations, First, due to the need for protection, the external wall is thick and tall. The earth building in Yongding, Fujian province, for example, has a wall of 1–1.5 m thick. The earth buildings in Zhao'an, Fujian, are even 2.4 m thick. In practice, bamboo tendons and pine branches are used inside the raw soil wall as reinforcement. After

Picture 3.20 Interior of a Hakka *tulou*

that, the soil is paired with rock blocks to make it extra firm after ramming. Second, because it is in the south of China, attention needs to be paid to protection from the sun. Therefore, cornices on the inner walls, patios, corridors, windows, and roofs are extended in order to create more shadow inside the building. Third, in the interior of the building, movable screen doors and fans were used so that the space became more open and transparent, making it more conducive to air circulation. Fourth, the building's windows on the outer ring are trapezoidal arrow windows, which are small on the outside and large on the inside, which makes them suitable for defense as well as daily use. Fifth, it is built in accordance with *fengshui* rules when selecting the site and keeping it south-facing like in northern China so that the house base "backs onto *yin* and faces *yang*". However, in special circumstances, for example, in order to restrict taboos from occurring and ward off evil, it can face east or west, but never north. It is built close to rivers or ponds, but never with its back towards the water. It can, however, be backed by mountains or hills, or face the mountains, and it can also be flanked by small hills.

3.2.2.11 Kaiping Watchtowers in Guangdong

The Kaiping watchtowers, also known as the Kaiping *diaolou*, located in Kaiping, Guangdong province, are multi-story tower-type buildings that are used for both defensive and residential purposes. These towers combine Chinese and Western architectural art (as shown in Picture 3.21), and are usually built much higher than ordinary residential buildings in order to be more suitable for mounting defenses. The windows are smaller and the walls are thicker and stronger than those of ordinary houses. There are iron bars and a window sash, with external iron windows and doors as well. At the four corners of the upper part of the tower, there are erected and fully

Picture 3.21 Kaiping
watchtower in Guangdong

enclosed or semi-enclosed hornworks (commonly known as "swallow nests") with gunports. There are also gunports on the walls of every story, which serve as extra defensive points.

The upper part of a Kaiping watchtower is the most distinctive. The builders go to great lengths to do each of them justice by using elements such as domes, pediments, and columns. As per the upper shape, Kaiping watchtowers can be divided into different types, namely, the colonnade type, the platform type, the set-back type, the cantilever type, the castle type, and the mixed type. Each different architectural style reflects the region's economic strength, aesthetic taste, and amount of influence on foreign architecture. This is the most fascinating aspect of a Kaiping watchtower. Watchtowers can also be categorized according to the main building material that is used, namely, stone, earth, brick, or concrete.

3.2.2.12 Shanghai Stone Gate Houses

Stone gate houses or *shikumen* are the most characteristic dwellings in Shanghai. A traditional stone gate house, adapted from Jiangnan dwellings, generally is as wide as three or five rooms and thus keeps the characteristics of traditional Chinese buildings with a symmetrical layout around a central axis. Entering the door, you can see a long horizontal patio, flanked by side houses, and the parlor with French windows is directly opposite you. The parlor is about 4 m wide and 6 m long and is the place for receiving guests and holding banquets. There are secondary rooms on both sides of the parlor, a wooden escalator leading to the second floor at the back, and a back patio, which is only half as deep as the front patio and contains a well.

Behind the back patio is an annex with a sloping roof, which is used as kitchen, storage room, or spare room. There are entrances and exits in the front and back of the house. The front facade is composed of a patio wall and side house gable, with the *"shikumen"* (stone gate) in the middle: a stone gate frame with a black thick lacquered wooden door. The rear wall is roughly the same height as the front wall and forms a circle around the nearly closed-off facade. Therefore, although stone gatehouses are located in the downtown area, their high walls can manifest a deep and quiet courtyard.

In the early twentieth century, the traditional *shikumen* were gradually replaced by the modern *shikumen* (as shown in Picture 3.22), most of which are as wide as single or double rooms, with the former completely getting rid of side houses and the latter keeping the front and rear side houses on one side. The biggest change in the internal structure is that the roof of the annex is changed to a flat roof, and a garret (a small bedroom) is built on top of it. The roof of the garret is made of reinforced concrete slabs surrounded by railing walls, which are used as a place for drying clothes. In order to reduce the occupied area and use less building materials, the modern *shikumen* have reduced the depth of the living room and the height of the floor and enclosing walls. Compared with the traditional *shikumen*, the appearance of the modern *shikumen* is also different. The outer wall is mostly made of ganged grey bricks, red bricks, or a combination of both, with lime for pointing instead of plastering the walls with white lime like the traditional one. The horsehead wall or gable in the shape of Guanyin *dou* (a kind of hood) commonly used in the traditional *shikumen* is no longer used in the modern one either. Another important difference is that the modern *shikumen* no longer uses stone as the gate frame, but uses clear water brick instead, and the decoration of the gate lintel has become more elaborate. Early gate lintels of *shikumen* were often traditional brick-carved, blue-tile-topped gate heads by imitating the front gate seen in traditional Jiangnan buildings. Influenced by western architectural styles, however, modern *shikumen* are often decorated with triangular, semi-circular, curved, or rectangular flowers, which resemble the mountain lintels on the upper doors and windows of western buildings. These ornaments have various forms and styles and are the most distinctive part of *shikumen* architecture. Some modern *shikumen* have classical western pilasters on both sides of the gate frame. In short, the architectural style of the modern *shikumen* is more westernized than the traditional one. The most typical feature of a *shikumen* residence is the combination of Chinese and Western architecture.

Picture 3.22 *Shikumen* in Shanghai

3.2.3 Palace Architecture

A "palace" often refers to a group of buildings, while a "hall" is often used to refer to a single building inside a "palace". In the early days of Chinese palaces, the design of three courts and five gates was implemented. According to the Zhou dynasty's system, a palace consisted of (1) the outer court, where national affairs were decided, (2) the middle court, where the emperor attended to business, and (3) the inner court, where the royal family lived and held banquets.

The Han dynasty pioneered the use of an East-to-West palace design, a system unequivocally adopted by the Jin, Southern and Northern dynasties (except the Northern Zhou dynasty). After the Sui dynasty, the preferred design reverted to that used in the Zhou dynasty. The five gates used by the Sui and Tang dynasties were called the Chengtian Gate, Taiji Gate, Zhuming Gate, Liangyi Gate, and Ganlu Gate. From then on, the three courts and five gates design was used until the Qing dynasty. The Forbidden City in Beijing is the most well-preserved example of a palace like this.

3.2.3.1 The Forbidden City

The current Forbidden City in Beijing was built in the fourth year of the Ming dynasty's Yongle period (1406) (as shown in Picture 3.23). It is surrounded by a moat and has an east-to-west width of 760 m and a north-to-south length of 960 m. The city wall has a gate on all four sides: (1) the main entrance in the south is the Meridian Gate, (2) the one in the north is the Shenwu Gate, (3) the one in the east

Picture 3.23 The Forbidden City

is the Donghua Gate, and (4) the one in the west is the Xihua Gate. Each gate has a double-eave arch over the gateway. The four corners of the city wall have resplendent turrets with 3 eaves and 72 ridges.

The palace is divided into two parts: the outer court and the inner court. The outer court includes the "Three Halls", Wenhua Hall and Wuying Hall. The "Three Halls" are (1) the Hall of Supreme Harmony, (2) the Hall of Supreme Harmony, and (3) the Hall of Preserving Harmony. They stand together on a white stone base with red walls and yellow glazed tiles. Among them, the Hall of Supreme Harmony uses a double-eave hipped roof, the Hall of Supreme Harmony uses a pyramidal roof, and the Hall of Preserving Harmony uses a double-eave gable and hip roof. There are dozens of clear distinctions between them.

North of the Qianqing Gate (Gate of Heavenly Purity) is the inner court, which has a road running straight through its center to the Palace of Heavenly Purity with the resting quarters of the concubines on the left and right. The Palace of Heavenly Purity is the emperor's main residence and the Palace of Earthly Tranquility is the empress' residence. In the Ming dynasty's Jiajing period, the Hall of Union was built between the two palaces, thus forming the layout of three outer halls and three inner halls. To the west and east of the Palace of Heavenly Purity are more palaces and residences: Six palaces to the east, six palaces to the west, as well as five residences to the east, and five residences to the west, where the concubines, princes, and princesses lived. This arrangement is based on astronomical symbolism: The Palace of Heavenly Purity symbolizes heaven, the Palace of Earthly Tranquility symbolizes earth, the 6 eastern palaces, and the 6 western ones symbolize the 12 Earthly Branches (*dizhi*), and the 5 residences to the east and the ones to the west symbolize the 10 Heavenly Stems (*tiangan*), thus forming the structure of stars guarding heaven. On the east side of the six eastern palaces, there are several groups of small courtyards running north to south, which are used for managing food and clothing, while their southern end, the Fengxian Hall, is used by the palace for ancestry worship. There are two courtyards on the west side of the six western palaces: The closer courtyard contains some small halls and gardens used for funerals or recreational purposes, and the farther courtyard contains a Lamasery (which used to be a Taoist temple in the Ming dynasty).

The Forbidden City is the embodiment of Chinese imperial architecture. It is the epitome of fully using a structure to highlight the sublimity and sacredness of imperial power. The most important feature is the way the continuous and symmetrical enclosed spaces are used on a 1.6-km axis to form a gradually unfolding architectural sequence that sets off the dignity, majesty, and grandeur of the "Three Halls".

From the Qing dynasty, the Imperial Palace in Beijing starts from the Daqing Gate (now the Zhonghua Gate) and goes through six enclosed courtyards before reaching the main hall: The 500-m-long "Thousand-Step Corridor" to the north of the Daqing Gate forms a long and narrow front yard, followed by another 300 sq.m. This long horizontal space forms a T-shaped plane, and the towering front gate of the Imperial Palace, the Gate of Heavenly Peace, better known as Tian'anmen, is equipped with ornamental columns and Jinshui (Gold Water) bridge, forming the first architectural highlight. Once through the Tian'anmen, there is a small courtyard with the Duan Gate at the end mirroring the size and form of the Tian'anmen. This repetition strengthens the overall image of Tian'anmen. Once through the Duan Gate, one can enter a long and narrow courtyard over 300 m long. The Meridian Gate forms the second highlight with its rich outline and overall grandeur. Past the Meridian Gate is the open and clear courtyard of the Gate of Supreme Harmony, which has a width of more than 200 m. Past the Gate of Supreme Harmony is an even larger courtyard, which is a nearly perfect square of more than four hectares. The Hall of Supreme Harmony on the high platform in the middle has more than 10 gates and surrounding arches, which is the main highlight of the palace.

In terms of architectural construction, the Forbidden City uses contrasting techniques such as small and large lining as well as low and high lining to really make the main buildings stand out. For example, Tian'anmen and the Meridian Gate both use the same style of towers, with the base building of more than 10 m tall; the Hall of Supreme Harmony has a three-story white marble Sumeru seat, equipped with luxurious and noble railings and *chishou* (legendary hornless dragon) gargoyles; on the other hand, the bases of the annex buildings are simpler and lower in order to set off the prominent position of the main buildings. The roofs are used in a certain order: double eaves, hip roof, gable and hip roof, pyramidal roof, overhanging gable roof, and flush gable roof, as seen on the Meridian Gate and the Hall of Supreme Harmony, which use double eaves, and the Tian'anmen, the Gate of Supreme Harmony, and the Hall of Preserving Harmony, which use double-eave hipped roof. The heights of the remaining halls are similarly lowered. Architectural details and decorations are also different in complexity and simplicity. For example, the upper eaves of the Hall of Supreme Harmony use a nine-step bucket arch of the highest grade, while the lower eaves use a seven-step bucket arch. In front of the main halls and gates, bronze lions, tortoises, cranes, sundials, and grain measuring vessels are used as foils to set off the scale of the halls and demonstrate imperial power. The buildings also use strong contrasting colors. The white bases, ochre walls, vermilion doors and windows, turquoise paintings densely covered with gold, plus the yellow, green, and blue glazed roofs, together make the Forbidden City look extraordinarily bright and brilliant against the blue sky and the backdrop of the large gray-tiled roofs of the buildings surrounding the palace. All of the above are good examples of how imperial rule can be safeguarded and prolonged through inspiring awe.

3.2.3.2   Kuai Xiang, Xiangshan and the Lei's Architectural Styles

What labeled as "major construction projects" are, in fact, highly comprehensive systematic projects on a big scale. In ancient times, people at all levels, ranging from various types of local workmen all the way to even the emperor himself wanted to be and were involved in them. Master craftsmen played a key role in the performance and success of these projects. With their excellent ideas and talents, along with their teams of fellow craftsmen, they were the designers and implementers of these projects. The magnificent project of the Forbidden City in the Ming and Qing dynasties, for example, was created by the craftsman Kuai Xiang, a descendant of Suzhou's Xiangshan style. Kuai Xiang was born in 1398 in Huanfan village, Xiangshan, Suzhou, and died in 1481. The Xiangshan style was created by a group of artisans that included different craftsmen, such as carpenters, stonemasons, plasterers, lacquerers, sculptors, rock stacking artisans, and painters. During the Ming dynasty's Yongle period, Kuai Xiang presided over the construction of the main palaces at Tian'anmen and the Meridian Gate, and also the main palace building of the Forbidden City. During the Zhengtong period (1436–1490), he also presided over the rebuilding of the Three Halls, the imperial garden and the government offices of the five military administrations, and six ministries, which laid the foundation for the palace architecture of the Ming and Qing dynasties. Kuai was proficient at construction, being pretty good at measuring. Once built, the new building resembled in all aspects without any difference from the original design. He had outstanding skills, being able to create blueprints for halls, towers, pavilions, and even corridors and eaves. Kuai Xiang also introduced Suzhou-style paintings and gold bricks to make the decoration of the palace more solid and resplendent. Therefore, his peers admired and praised him and his extraordinary workmanship by dubbing him "Lu Ban incarnate". Kuai Xiang can be called the chief designer of the Forbidden City. He was a leading authority of the Xiangshan style and the representative of the architectural skills of that era. Xiangshan architectural skills, which are an important part of traditional Chinese construction practices, were added to the *List of the First Batch of National Intangible Cultural Heritage* in 2006, and are therefore expected to be better protected and passed on.

During the Kangxi period (1662–1722) of the Qing dynasty, Lei Fada, who was originally from Jianchang, Jiangxi (now Yongxiu county, Jiangxi), came to Beijing from Nanjing to participate in the construction of the imperial palace. He was the creator of Lei-style architecture. However, it was Lei Jinyu, the son of Lei Fada, who pushed the style of the Lei family to its peak. During the Kangxi and Yongzheng periods, his prowess stood out during the construction of the Changchun Garden and the Old Summer Palace. After that, Lei Shengwei, Lei Jiaxi, Lei Jingxiu, Lei Siqi, Lei Tingchang, and Lei Xiancai, seven generations of the Lei family, successively presided over the construction of projects at Longevity Hill, Jade Spring Hill, Zhongnanhai, the Summer Palace, Rehe Summer Resort, the Xingling and Dongling Tombs, the reconstruction of the Hall of Prayer for Good Harvest, the Zhengyang Gate and many more, spanning over 200 years. Their works account for one-fifth of China's world cultural heritage, which is unique in architecture everywhere. With the collapse of the imperial dynasty, the Lei family's success declined and their blueprints began to circulate on the market. At that time, Zhu Qiqian presided over the China Architectural Society. He followed the suggestions of Liang Sicheng, Liu Zhiping, and Liu Dunzhen, and invested in collecting the Lei family's drawings and stereomodel. In 1930, more than 10,000 Lei-style drawings were collected and stored in Beiping Library. In 1964, descendants of the Lei family sent some of the drawings to the Beijing Municipal Bureau of Culture. The rest were burned during the Cultural Revolution and their ashes were thrown in the city moat. There are currently about 20,000 patterns of Lei drawing files held by the National Museum of China, the First Historical Archives of China, and the Palace Museum in Beijing. These precious cultural relics contain rich traditional architectural concepts, and techniques and have significant humanistic connotations. Some of them are comparable to modern architectural design theories and methods. They also fully demonstrate that these major construction projects in China's history were under a strict management system that was carefully regulated, planned, designed, and constructed, and that, through this, greatness was achieved. Acknowledged as a cultural treasure, the Lei style Archives were added to the *Chinese Archives Document Heritage List* in 2003 (as shown in Pictures 3.24 and 3.25).

3.2.4   Religious Buildings

In ancient China, there were many different kinds of religions. Buddhism was the most influential, but there were also Taoism, Islam, Manichaeism, Zoroastrianism, Catholicism, Christianity, and Bon.

Buddhism originated in India and initially opposed idolatry. Therefore, no statue of Sakyamuni was erected and pagodas became regarded as the object of worship. Early Buddhist temples were square, with a pagoda in the middle and monks living on all sides. The roof, eaves, and courtyard floor of the temple were covered with special materials, made of walnut-sized pieces of bricks and clay. After being ground flat, these pieces were coated with lime mixed with oakum, hemp offscourings, and stem fiber, and covered with grass. Before fully dried, they are polished with talc, coated with terracotta mud, and painted to make them shine like a mirror. After such treatment, the temple floor becomes solid and is able to withstand many people treading on it for many years.

Picture 3.24 Vertical design of the Zhengyang Gate, now held in the Beijing Palace Museum

Picture 3.25 Stereomodel of the Old Summer Palace

Buddhism was introduced to China in the early Eastern Han dynasty. The oldest Buddhist temple in China is the White Horse Temple, also called Baima Temple, in Luoyang, Henan province, which is built on a square plane. Its main pagoda is located at the temple's center, just like temples in India. This type of design is called the "Tianzhu style". As a result of the continuous spread of Buddhism and its integration into Chinese culture, the shapes and design of Buddhist temples changed as well, transforming from the Tianzhu style that is mainly composed of pagodas to the courtyard style that is mainly composed of temple buildings.

Buddhism developed rapidly during the Jin and during the Southern and Northern dynasties, exemplified by the construction of most of China's famous Buddhist grottoes (as shown in Picture 3.26) as well as the construction of over 1,200 temples in Luoyang alone. One temple, called Yongning Temple, was famous in the Northern Wei dynasty when it was built by the imperial family. The main part of the temple is composed of a pagoda, a hall, and a gallery, which are symmetrically arranged along the central axis. At its center is a square, nine-story pagoda on a raised three-story platform, north of which lies the hall and a rectangular courtyard surrounded by walls. The east, south, and west sides of the courtyard have gates in the middle, with gate towers built on top. The north side has a simpler *wutoumen*, which is a gate with a black column head. The auxiliary buildings, such as the monks' quarters, are at the back and west of the main building. For protection, each of the four corners of the surrounding walls has a turret and the periphery outside the walls contains trenches and locust trees. The top of the walls is roofed with short rafters and tiling. Yongning Temple is a typical example of the "pagoda in front, hall in the back" style of the late Eastern Han dynasty. During this period, many residences were converted into temples. The original buildings were left intact, yet they were changed to fit the "hall in front, sutra hall in the back" style. Grotto temples at that time were still like Indian Buddhist temples. In addition to the Buddha statues, they had pagodas, flame-shaped arches, beam lotus columns, and swirl columns. However, the layout of the temple, the wooden beam-column frame, and many architectural symbols started to show Chinese characteristics.

The larger Buddhist temples of the Sui and Tang dynasties adopted an axis-symmetrical layout, with gates, lotus ponds, platforms, Buddha towers, auxiliary halls, and main halls arranged in sequence. The core of the building complex changed from a pagoda to the main hall, with the pagoda generally moved to the side or another dedicated courtyard. Esoteric Buddhism prevailed in the late Tang dynasty. Statues of Guanyin with eleven faces and Guanyin with a thousand hands and thousand eyes appeared, as well as stone Dhanari columns engraved with the *Usnisa Vijaya Dharani Sutra*. Clock towers were added and generally built on the east side of the axis, and drum towers built on the west side began to prevail in Ming dynasty. In the Five Dynasties period, Arhat Halls in the shape of the character "田" (*tian*) appeared. The Southern Dynasties period was the first to witness the appearance of prayer wheels. The Song dynasty saw the addition of a special altar used for taking Buddhist vows in the Vinaya School temples. Tibetan Buddhism prevailed in Tibet and Mongolia during the Yuan and Qing dynasties, but it had little influence

Picture 3.26 Interior of
Yungang Grottoes

on the Buddhist architecture in the Central Plains. Small Buddhist temples in Han-dominant areas are called Buddhist convents (or Bhikshuni temples), larger ones are called temples, and the ones that are even larger have the character "大" (*da*), which means "big", added in front of their names, such as Daci'en Temple, Daxiangguo Temple, and Dayuanman (Dzogchen) Temple.

Taoism originated quite early in China and formed a religion in the Eastern Han dynasty. Taoist architecture follows the traditional design of palaces and ancestral temples. They are generally laid out on a central axis with mostly halls and pavilions (as shown in Picture 3.27), and without pagodas or stone columns with scriptures.

Picture 3.27 Sanqing Hall
of the Ancient Changdao
Taoist Temple at Mount
Qingcheng

3.2.4.1   East Hall of Foguang Temple at Mount Wutai, Shanxi Province

During the Tang dynasty, a time when many Buddhist temples were being built, Mount Wutai was already an established center for Buddhism. Foguang Temple is located on the mountainside of Foguang Mountain, 5 km northeast of Tainandou village, Mount Wutai. The existing buildings in the temple include the main hall, built in the 11th year of the Dazhong period (857) in the Later Tang dynasty, the Manjusri Hall, built in the Jin dynasty, the Sakyamuni pagoda, and two stone Dhanari columns, built in the Tang dynasty. The main hall is the largest existing wooden structure of the Tang dynasty (as shown in Picture 3.28).

It is seven rooms wide, measuring over 34 m, four rooms deep, measuring 17.66 m, and covered with four hip roofs with single eaves. There are inner columns (golden columns) in the temple, which divide the space into an inside part and an outside part. There are three walls along the back of the inside part which surround the Buddhist altar, on which there are more than 30 Later Tang dynasty colored statues. All along the back, left and right walls, there are 500 arhat statues which were added in the Qing dynasty. The inside part is high-ceilinged to strengthen the position and impression of the Buddhist altar. On the altar, wooden bars form a square shape and steep rafters around it form a high inverted-bucket-shaped ceiling. The space is divided into five smaller parts by the four girders exposed under the ceiling and the upper and lower bucket arches. The space of the outside part is narrow and low-ceilinged, which serves as the foil for the inside part. Its ceiling, exposed beam frame, and bucket arches are coherent with those of the inside part, which gives it all a strong sense of order and integrity. The architectural space is in harmony with the sculpture. The three statues in the center are the largest, which are three sitting Buddhas, while another two statues, one on each end, are a bit smaller, which are Manjusri riding

Picture 3.28 The east main hall of Foguang Temple, Mount Wutai

a lion and Samantabhadra riding an elephant. There are also five groups of small statues, each group surrounding one of the five larger statues. The height and volume of each statue correspond to the space in which it is located, without congestion or emptiness. At the same time, the perspective is also taken into account. When a person stands at the temple gate, the inner columns surrounding the inside part do not block the integrity of the statue nor the backlight of the sitting Buddhas; when standing at the inner column line, the connection between the Buddhas' top and one's eyes is still within the normal vertical visual angle. Several of the temple's murals and architectural colored paintings dating back to the Tang dynasty are still intact. The facade of the main hall is vertically divided into three parts: the base, the body, and the roof. The base is plain and unpretentious. The vertical columns of the building have sided footholds and raised eave columns, which imbue the building with stability as well as charm. The bucket arch on each of the columns is large: its height accounts for half of the column's height. The eaves reach far as well, peaking out up to 4 m, which is about half of the height from the eaves to the bottom of the column. There is only one mending bucket arch: its layout is sparse, and its most pleasing feature lies in its structural strength and simplicity. The roof slopes gently. The eaves rise slightly from the middle of the facade to both ends of the facade with flexible curves, and the entire roof stretches gently. The main ridge on a whole is an arc low in the middle and high on the sides, and the two sides end with involuting *chiwen* (mouth of *chi*, a legendary animal in China) with a large scale and a simple outline, and the upright column is right under it, which strengthens the organic character of the overall structure.

3.2.4.2 Dule Temple's Guanyin Pavilion and Temple Gate

The temple is located in Xidajie Street, Jixian county, Tianjin city. It was first constructed in the 10th year of the Tang's Zhenguan period (636) and rebuilt in the second year of the Tonghe period (984) in the Liao dynasty. The current temple gate and Guanyin pavilion date back to the Liao dynasty.

The temple gate faces south, measuring three rooms in width and two in depth, and has a single-eave hip roof. Its columns are not high, and their sloping angle is clearly visible. The bucket arches are large and sparsely arranged, with their height taking up about half the height of the columns. The *chiwen* at both ends of the roof's ridge, with their tail wings turning inward, are different from the *dawen* dragon tails turning outward seen on the temple buildings of the Ming and Qing dynasties.

The Guanyin pavilion, which is the main building, is a pavilion with a three-story wooden structure. Because the second floor is a dark room and it is separated from the third floor without eaves, it appears like a two-story building. The pavilion is 23 m high, surrounded by middle eaves and railings, with a single-eave gable and hip roof on the top, and far-reaching cornices, which are both beautiful and spectacular. On the Sumeru seat, which is an elevated platform in the center of the pavilion, stands a 16-m-high clay statue of Guanyin (as shown in Picture 3.29) with its head reaching all the way up to the roof. It is also called "Guanyin with eleven faces" because there

Picture 3.29 Interior of the Guanyin Pavilion at Dule Temple

are ten small heads on Guanyin's head. Guanyin's body leans forward slightly, and its stance is dignified. There is a Bodhisattva statue on each side of the Guanyin statue, which was originally made in the Liao dynasty. The pavilion is centered on the Guanyin statue, surrounded by two rows of columns, with bucket arches on the columns and beams on them, on which wooden columns, bucket arches, and beams are erected. The interior is divided into three layers so that people can admire the Buddha from different heights. The wooden structure is set around the statue, and the middle part forms a patio, which runs from top to bottom and accommodates the statue's body as if it is clothed by the building and capped by the ceiling. The whole internal space closely corresponds to the Guanyin statue.

3.2.4.3 Pagodas

China's pagodas can be roughly divided into pavilion-style pagodas, dense-eave pagodas, single-story pagodas, Lama pagodas, and Vajrasana pagodas of Mahayana, and Buddhism pagodas of Hinayana.

Pavilion-style pagodas are the oldest and largest in number. They are considered the mainstream of Chinese pagodas. Their most distinctive feature is that each floor of the pagoda can be seen as one architectural whole, with eaves, body, and base. The whole pagoda construction is like several single-story buildings stacked on top of each

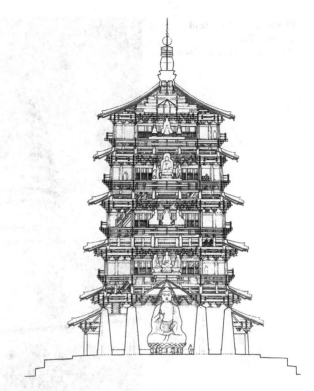

other. Among these types of pagodas, the most famous one is the Sakyamuni Pagoda at Fogong Temple in Yingxian county, Shanxi province (as shown in Picture 3.30).

The bottom layer of the dense-eave pagoda is high, with 5–15 dense layers of eaves (usually, it's an odd number with 7–13 layers). Each floor of this pagoda does not architecturally mirror the preceding floor completely. The most visually prominent feature is the eaves. The whole pagoda seems to be formed by overlapping layers of eaves, hence the name.

Pagodas are generally built from brick and stone. The Pagoda of Songyue Temple in Dengfeng, Henan province is the oldest existing brick pagoda with dense eaves (as shown in Picture 3.31). Single-story pagodas are mostly used as tomb pagodas or contain Buddha statues for worship. The former first appeared in the Northern Qi dynasty, the latter in the Sui dynasty. By the Tang dynasty, the structure of single-story pagodas had imitated the wooden structure, using the components such as columns, square columns, and bucket arches. The plane of the pagodas can be square, round, hexagonal, or octagonal.

Lama pagodas are mostly found in Tibet and Inner Mongolia. They are often used as the main pagoda of a temple or as a monk's tomb; however, some of these pagodas are in the shape of a gate for people to pass through. The Lama pagodas found in the mainland were built in the Yuan dynasty. These pagodas became taller and thinner in the Ming dynasty. Only a few Vajrasana pagodas were built. These pagodas have a

Picture 3.31 Pagoda of
Songyue Temple in
Dengfeng, Henan province

Picture 3.31 Pagoda of Songyue Temple in Dengfeng, Henan province

base shared by five pagodas and were only built in the Ming and Qing dynasties. The one at Dazhengjue Temple in Beijing, built in the Ming dynasty's Yongle period, is a typical example of a Vajrasana pagoda.

Pennants or streamers were banners and flags carried by a guard of honor in ancient China and were made of silk fabric tied to the top of a pole. After the introduction of Esoteric Buddhism in the mid-Tang dynasty, Buddhist scriptures that were originally written on silk fabrics or Buddha images that used to be painted on silk fabrics were now engraved on stone columns to make them last longer and keep them from being destroyed. Therefore, it is called columns with scriptures, which generally consist of three parts: the top, the body, and the base. The main body of the column, which is usually hexagonal or octagonal, is engraved with mantras or scriptures and Buddha images of Esoteric Buddhism. The scripture columns in the Tang dynasty were robust and simple in decoration. In the Song dynasty, the columns became taller and slenderer. The column body was divided into several sections and the decorations became more elaborate. The Dharani scripture column (as shown in Picture 3.32) in Zhaoxian county, Hebei province is the most representative building of its kind.

Picture 3.32 Dharani
scripture column in Zhaoxian
county, Hebei province

3.2.5 Sacrificial Architecture

Offering sacrifices to gods and worshipping ancestors were important activities in ancient China. As *Classic of Rites* says: "a man of noble character who wants to build a palace must first build an ancestral temple". These sacrificial architectures are also called temples and altars. The original meaning of "庙" (*miao*, temple) refers to places where ancestors are given offerings, such as family temples, ancestral temples, and the imperial court. The "坛" (*tan*, altar), on the other hand, is used to worship and give offerings to gods. For example, the altar worshipping heaven is called the *Tiantan* (Temple of Heaven), and the altar worshipping the earth is called the *Ditan* (Temple of Earth).

Altars and temples have different architectural forms due to their different functions. Temples require a place with ancestral tablets and an area to practice such activities including offering sacrifices to ancestors, so they consist of a building or a group of buildings. Altars are places to communicate with gods, so they are often just high platforms.

3.2.5.1 Beijing's Temple of Heaven

Built in the 18th year of the Ming dynasty's Yongle period (1420), the *Tiantan* (Temple of Heaven) in Beijing is located to the south of the palace city. It is an altar where emperors worshipped heaven (as shown in Picture 3.33). The altar is divided into inner and outer parts, with the main buildings in the inner part. The Circular Mound Altar and the Imperial Vault of Heaven are to its south, and the Hall of Prayer for Good Harvests and the Huangqian Hall are to its north. Both areas are connected by a 360-m-long paved path.

As the early Ming dynasty established Nanjing as its capital, they offered sacrifices to heaven and earth simultaneously. During the Jiajing period (1522–1566), when the capital had been shifted to Beijing, the Circular Mound Altar was set up as an altar for offering sacrifices to heaven, and an altar of Earth was added to the north of the city, making it possible to offer sacrifices to heaven and earth separately. The Sun Altar and the Moon Altar were built in the eastern and western suburbs. In the Qianlong period (1736–1796) of the Qing dynasty, the Temple of Heaven was rebuilt, and the Circular Mound Altar was enlarged. The stones used for the altar surface, steps, and railings all became multiples of 9.

Located just north of the Circular Mound Altar is the Imperial Vault of Heaven, which is a single-eave circular building with a gold-gilded roof. There is a circular wall outside the Vault. Because the wall is hard and smooth and the circumferential curvature is very precise, sound waves can be continuously reflected along the inner wall. If you stand closely against the wall, you can hear people talking on the opposite side, which is why it is called the "echo wall".

Picture 3.33 Temple of Heaven in Beijing

3.2.5.2 Beijing's Imperial Ancestral Temple

Beijing's Imperial Ancestral Temple is located on the east of the central axis and south of the Forbidden City. This layout conforms to the design of having "the ancestral temple on the left (east) and the state altar on the right (west)" as written in the *Book of Diverse Crafts* in *Rites of Zhou*. This is the family temple where emperors of the Ming and Qing dynasties offered sacrifices to their ancestors (as shown in Picture 3.34). The temple was built in the 18th year of the Ming dynasty's Yongle period (1420). It has 15 auxiliary halls on both sides, with the eastern auxiliary halls dedicated to the ancestors of the royal family and the western auxiliary halls dedicated to the heroes with different surnames. The middle and back halls are both nine rooms wide, with yellow-glazed tile hip roofs, in addition to the kitchen of gods, the storehouse of gods, the sacrifice slaughter pavilion, and the sacrifice treatment house. the Imperial Ancestral Temple is famous for its ancient cypresses, which are hundreds of years old.

The Sacrificial Hall is at the heart of the temple, located on a three-story white marble Sumeru seat, with yellow-glazed tiles and double eaves. Inside the temple, the beams are decorated with gold, and the floor is paved with gold bricks. The 68 large columns and main beam bridges are made of Phoebe Zhennan wood. There are offerings displayed on the incense table and sacred seats for the emperor and empress.

Picture 3.34 Halberd gate of the Imperial Ancestral Temple in Beijing

3.2.6 Landscape Architecture

Chinese gardens have experienced more than 2,000 years of development, going from being mostly practical to be sublimated into an architectural form rich in connotations. Before the Han dynasty, the imperial and aristocratic hunting grounds were the closest thing to a garden they had; the actual foundation of landscape gardening only came in the period of the Wei, Jin, Southern and Northern dynasties, when the construction of gardens became a form of art. In the Tang dynasty, gardening developed further, and it became fashionable in the Song dynasty. However, the peak of landscape art was during the Ming and Qing dynasties. The characteristic of Chinese gardens is that they are good at artificial landscaping that has rich cultural connotations in which, counterintuitively, buildings account for a large proportion of the garden. The imperial gardens of the Qing dynasty and the literati gardens in Jiangnan area are the most representative traditional Chinese gardens.

3.2.6.1 Imperial Gardens of the Qing Dynasty

Beijing's Jingyi Garden (now Xiangshan Park), Jingming Garden (now Jade Spring Hill), Changchun Garden, Qingyi Garden (now Summer Palace), and Chengde Mountain Resort were all built or rebuilt during the Qing dynasty's Kangxi and Qianlong periods and consisted of two types: one was a royal palace for living and having audiences, while the other was a garden for amusement. The guiding ideology is to imitate famous scenic spots in China and divide them according to the garden's topographical features, each area containing different scenic spots.

Compared with the royal buildings, regular garden buildings are arranged according to the needs, with varied architectural styles. They are integrated with rocks, flowers, trees, and pools of water. The buildings are small with roofs mostly covered with grey tiles. They don't or seldom use colorful paintings or a lot of decoration, keeping it simple, elegant, and light. However, compared with the literati gardens, royal gardens are magnificent, with the sheer amount of large wooden structures basically serving as an official mandate. Temples are often arranged in these gardens, which become important scenic spots or the center of the gardens, such as at the Chengde Mountain Resort, the Summer Palace, or the Old Summer Palace. As central buildings, some pagodas and pavilions are very large, such as the White Pagoda in Beihai Park and the Foxiang Pavilion in the Summer Palace. In traditional gardens, the technique of stacking stones is commonly used (as shown in Picture 3.35). The stacked stones are usually used to create a small garden in a large one.

The "Sanshan Wuyuan", also known as the "Three Hills and Five Gardens", is the general name of the Imperial Palace Gardens in the western suburbs of Beijing. They were built one after the other from the Kangxi to the Qianlong period and are typical examples of imperial gardens in the Qing dynasty. The Summer Palace (as shown in Picture 3.36), Jingming Garden, and Jingyi Garden are all large in scale, and

Picture 3.35 Stone stacking technique

their overall layout is relatively open. Each has their own characteristics simply due to their own local topographical features. In planning and layout, a lot of attention is paid to the relationship between the landscape and the axis. Gardens within the garden are a major feature of the gardens of the Qing dynasty, in which they use the contrast of the landscape to emphasize the different features.

Picture 3.36 The Summer Palace

3.2.6.2 Literati Gardens

The private gardens in Jiangnan area are the most representative examples of Chinese classical gardens. The Jichang Garden, the Garden for Lingering, the Humble Administrator's Garden, and the Lion Grove Garden are the most famous.

The Jichang Garden is located at the foot of Huishan Mountain in Wuxi, Jiangsu province, covering an area of 14.85 *mu* (about one hectare). The garden is centered around the mountain pool, which is skillfully incorporated into the natural elements. The rockery is shaped as a continuation of the mountain's rocks and is injected with the spring current. The Yupan Pavilion Gallery, the Zhiyu Sill, the Qixing Bridge, and the Hanbi Pavilion are built around the water in harmony with the rockery. The trees in the garden are towering, creating a simple and quiet atmosphere. The garden is unique among the Jiangnan gardens as it blends in well with the scenery, stacked stones, water modulation, and architecture.

Located outside the Changmen Gate of Suzhou city, Garden for Lingering (*Liuyuan*) combines residence, ancestral hall, Buddhist convent, and garden into one. It is known for its characteristic Jiangnan gardening art and its architectural structure. It uses contrastive techniques in size, curvature, light, and height, and integrates it into the surrounding scenery to form a spatial system with rich layers, both scattered and connected, rhythmic and colorful (as shown in Picture 3.37). The halls, corridors, lime walls, cave gates, rockery, pools, flowers, and trees are combined into dozens of garden ornaments, which fully showcase the wisdom of the gardeners and the architectural style of the Jiangnan gardens. There are various doors and windows in each building, which also blend with the scenery of each garden element. When watching the outdoor scenery from indoors, the pictures composed of landscapes, flowers, and trees can be seen at first glance.

The Humble Administrator's Garden (*Zhuozhengyuan*), located in the northeast of Suzhou city, was built in the Ming dynasty's Zhengde period (1506–1521). It is characterized by its vast water surface, and its plain, innocent, and natural scenery. Pavilions and terraces are built around the pool, which are connected by traceries and cloisters (as shown in Picture 3.38), creating winding, flowing water that is quite the sight to see. Different scenic spots reproduce the artistic concepts of landscape poetry and landscape paintings in the garden, which creates its own poetry and picturesque splendor. The whole landscape seems to float on the water, creating different artistic tastes from different realms.

The Lion Grove Garden is famous for its rockery (as shown in Picture 3.39). The garden can be categorized into three parts: the ancestral hall, the residence complex, and the rockery garden. The Yanyu Hall is the main hall of the whole garden, located in a residential complex and characterized by being built tall and spacious, and equipped with elegant furnishings. The traceries have various styles with exquisite workmanship. Those themed with "*guqin*-playing", "go-playing", "calligraphic work" and "painting" and clay sculptures themed with flowers are regarded as the top grade ones. The Shanting Pavilion, Wen Tianxiang Monument Pavilion, and Royal Monument Pavilion are built along a long corridor, which reduces the straightness and height of the south wall.

Picture 3.37 The Quxi Tower in Garden for Lingering

Picture 3.38 The Yuanxiang Hall of the Humble Administrator's Garden

The skill of stacking rocks holds an important position in Chinese landscape architecture. At the end of the Ming dynasty, Ji Cheng wrote *Art of Garden Building*, elaborating on the details of creating gardens, followed by Wen Zhenheng and Li Yu. Modern rockery garden styles are the "South Han" and the "North Zhang", and the latter also originated Jiangnan area. The South Han style was founded by Han Hengsheng, known as "Shanshi Han" or "Rockery Han" in English. Han Buben, his son, once repaired gardens such as the Garden for Lingering and the Yuyuan

Picture 3.39 Lion Grove Garden

Garden. The three sons of Han Buben all inherited their ancestors' trade, especially Han Liangshun, the second son. He worked with his father and brother to renovate many classical gardens and rockeries. He also built the Wang Villa and the Liu Villa at the West Lake, as well as the Ming Hall (the Astor Court), a re-creation of a Ming dynasty-style, Chinese-garden courtyard, at the Metropolitan Museum of Art in New York. In 1979, Han Liangshun was transferred to Beijing to engage in the creation of rockery, and his book *Rock Stacking Techniques of Rockery Han Style* is considered a high-level academic monograph in his field. His three children, Jianzhong, Jianwei, and Xueping, also inherit the family's art and make many innovations in scenic courtyard scenery and greening. They once cooperated with Ieoh Ming Pei to build the indoor garden of the Bank of China Tower.

3.2.7 Bridges

Bridges have become an important category of architecture due to their unique composition and functionality. Traditional bridge construction techniques in China vividly reflect the development of the nation's technology and art. They can be categorized by material into wooden bridges, stone bridges, and iron bridges, or by type of structure into beam-column bridges, arch bridges, and suspension bridges.

3.2.7.1 Wooden Arch Bridges

Wood has strong tensile and flexural capabilities, but weak compressive capabilities, which is why it is mostly used as a structural material for beam-column systems. With its elegant appearance and ingenious structure, the wooden arch bridge has become one of China's most favored wooden bridges (as shown in Picture 3.40). The basic assembly unit of this kind of bridge is six rods, four in the longitudinal direction and two in the transverse direction, and the plane is in the shape of a "*Jing*" ("井" shaped). Due to the friction generated by pressure, the more the components are pressed, the tighter they become. The bridge doesn't need to be riveted. The rods only need to be inserted and pressed in the same specific way. The upper longitudinal beam is pressed onto the cross beam, and the cross beam is pressed onto the opposite longitudinal beam; this way the two longitudinal beams clamp the cross beam and friction prevents the cross beam from sliding. The whole structure is arch-shaped, which is compressed along the arch's central line, preventing bending force from being generated. As far as the other rods are concerned, they simply function as additional support beams. The components of the bridge all fit as one, requiring no special-shaped components. Felled trees can be made into the required components with minimal processing. It is even convenient to load and unload. If disassembled, it can be done without damaging the components and everything can be reused.

An image of a wooden arch bridge was first seen in the painting *Along the River During the Qingming Festival*. For a long time, people thought that this type of bridge and its construction method had fallen into oblivion. In fact, in October 1980, the

Picture 3.40 Xiaqiao Bridge in Sixi, Taishun county, Zhejiang province

conference on compiling the *History of Ancient Chinese Bridges* held in Hangzhou released that dozens of wooden arch bridges shaped like the Chinese character "八" had been found in the archaeological survey in Taishun county, Zhejiang province, which attracted great attention from the participating experts. Through field investigation, it is confirmed that these wooden arch bridges belong to the same bridge type as the long-lost Rainbow Bridge, which is a major discovery with world influence in the study of Chinese bridge history. Because the wooden arch bridges in Taishun are built with a corridor, they are also called timber-arched corridor bridge. This kind of bridge is common in Yunnan, Gansu, Hubei, Sichuan, Hunan, Fujian, and Zhejiang provinces, especially in southern Zhejiang and northern Fujian, which have more than 100 of these bridges. The local people also call them "*cuo* bridges".

"*Cuo*"(厝) refers to the corridor house above the bridge. The construction of the wooden arch bridge makes it have an upward rebounding force. The weight of the corridor house offsets it, which ensures the stability of the bridge. In Taishun, Dong Zhiji, currently in his eighties, is the most famous inheritor of the techniques to construct timber-arched corridor bridges. As the "*shengmo*", that is, the master craftsman of bridge building, his name is engraved on the Tongle Bridge: "*Shengmo, Dong Zhiji*" (as shown in Picture 3.41). There is also a bridge-building family, the Xu family, in Shouning county, Fujian province, which is close to Taishun, whose descendants can be traced back to the Qing dynasty's Jiaqing period two centuries ago. In the sixth year of the Jiaqing period (1801), Xu Zhaoyu, a bridge builder in Xiaodong village, presided over the construction of the Xiaodong Shangqiao Bridge. These skills were passed down to his great-grandson Xu Zechang, making him the fifth generation of bridge builders. Xu Zechang, who had outstanding skills, built many corridor bridges in Fujian and Zhejiang provinces. Zheng Huifu from Dongshanlou village learnt most from him and went on to build 11 corridor bridges. Zheng Duojin, his eldest son, inherits his father's business and independently built Yangxitou Corridor Bridge in 1967. In 2006, he presided over the relocation project of the Zhangkeng Bridge (as shown in Picture 3.42), which was originally built in 1828.

Like other construction projects, bridge construction also follows a series of folk customs, which have been passed down by bridge craftsmen from generation to generation and are still used today. When directors and facilitators prepare to build a corridor bridge, they first select a bridge maker. Then they also hire a geomancer to determine the direction of the bridge's abutment and choose an auspicious day for erecting the main beam. The beam at the top of the roof in the middle of the bridge is commonly known as the "*xiliang* beam" (happy beam), which is made of a Chinese fir from the same area and with lush foliage. An auspicious day will be chosen for logging. Incense, candles, tea, and wine are also prepared to worship the mountain, and only people who lead a charmed life with three generations living together and both parents alive may conduct the logging. The *xiliang* beam should be covered with red cloth and carried to the bridge site while firecrackers are set off (as shown in Picture 3.43). Construction starts mainly during the dry season. Offerings to worship the river, like incense, candles, tea, wine, fruit, pastry, vegetable dishes, and three sacrifices (pigs, sheep, and cattle) must be prepared as well. Taoist priests will then invite the gods, read scriptures, and send off the gods after the offerings. The

Picture 3.41 Dong Zhiji, the *shengmo*, quoted from *Taishun: An Old Man and His Bridges* by Sun Xiaoning

Picture 3.42 Zheng Duojin and the relocated Zhangkeng Bridge

ceremony of offering sacrifice to the beam is presided over by the master carpenter in charge of the marking line. When the corridor bridge is completed, a banquet shall be prepared to thank the bridge workers. In addition to the corridor house, many shrines are built on the bridge for the villagers to offer sacrifices to the gods and help them pray for good weather and a happy family (as shown in pictures 3.44 and 3.45). The 37 *Bridge Contracts* for building corridor bridges found in Dongpan village, Zhouning county, Fujian province in 2002 are also quite valuable. The earliest one was signed in 1800.

Picture 3.43 *Xiliang* beam

Picture 3.44 Interior
structure of a corridor bridge

Picture 3.45 The Puxiu Bridge in Dong township, Tongdao county, Hunan province was built during the Qing dynasty's Qianlong period. The names of thousands of donors are carved on both sides of the corridor

The ridge of the Houkeng Bridge has the words "Reconstructed in the 11th year of the Guangxu period" and "master carpenter Zhang Maochun, Zhang Maoxiu, deputy carpenter Zhang Yunliu, Zhang Maozhao", and the newly discovered bridge contract shows approximately the same features of the one found in Fujian in 2002.

3.2.7.2 Stone Bridges

Due to its excellent compressive properties, stone is widely used in traditional bridges. Smaller stones are combined to form an arch, making full use of the properties of stone to achieve a huge span. Therefore, although stone buildings are rare in ancient China, the arching techniques they used were excellent. The Anji Bridge, also called Zhaozhou Bridge, in Zhaoxian county, Hebei province and the Lugou Bridge, also called the Marco Polo Bridge, in Beijing are famous stone bridges in China.

The Zhaozhou Bridge has the largest span and is the earliest single-hole open-spandrel stone arch bridge ever built in China (as shown in Picture 3.46). There are two small holes on the shoulders at both ends of the bridge. This is a first in the history of bridge building. It came 1,200 years earlier than the construction of the same type of bridge in Europe. This method of making the two ends into hollow arches does not only reduce the weight of the bridge but also reduces the impact of flooding.

Picture 3.46 Zhaozhou Bridge

The Lugou Bridge is a combined arch stone bridge with a total of 10 piers and 11 segmental arches with corresponding holes. The key parts are all connected by silver-ingot-shaped iron tenons. The most distinctive feature is the construction method of the bridge's piers. The piers are in the shape of a boat, and the water-facing surfaces are built as water-dividing points. Its shape is like a pointed bow, which can withstand and divert the impact of running water.

3.2.7.3 Suspension Bridges

Most ancient Chinese suspension bridges have decks resting on the suspension cables. This is different from modern suspension bridges, but the distribution of force is the same. The Luding Bridge on the Dadu River in Luding, Sichuan province (as shown in Picture 3.47) has a span of 100 m and a width of 2.8 m. The whole bridge uses 13 iron cables, of which 9 are used as load-bearing bottom cables, the overlying planks are used as bridge decks, and 4 are used as handrails. Each cable is interlocked by 862–997 iron rings, weighing over 21 tons.

Picture 3.47 Luding Bridge

3.3 The Craft of Chinese Construction

3.3.1 Timber Frame Architecture and Its Construction

It is not the walls that make up the traditional timber frame of Chinese architecture, but the system is composed of column grids and beams. Each component is connected by mortises and tenons, which makes it extra sturdy. Depending on the climate and living habits in different regions, timber will be treated and used in different ways, thus resulting in different frame systems, such as (1) a column-and-tie system, (2) a post-and-lintel system, and (3) log cabins.

To use the column-and-tie system, you need to set up rows of columns according to the number of purlins along the depth of the house, with one purlin on each column and rafters on the purlins. The roof load is directly transmitted from the purlins to the columns. It is characterized in that the distance between the columns is small and that each row of columns transversely penetrates through the column body to form a frame. Every two frames are connected by an arched square column and towing wood to form the truss (as shown in Picture 3.48). Arched square columns are used between thechapters of eave columns, which are shaped like the architrave of post-and-lintel frame, and the towing wood is used between inner columns. Arched square columns and towing wood often serve as the joists of the attic. This kind of structural system requires less construction materials. It is first assembled on the ground to form a complete roof truss and then erected, which has the advantage of saving both labor and materials, making it both convenient and economical. At the same time, the

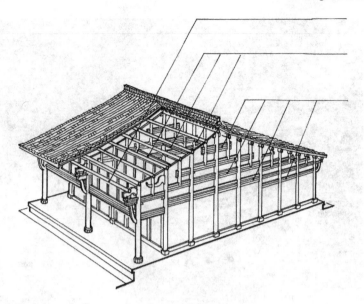

Picture 3.48 Schematic diagram of a column-and-tie system

densely arranged columns also facilitate the installation of siding and wall masonry. The techniques for this kind of system were already quite mature in the Han dynasty and are still widely used in China's southern provinces.

The Post-and-lintel system is the most commonly used timber frame in ancient Chinese architecture. A characteristic of this kind of system is that on the top of the column or the horizontal paving layer of the column grid, a number of stacked beams are laid down perpendicularly. The beams are shortened layer by layer, with short columns or wooden blocks between each layer, and small columns or triangles erected in the middle of the top girder, forming a triangular roof truss. Between adjacent roof trusses, purlins are placed on both ends of each layer of beams and small columns in the middle of the uppermost beam. To form the spatial framework of a double-slope roof house, rafters are placed between the purlins. This way, the weight of the roof is transmitted to the foundation through the rafters, purlins, girders, and columns (as shown in Picture 3.49). This system already existed in the Spring and Autumn period, but really matured in the Tang dynasty.

The above-mentioned interconnecting grids of columns and beams are not used in the log cabin system. This frame is made of round logs or rectangular and hexagonal timbers stacked upwards in parallel, and at the nook, the ends of the timbers intersect to form four walls shaped like the wooden fences on ancient wells. Then, short columns are erected on the left- and right-side walls to underpin the ridge purlins and form a house. Log cabins need a large amount of wood and are limited in scale, as well as door and window openings, so it is not as popular as the post-and-lintel system and column-and-tie system.

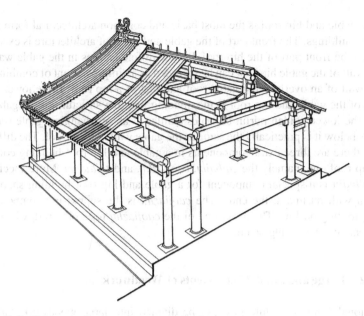

Picture 3.49 Schematic diagram of a post-and-lintel system

3.3.1.1 Architecture with Timber Frames

Traditional Chinese architecture made of timber frames mainly includes the flush gable roof, overhanging gable roof, and gable and hip roof.

The flush gable roof is the most used timber frame in ancient buildings, especially small ones. Their size and style depend on the number of purlins, often ranging between five and nine purlins, but most garden buildings, for example, have less than seven. The five purlin buildings are the simplest, and the seven purlin ones are the most luxurious.

The overhanging gable roof can be divided into two types depending on the structure's shape and size: (1) a big overhanging gable roof and (2) a round ridge overhanging gable roof. There is a main ridge at the intersection of the roof in front of and at the back of the big overhanging gable, which divides the roof into two slopes. The most common big overhanging gable roofs are the five- and seven-purlin overhanging gable roof, and the five-purlin middle-column and the seven-purlin middle-column overhanging gable roof. The latter two are mostly used for annex rooms by a main entrance. Double purlins are placed on ridge of the round ridge overhanging gable roof'. Such a roof does not have a main ridge. The front and back sloping roofs form a long-raised ridge. The common ones are the four-, six-, and eight-purlin round ridge overhanging gable roof, and there is another type which combines a big overhanging gable roof and a round ridge overhanging gable roof. They hook and overlap each other and are referred to as "one hall one round-ridge roof". This is often used in floral-pendant gates.

The gable and hip roof is the most basic and common architectural form seen in ancient buildings. The front part of the gable and hip roof architecture is exactly the same as the front part of the hip roof, but the difference lies in the gable wall. The gable wall of the gable hip roof can be regarded as an improvement of combining the gable wall of an overhanging gable roof and that of a hip roof. If the lower golden purlin of the gable and hip roof is the dividing line, the structure of the gable wall above the lower golden purlin becomes similar to an overhanging gable roof, and what is below it is practically the same as the gable wall of a hip roof. The difference is that there are three extra components in a gable and hip roof building compared to a hip roof one, namely the *caibujin*, *caojiazhu*, and *tajiaomu*. More specifically, the *caibujin* is a special component for a gable and hip roof building shaped like a beam with purlins at the ends. The *caojiazhu* is the supporting component for the protruding purlins. The lower end of the *caojiazhu* rests on a rod, which is the *tajiaomu*, and a crossing beam.

3.3.1.2 Large and Small Components of Woodwork

Traditional Chinese architecture can be divided into large woodwork and small woodwork. Large woodwork refers to the production of the load-bearing parts of a building, such as columns, beams, purlins, rafters, tenon-and-mortise structures, and bucket arches, while small woodwork refers to the production and installation of non-load-bearing parts.

As Chinese ancient architecture used timber frame and systems for the backbone and support of a house, large woodwork is used for the overall design of a house. The large woodwork of traditional buildings includes columns, beams, square columns, purlins, *chashou* (inverted V-shaped braces), short columns, camel's humps (the component to hold beams together with bucket arches), and *timu* (the component fixed on the arch to strengthen the connection of different parts). Chinese frame structures eliminate the load-bearing requirements of walls, saving the trouble of calculating how and where to arrange the position of the walls, doors, and windows, giving more freedom in the design of the building's facade. The regulations on small woodwork are not too strict. Artisans can make them however they wish and on demand, which leaves room for them to be creative with it.

Those most categorized under small woodwork are doors, windows, top ceilings, flat ceilings, check-shaped ceilings, caisson ceilings, and balustrades.

3.3.1.3 Bucket Arches

Bucket arches played a very important role in the development of Chinese architecture. Its evolution goes hand in hand with the evolution of all traditional timber frames and can even be used to distinguish the different architectural time.

A bucket arch consists of a bearing bucket, an arch, and a cantilever (as shown in Picture 3.50). A bearing bucket is a squared piece of timber that directly undertakes

Picture 3.50 Bucket arches

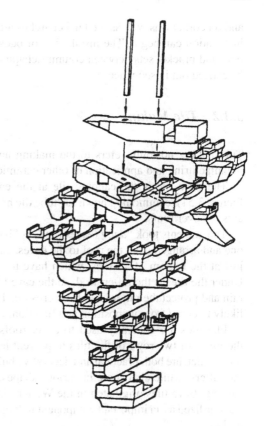

a transverse arch, a joist, or a beam. An arch refers to a piece of wood whose mouth extends out to bear the weight of the bearing bucket. A cantilever is a tilted, wooden structural component, and its end is pressed at an angle under the beam or purlin, which uses leverage to support the eaves.

To make a bucket arch, you first need to use the original blueprint and create a reproduction sample. The original blueprint is used to draw a 1:1 full-scale sample according to the specifications in the design. Once completed, you separate the sitting bearing bucket, tilting, cantilever, decoratively nosed timber, small tie beam and truss bowl, oval arm, long arm, regular arm, interactive bearing bucket and *sancaisheng* (a kind of bucket arch component used in the Qing dynasty placed at the end of a single arch to support the last layer of an arch or square-column component) to be used, one by one, as reproduction models on their own. These models serve as the basis to produce single-use line drawings, which are then used to outline the drawing directly on the material and finally made according to that outline.

In order to ensure the smooth assembly of the bucket arches, it is necessary to "run a test" before installation, a trial installation. During the actual installation, the assembled bucket arches are transported to the site and placed in the predetermined positions. Once the bracket sets designed for placement between and on the columns, and on corners, as well as the bucket arches are all manufactured and delivered, the installation can begin. The installation of bucket arches should be finished one by one, and bracket sets between columns,chapters, and bracket sets on corner should be carried out in sequence.

3.3.2 Tile Making

Tile making not only refers to the making and use of tiles but also includes the manufacturing and application of other ceramic building materials, such as bricks.

The eave tile is the capping tile at the eave end of the palace (as shown in Picture 3.51), commonly known as "tube tile head" or "tile head". Tiles have circular arcs to cover roofs.

The ancients took "*dang* (eaves tile)" as "base", because tiles are pressed one by one and arranged from the roof to the eaves, and the tubular tiles with tile heads are just at the bottom of all tiles, which have the function of blocking and shielding. Under the tiles is the rafter head, so the eave tiles can help resist the wind, sun and rain and protect the rafter head from erosion. The name of "*wa dang*" (eaves tile) is likely to come from its position and function.

Flat tiles are generally used to cover roofs, while tube tiles are used to cover the joints of two rows of flat tiles to prevent leaks. Eave tiles are the heads of tube tiles, which are both practical and decorative building components, since they protect the rafters, while also come in various shapes, combining the arts of painting and carving. Eave tiles originated in the Western Zhou dynasty but were perfected and standardized as an important component in large buildings in the Spring and Autumn period.

Picture 3.51 Eave tile from the Eastern Han dynasty

Tiles and eave tiles have been used for a long time. Tiles are used for roofs of palaces and residential buildings alike, but glazed tiles are used only in palaces, while regular people use black tiles. Eave tiles can be divided into two types: One is an eave tile with an upward arc, called *dishui*, which means dripping water, while the other is an eave tile with a downward arc, called *yantou*, which means cornice.

When categorized by material, eave tiles can be divided into three types: (1) grey pottery eave tiles, (2) glazed eave tiles, and (3) metal eave tiles. Among them, grey pottery eave tiles are the oldest and most common. Glazed eave tiles first appeared in the Tang dynasty, and were made by glazing and firing clay tiles. They were used in higher-grade buildings and came in various colors, such as black, green, blue, and yellow.

During the Song, Yuan, Ming and Qing dynasties, metal eave tiles were used in individual buildings to show their hierarchical status. There are three types of metal tiles: (1) cast iron, (2) brass, and (3) gilded.

There are three shapes of eave tiles: semi-circular large semi-circular, and circular. From a construction point of view, circular eave tile is more advanced than the semi-circular one. The former is twice as large as the latter and can completely cover the rafters. Although its production is laborious, it is better at protecting the building. From an aesthetic point of view, the surface area of the circular eave tile is twice that of the semi-circular one, which makes more elaborate and beautiful designs possible. Early round eave tiles were made of the coil method, but in the Western Han dynasty, it was replaced with the molding technique, which means that eave tiles and tube tiles were made separately and jointed together. This technique is used to later generations. Eave tiles are categorized into three types according to ornamentation: (1) those with patterns, (2) those with images, and (3) those with characters. Among them, eave tiles with characters prevailed in the Eastern and Western Han dynasties and declined after the Eastern Han dynasty.

3.3.2.1 Tile Making

Tiles are mainly made with clay, and processed with mud, shaped, dried, and finally roasted. The clay requires a pure yellow color, good viscosity, and contain as little sand as possible.

The process of making tiles in the Ming and Qing dynasties was similar to that of making bricks. To avoid water seepage, fine clay should be mixed with water, pressed, and then moistened. The next day, the clay is wedged well, and a cloth tube is put over it. Then, mud is put on the cloth tube with water; the tile roller is used to pat and press the mud onto the tile until it is smooth and level. Once done, the muddy tile tube is taken from the wedged tile field and the cloth tube removed to dry the base. Once dry, the base is cut into four pieces with a knife to form four flat tiles. The diameter of a tube tile is small, and a tenon is made at the upper end. When the tile base is slightly dry, it is cut in half with a knife, to get two symmetrical semi-circular tiles.

3.3.2.2 Brickwork

Brickwork is the trade of using brick materials to construct buildings. In Northern Song dynasty, Li Ji's *Construction Skills* described various specifications and usages of bricks, which were used to build bases, Sumeru seats, steps, walls, arches, waterways, hearths, wells, and to pave floors, pavements, and ramps.

Bricks were called *pi* (甓) and *ling* (瓴) in ancient China. Bricks and tiles have improved the quality of civil wooden structures, and the further application of bricks laid the foundation for a wider variety of brick structures.

Baking grey bricks from clay has a long history in China. Records about bricks have been found going back as early as the Zhou dynasty. Actual bricks were produced during the Warring States period and used to pave floors. After being fired in a kiln, water would be poured onto bricks and dried by heat, turning them grey. Later, in the Qin dynasty, bricks were used as bearing bricks. Hard brick walls were found in the pit on the east side of Mausoleum of the First Qin Emperor. Bricks were widely used in the Han dynasty, and many floored quadrel and hollow bricks in the Qin and Han dynasties were printed with patterns and used as decoration. However, although solid bricks were used for masonry, decorative floor bricks and large hollow bricks dominated during the Warring States period and the Western Han dynasty, and they gave way to narrow bricks in the Eastern Han dynasty.

As ancient architecture evolved, so did the application of bricks. Bricks became indispensable for the construction of fortresses. Construction of brick tombs and brick pagodas were also on the rise. Most of the pagodas in the Tang dynasty used bricks and stone. Masonry reached new heights in the Song dynasty, mainly in regard to the construction of pagodas and bridges. The firing techniques of bricks began to be systematically standardized during the Five Dynasties period, and the Song and Yuan dynasties. *Construction Skills* recorded a quite scientific summary of the brick-firing process and regulations on the size, raw materials, forming, drying, stacking in the kiln, and firing of bricks and tiles, as well as the specifications and construction of brick kilns.

The rulers of the Ming dynasty attached great importance to the construction of city walls and palaces. They set up factories in Linqing and Suzhou to fire bricks, including arch bricks for making arches, flat-boarded bricks for walls, roof boarding bricks for roofing, and axe-blade bricks for pavements. As brick walls gradually replaced earth walls, their widespread use created the conditions for the development of the flush gable roof.

The so-called metal bricks were produced in Suzhou (as shown in Picture 3.52). They were used for paving the floors of palaces in the Ming dynasty. They were available in sizes of 2.4, 2.2, 2, and 1.7 sq. *chi* (1 sq. *chi* is about 0.11 sq.m). Its texture was extremely dense and fine. To manufacture them, in addition to the various, more careful procedures, the clay had to air-dry from winter until summer, then the bricks had to be tightly sealed with oil paper and dried in the shade for another year, before being put into the kiln. After firing, it had to be inspected piece by piece, to make sure that the surface was smooth and flawless. When this metal brick is struck, it will sound like metal, hence the name.

3.3.3 Stone Masonry

The feet of the wooden columns used in the architecture of the Shang dynasty were supported by stone plinths to prevent moisture and corrosion. The use of bricks gave birth to arch and dome structures and pushed stone architecture to develop further as well. For example, single-layer and multi-layer arches construction techniques were used by stone arches. Stone architecture developed by leaps and bounds in the Eastern Han dynasty, with the appearance of buildings built entirely of stone, such as stone temples, stone Que-towers, beam-slab stone tombs, and arch-type stone tombs. These buildings were often carved with character stories and patterns. The quality and detail of these stone carvings steadily kept improving over time.

Grottoes, stone towers, and stone arch bridges, displaying a high level of architectural skill, were built on a large scale during the Wei, Jin, and Southern and Northern dynasties. Due to the wide spread of Buddhism, many temples were built on cliffs and mountainsides, creating religious architectural treasures such as the Yungang Grottoes in Shanxi province and the Longmen Grottoes in Luoyang, Henan province. During this period, the use of stone pillars, stone plinths, stone steps, and stone railings increased.

Stone architecture really matured in the Sui and Tang dynasties. The large stone pagoda, called the Four Gate Pagoda of Shentong Temple, that still stands in Liubu town, Licheng county, Shandong province, is a tower-like stone structure built in the seventh year of the Daye period (611) in the Sui dynasty, measuring about 15 m in height and built on a square plane. Every side of the walls, which is made of large bluestones, is about 0.7 m thick and contains a small arch in the middle. There is a square core column in the center of the pagoda. The spire roof is made of stacked stone slabs. The eaves are stacked with five layers. Except for the spire, the rest of the building is not decorated with carvings, making it simple and serene.

After the Tang dynasty, the number of stone pagodas gradually increased, and their styles continued to change. During this period, the combination of stone carvings and architecture attained a high level of proficiency. The exquisite art of stone carvings was displaced in the construction of grottoes. Stone steps were decorated with carvings or colored paintings, and stone railings were also used on the bases of important buildings.

The Song dynasty was a period of great transformation for traditional architecture, and masonry pagodas and tombs in the form of wood-like buildings became popular. With the promulgation of *Construction Skills*, specialized stone masonry emerged and became more standardized. The order in which the construction of stone buildings was done, was (1) stripping, in which the raised part of the stone surface is chiseled away with a small chisel, (2) rough beating, where the surface of the stone is roughly beaten with a small chisel to make the depth of the chisel marks even, (3) fine percolation, where the stone surface densely covered with chisel marks is finely processed so that the surface's unevenness gradually becomes shallow, (4) rugged edges, where the edges and corners of the components are cut with a narrow chisel to make the four sides straight, (5) ballasting, where the components to be carved are sculpted according to their different requirements and then cut or hacked one to three times with a knife and axe to make the stone surface more level, and (6) grinding, which is the final step in which the surface is smoothened with sand and gravel, and sometimes a bit of water, which is only necessary when the surface of the carving has little undulation. At this time, different pagoda structures were built, such as double-sleeve structures, single sleeve with a central column structures, thin-walled whole-body structures and flower pagodas. The multi-story and multi-sleeve stone pagoda is represented by the twin stone pagodas of Kaiyuan Temple in Quanzhou, Fujian province (as shown in Picture 3.53). The existing twin pagodas were built in the Southern Song dynasty's Jiaxi period (1237–1240) on an octagonal plane and have five floors and a height over 40 m. The outer ring is made of large stones, and the middle consists of a core pillar made of stones. There is a door on each side of the outer walls, and the four doors are staggered to ensure the integrity of the pagoda. Its core pillar and its outer cylinder walls are tied together with strips of stone, layer by layer, to ensure the structural stability of the pagoda.

In the Song dynasty, the variety of stone bridges increased, such as stone beam bridges and stone arch bridges. The Luoyang Bridge in Quanzhou, also known as the Wan'an Bridge (as shown in Picture 3.54), was built in the fifth year of the Northern Song dynasty's Huangyou period (1053). The bridge is 1,200 m long, 5 m wide, and has 46 piers. It is one of the famous ancient Chinese stone beam bridges.

During the Yuan, Ming, and Qing dynasties, besides being used for bases, railings and flooring, stone was also widely used in the construction of pagodas, bridges, memorial archways, and underground tombs. The tombs of the Ming and Qing dynasties are all made of high-grade stone, and several rooms are connected with each other to form magnificent underground palaces. In front of these mausoleums, there is usually a general sacred way flanked by stone statues, dotted with stone pavilions housing tablets, big red gates, and memorial archways, creating a dignified atmosphere. Stone masonry reached its peak in the Qing dynasty. *Examples of Engineering Practices*, published during the Yongzheng period (1722–1735), summed

Picture 3.53 One of the twin stone pagodas of Kaiyuan Temple in Quanzhou

Picture 3.54 Wan'an Bridge

up a set of completed technical specifications for stone masonry and construction procedures that included cutting waste, roughing, chiseling with an axe, flattening, picking and chiseling, laying joints, grouting and numbering swinging rollers. Just like the tile, wood and oil industries of that time, stone masonry became a highly technical industry.

3.3.4 Earthen Architecture

The earliest materials used in ancient architecture were earth and wood, which laid the structural foundation of civil construction. The main methods of building city walls, towers, mausoleums, and walls were to make adobe, build walls with them, as well as use the local soil. Therefore, in ancient China, large-scale construction activities were referred to as activities in which "a great bustle of masons and carpenters".

Rammed earth technology can be traced back to the Neolithic Age. The earth in China's Central Plains is dry and hard, and the particles are fine, which can adhere together with a little bit of ramming. Later, glutinous rice juice was poured into layers as an adhesive for ramming the earth. Once dried, it resulted in a product as hard as stone. The Erlitou site in Yanshi, Henan province has the earliest known palace foundation in China, which was built of rammed earth. The mastery of rammed earth technology promotes the emergence of "high terraces and beautiful palaces". In order to build these ancient, raised buildings, a platform first needed to be built on the ground and then a building built on that platform. Another way was to use the naturally occurring highland or earthen platforms by landscaping it to make it suitable and then put a building like a house or palace on it. During the Spring and Autumn period, various vassal states competed in building raised buildings, such as the Zhanghua Palace by King Ling of Chu, and Huangjin Terrace by King Zhao of Yan. The Chinese idiom "the lack of one basketful of earth spoils the entire effort of building something measuring nine *ren* (one *ren* equals about 158 cm)" shows that the use of earth to construct raised buildings was quite common in ancient times. Since wooden architecture was not very tall, it was necessary to build a house layer by layer with a trapezoidal rammed earth platform as the base in order to obtain a larger building volume.

The invention of board building, which is used to construct rammed earth walls, laid the foundation for palace buildings to be turned into tall buildings and played an important role in establishing the earth-timber hybrid structure system of Chinese ancient buildings. Rammed earth walls were used in the Neolithic buildings found in Dadiwan, Gansu province. Mud walls with a timber frame were used as load-bearing walls in the Shang dynasty and adobe-built load-bearing gables were found in late-Shang site. In the Western Zhou dynasty, rammed earth walls only served as enclosures, but even the timber frame houses that used a purlin frame as the main beam frame, still built its base with grass mud adobe and used a rammed earth base and rammed earth walls for a long time.

China's traditional city walls originated from the Yangshao Culture. Due to the demands and pressure of the clan wars, ramming earth technology developed rapidly. Layered ramming of loess greatly improved the compactness and strength of the earth layers. The rammed living surface reduced humidity and did not easily show signs of wear, while still remaining firm and stable after being trimmed into steep walls. This is why the rammed earth technology was first applied in defense engineering. With the improvement of the range of the bow and arrow, the defensive function of a moat became moot, thus inciting the construction of city walls. In the early stages, city walls were rammed and cut simultaneously. The earth ridges were rammed first and then the outer slope was trimmed to make it stand straight as walls on that side. When a city was built, a moat was formed parallel to the city wall, thus the earth was simultaneously taken and used in the same area.

These rammed earth walls were modeled with wooden boards, filled with clay or limestone, and hammered and compacted layer by layer. During the Spring and Autumn period and the Warring States period, due to the high frequency of wars, rammed earth was widely used to build cities. One of the outstanding contributions was the Qin Great Wall. According to the existing Qin Great Wall site in Lintao (as shown in Picture 3.55), the lowest layer consists of raw earth with a layer of loess laminated firmly on it. A rammed earth city wall is built on top of the loess with yellow clay and gravel. Even now, rammed earth walls built after the Qin dynasty can still be seen along the Great Wall, some of which are rammed with clay and sand, pressed with red willow or reed branches (as shown in Picture 3.56), while others are rammed with earth, sand, lime, and gravel. Generally, the height of rammed earth walls is twice the bottom thickness, and the width at the top is 1/4 to 1/5 of its height. This kind of wall can be made with local materials and is easy to construct. Before the Sui dynasty, most of the Great Wall was built with rammed earth.

3.3.5 Decoration on Buildings

Decoration plays a very important role in Chinese traditional architecture. Buildings cannot only separate spaces, bring in light and ventilation, protect against moisture and the elements, but they can also be aesthetically pleasing. The combination of decorating with calligraphic work, painting, embroidery, and inlaying results in a gorgeous artistic style. The *queti* (sparrow brace), lintels, hanging fascia and wooden railings under the eaves-purlins form a sharp contrast with reality, setting off the lines, the rigidity, with softness against the bases, roofs and walls, giving it all an overall sense of rhythm and harmony. The partition doors, wooden partitions, antique shelves, and bookcases used to divide the interior space are not only cleverly laid out but also finely carved, which makes the interior feel both separated and connected. Antique bed cases, tables, chairs, makeup tables, cabinets and boxes can range from rough to exquisite and small, and the sheer variety of Chinese furnishings, exquisite skills and rich techniques are rarely seen in the rest of the world.

Picture 3.55 Qin Great Wall site

Picture 3.56 Han Great Wall in Dunhuang

The modeling beauty of Chinese traditional architecture is, to a great extent, expressed as structural beauty. From a single house to a group of buildings, decoration plays an extremely important role in shaping the artistic image of architecture. It not only increases the artistic charm of architecture but also makes the architectural art have the expressive force of an ideal. Components such as columns, beams, joists, purlins, and rafters have often been given an artistic twist during manufacturing. For example, columns that are made into spindle columns with slightly smaller upper and lower ends, with the centers of the beams slightly arched upwards, short columns on the beams are made into chapters, the bolsters on both sides of the short columns become curved joists, the skids between the upper and lower beams are made into various hump shapes, the raking shore supporting the protruding eaves from underneath is mostly turned into animal-shaped and geometric arch supports, and the coupling beam square-columns are made into shapes such as chrysanthemum heads, grasshopper heads and hemp leaf heads. The creation and addition of these elements is done on the premise of not damaging or endangering the column's structural function, which makes it feel natural and appropriate.

The ancient craftsmen used the characteristics of wooden frames to create different roofs, such as hip roofs, gable and hip roofs, overhanging gable roofs, flush gable roofs, single eaves, and double eaves, and decorated the roofs with *chiwen*, treasure tops and animals, to make them more beautiful. Lattice patterns and apron board decoration patterns carved on doors, windows and partitions enhanced the artistic effect of the facade of a house. Even things like door nails, door knockers and corner leaves became important decorative parts. Bases and steps serve as the base of a house and the steps to enter a house, which become especially dignified and majestic with decorations and carvings on the railings. In addition, small woodwork also incorporates decoration, namely on the exterior and interior of the eaves (as shown in Picture 3.57). The former requires sturdiness and durability, while the latter requires finer materials and workmanship.

3.3.5.1　Caisson Ceilings

Caisson ceilings serve as a decoration of the indoor ceiling of a traditional building. The caisson ceiling (as shown in Picture 3.58) is decorated with patterns, carvings, and colored paintings, hence the name. It symbolizes nobility and rank and is generally only used in official buildings. It comes in a variety of shapes, such as a square, a rectangle, an octagon, and a circle. The use of caisson ceilings began in the Han dynasty; however, they are also frequently seen in grottoes dating back to Jin, Southern and Northern dynasties. Later generations of ceilings continued to use this style, but the level of painting and carving improved. *Construction Skills* contains a special discussion and stipulation on the method and scale of caisson ceilings as well.

Picture 3.57 Pillar and
square-column decoration on
the Daqing Gate

Picture 3.58 Caisson
ceiling

3.3.5.2 Decoration Techniques

There are three main types of traditional architectural decoration techniques: Color painting, oil decoration, and carving.

The various paintings, patterns, lines, and colors used to decorate buildings are collectively referred to as color paintings (as shown in Picture 3.59). Lin Huiyin, a contemporary architect, pointed out that the use of decorative oil paintings in buildings was "in the beginning for practical purposes to meet the practical needs of anti-corrosion and anti-moth on wooden structures, mineral raw materials such as cinnabar or vermilion (red), and black lacquer tung oil were commonly used to decorate the wooden structure; later, it was gradually unified with the requirements of fine arts, becoming more complex and rich, and it became a unique method of Chinese architectural art".

Color paintings were mainly made of mineral pigments, supplemented with plant pigments, which were mixed with glue and powder. Mineral pigments have a strong color payoff, do not fade for a long time, and some are even toxic, which plays a role in warding off insects.

Picture 3.59 Color painting of the Chongzheng Hall

Although the patterns, colors, and practices of color paintings differed between past dynasties, some techniques overlapped and evolved throughout, such as the use of the overlapping halo, secondary color, embossing, and gold foil.

There are three working procedures when creating oil decorations on traditional wooden buildings: Cleaning the bottom layer, applying plaster, and painting the decoration. The surface of the bottom layer is first cleaned to prepare for the next step, helping the plaster closely adhering to the component. There are different procedures in this step, namely cutting, scratching, washing, burning, shoveling, tearing, picking, carving, scraping, grinding, and coating. The plaster is necessary to provide a smooth surface for the oil decoration. It is wrapped on the surface of the component mixed with oil putty and hemp, going through the procedures of applying a roughing-cast, mopping-up oil putty, applying hemp, pressing the hemp and oil putty together, medium thick lime, fine and thin lime, grinding and drilling.

The last step is to apply oil, which is made of gloss oil mixed with the colors such as cinnabar, vermilion, and *guanghong* (rust red), on the plaster with hemp thread ends. When the oil is dry, it becomes bright and full, durable, and its colors are preserved.

Some of the procedures required include starching mortar, using fine putty and applying the first, second, third, and final layer of oil.

The carvings in the decoration include wood carving, stone carving, and brick carving, which have been introduced in detail in the "Sculpture" chapter of this book.

* * * * *

Traditional Chinese construction ideas, techniques, and culture, brimming with distinctive features and meaning, are an important part of Chinese civilization and have had a far-reaching influence on the entire East Asian region. Liang Sicheng and Liu Dunzhen are the pioneers and founders of modern Chinese architectural history research, followed by famous scholars such as Chen Congzhou, Fu Jianian, and Luo Zhewen. Thanks to their unremitting efforts, the value and connotation behind many ancient buildings have been excavated and interpreted, thus properly evaluating and promoting the history of Chinese architecture in international academic circles. So much so, that it has become a compulsory subject for practitioners of the history of astronomy and physics. Under today's new historical conditions, finding ways to inherit and carry forward the essence of traditional architecture and combining it with modernization is still a major issue with practical significance.

Chapter 4
Spinning, Dyeing, and Embroidering

Wang Lianhai and Zhao Hansheng

Clothing, food, shelter, and transportation are the four basic needs of human life. Clothing, which can cover our body and protect us from the cold, is a symbol of civilization because it is one of the main differences between human beings and animals. Therefore, it is clear that spinning and weaving are of great importance to human life.

Agriculture was the foundation of ancient China. The agriculture and textile industry were always the two pillars of its national economy, and the most important social division of labor was that men were engaged in farming while women were weaving.

As early as the Warring States period, female workers were listed as one of the "six posts" of the country. According to *Treaties of Foods and Commodities* in *History of Han*, "Women in the same neighborhood spun and wove together from dusk until midnight and worked for 45 portions of work in one month including 30 whole days and 30 half nights". It's also recorded in *Peacock Flies Southeast*, a Yuefu poem in the Han dynasty, "She started weaving at the dawn of day and worked at the loom until the midnight hour". Yin Huiyi (1691–1748) writes in *Proposal on Farming and Sericulture* to the Emperor, "Girls (in Songjiang city, nowadays Shanghai) can spin wool at seven or eight, and weave cloth at twelve or thirteen". During the long history of China, women have always been the main force in the textile industry. Since childhood, they were asked to spin, weave, and do some needlework. It was the traditional way of life for the vast number of rural women in China to be engaged in various labors during the day and to weave in the light of oil lamps until late at night. The practice still exists among women in remote areas in China even today.

W. Lianhai
Academy of Arts and Design, Tsinghua University, Beijing, China
e-mail: wanglianhai327@163.com

Z. Hansheng (✉)
The Institute for the History of Natural Sciences, Chinese Academy of Sciences, Beijing, China
e-mail: hshzhao@ihns.ac.cn

© Elephant Press Co., Ltd 2022
H. Jueming et al. (eds.), *Chinese Handicrafts*,
https://doi.org/10.1007/978-981-19-5379-8_4

As the mother country of sericulture and silk weaving as well as a big country of cotton production and consumption, China's textile technology, which is characterized by its exquisite craftsmanship and great varieties, occupies a prominent position in global textile history. Spanning over 3,000 years, from the Xia and Shang dynasties to the Ming and Qing dynasties, the textile industry continued to develop, making great contributions to China's economy and cultural prosperity and affecting all aspects of people's social life. This chapter will mainly talk about the history of the Chinese textile industry, spinning and weaving of silk, cotton, hemp and wool, printing and dyeing skills, fabric varieties, and embroidery.

4.1 History of Weaving, Dyeing, and Embroidery

4.1.1 Pre-Qin Period

Textile originates from weaving. The ancients wove coarse fabrics from wild hemp, kudzu, or animal hair inspired by reed weaving and bamboo weaving. It was described in *Fan Lun Xun* in *Huang Nan Zi* as "weaving with hands and fingers", which is the beginning of textile.

Archaeological excavations show that spindles were used to twist yarns and weave hemp cloth during the Yangshao culture 7,000 years ago. Some parts used in the backstrap loom such as the beating-up bone knife, bone shuttle, dividing rod, and warp beam (as shown in Picture 4.1) were unearthed at the Hemudu site (6,000–7,000 years ago) in Yuyao county, Zhejiang province.

Sericulture and silk weaving technology were invented by the Chinese very early. The silk fragments unearthed at the Qingtai village site in Xingyang, Henan province were made about 5,500 years ago. Red-dyed silk pieces and ribbon fragments were unearthed from the Qianshanyang site in Wuxing district, Huzhou city, Zhejiang province, which were identified as made of cultivated silk. It shows that the ancestors

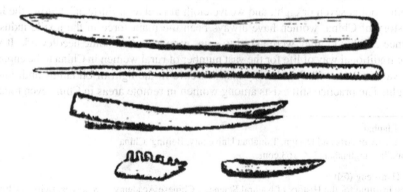

Picture 4.1 Parts of the backstrap loom, unearthed at the Hemudu site

in the Yellow River Basin and the Yangtze River Basin already mastered the skills of artificial silkworm rearing, silk reeling, and silk weaving in the middle of 4,000 B.C. at the latest. Those skills have a far-reaching influence on Chinese civilization.

Ancient legend has it that Lei Zu, the concubine of the Yellow Emperor and the daughter of the Xiling clan, was the originator of sericulture and silk weaving, and was honored as "the first silk farmer" or "the silkworm goddess". In ancient time, the royal families performed the "sericulture ceremony" every year (as shown in Picture 4.2), and the common people also built temples to worship and offer sacrifices to the silkworm goddess regularly (as shown in Picture 4.3).

With the Xia dynasty turning a new page in the Chinese civilization, textile technology made greater progress. A place named "Bo" (薄), which was rich in *zuanzu*, the ancient name of colored ribbon, was a city famous for dyeing and weaving. The Chinese character "薄" is interchangeable with "亳" (both have the same pronunciation), so Bo city might be "Xibo" (西亳) in history books, which is located in the west of Yanshi city, Henan province today. The "*can bo*" is a bamboo utensil with a flat bottom used to rear silkworms; therefore, it can be concluded that the name of Bo city might be related to sericulture, and it might be the earliest textile production center in China.

There are many oracle bone scripts in the Shang dynasty like "silkworm" (蚕), "mulberry" (桑), "silk" (丝), and "silks" (帛). The Chinese character "*sang*" (桑), meaning mulberry, is also used as place names, such as "Zaisang" and "Tiansang". Oracle bone inscriptions have records of offering sacrifices of three cattle to the silkworm goddess. Some surviving or unearthed bronze wares of the Shang and Zhou dynasties, such as the dagger-axe of Shang in the Palace Museum in Beijing (as shown in Picture 4.4), several ritual vessels unearthed from the Fuhao tomb in Yin Ruins, as well as the *fanjin*, fan bronze ban in Metropolitan Museum of Art in New York, have traces of having been wrapped in silk cloth. It was identified and confirmed to be damask jacquard cloth with mountain patterns, cloud, and thunder patterns, which shows that this kind of jacquard technology was quite common in the Shang and Zhou dynasties. It is recorded in *Genealogical Annals of the Emperors and Kings* that there were "more than 300 women dressed in silk" in the imperial palace of Shang, which is just like what Mo Zi (480–390 B.C.) describes, "The imperial halls were splendid, filled with gold and jade decorations; women were especially extravagant; the situation couldn't be forbidden and the country gradually exhausted. Therefore, the soldiers died, and the country was defeated. Wasn't it because of the of the wasteful use of splendid silk clothing?".

The successive dynasty of Shang was the Western Zhou, with its capital in *Bin* (nowadays Xunyi) in Shaanxi province. It attached importance to farming and sericulture. The ladies held sacrificial ceremonies during the silkworm season, which was the first of its kind followed by later generations. "*Lessons of the Capital*" and "*Lessons of Chen State*" in the *Book of Songs* contain such lines as "there's hemp around the hill" and "endlessly spread the kudzu and vine plants along the banks of the river". Hemp and kudzu are the most widely used raw materials for weaving, and this is why the common people were called "cotton garment" (*bu yi*) in ancient times. Wool was used to weave coarse clothes for commoners, which is described in

Picture 4.2 Painting *Empress Offering Sacrifices to the Sericultural Ancestor* in the Qing dynasty

the *Book of Songs* as "How can I celebrate the Spring Festival without fine clothes or coarse clothes". Refined wool fabric was called "*cui*", which was used to make clothes for nobles.

To strengthen the management of the government-run textile industry, the government of the Western Zhou dynasty set up some professional posts such as "*dian fu*

Picture 4.3 Folk activities of offering sacrifices to silkworm goddess

Picture 4.4 The dagger-axe of Shang in the Palace Museum in Beijing

gong" (officials in charge of women's work), "*dian si*" (officials in charge of silk quality), "*dian (ma)*" (officials in charge of hemp quality), "*ran ren*" (officials in charge of dyeing), and "*zhang ge*" (officials in charge of levying kudzu and ramie). *Qi* (damask) and *jin* (brocade) are two important varieties of silk fabrics. The former is a twill jacquard fabric, while the latter is the fabric woven into patterns out of dyed silk, which is stated as "weaving the colored silk into patterns" in *Quick Approaches of Chinese Characters*. The word "*jin*" (锦) first appeared in verse *Xiangbo* in *Minor Odes* in the *Book of Songs*. "The colorful silk is woven into *jin* (brocade) like the patterns on the shell". Expressions like "brocade clothing", "brocade food" and "wear brocade" are also common in *Book of Songs*. They proved that brocade was not uncommon at that time, and many objects made of brocade were also unearthed in tombs of the Zhou dynasty in Liaoning, Shandong, and Shaanxi provinces. Embroidery was called "*zhi*" (黹) in ancient China. On a piece of cloth unearthed from tombs

Picture 4.5 Embroidery marks of braid stitches

of the Western Zhou dynasty in Rujiazhuang village, Baoji city, Shaanxi province, there were embroidery marks of braid stitches, showing quite mature skills (as shown in Picture 4.5).

During the Spring and Autumn period, as well as the Warring States period, when the various states were in dispute, they made it an important policy to reward farming and sericulture in order to enhance national strength. According to *Tribute of Yu* in the *Book of History*, such states as Yan, Qing, Xu, Yang, Yu, and Jing all paid tribute to the emperor with silk and silk cloth; and the textile industry was particularly developed in Qilu area (nowadays Shandong province) with Linzi as the center. It is recorded in *Biography of Merchants* in *Records of the Historian* that "Jiang Shang, a politician who helped build the Western Zhou dynasty and was the first king of the Qi State, took many measures to promote economic growth, such as encouraging women in his fief to spin and weave cloth with great skills and encouraging people to sell fish and salt to other states". Gradually, people and wealth accumulated in the Qi State, making it more and more prosperous. Its hats, sashes, clothes, and shoes sold well in many states.

The textile technology of the Chu State and the Wuyue area in the south was not inferior to that in the north. A large number of exquisite fabrics, which were unearthed from tombs of the Warring States period in Jiangling, Hubei province, and Changsha, Hunan province, were physical evidence. After being defeated by the Wu Sate, Goujian, the King of Yue State, spent 10 years accumulating wealth and another 10 training his people. He also "encouraged farming and sericulture", and "tilled land himself with his wife spinning and weaving". In this way, he eventually avenged himself and destroyed Wu.

During this period, the government-run and private workshops, as well as individual craftsmen in various vassal states, continued to develop, and household spinning and weaving became more popular, as *Anti-music* in *Mo Zi* describes, "Women rise early and go to bed late, spinning and weaving hemp, silk and kudzu, which is their allotted task". The need for daily clothing and market trade urges the emphasis on the quality and specifications of cloth. There is an example in *A Collection of Sayings I* in *Han Fei Zi*, which says Wu Qi of the Zhao State divorced his wife because of the "narrow width" of her woven cloth. There are various types of traditional Chinese fabrics including *mabu* (hemp cloth), *zutao* (sash), *jin* (brocade), *juan* (silk tabby), *luo* (gauze), *sha* (plain gauze), *wan* (pure white fine silk), and *qi* (damask), represented by objects unearthed from Warring States tombs in Jiangling county, Hubei province and Zuojiatang, Changsha city, Hunan province. Depictions of the dragon and phoenix, flowers, and geometric patterns, which are vivid and splendid, are embroidered on quilts, robes, clothes, and trousers, either with braid or flat embroidery. There are 12 colors of embroidery threads such as dark brown, vermilion, golden yellow, and cobalt blue (as shown in Picture 4.6). Vassal kings at that time often gave beautiful embroidery as gifts to each other. For example, it is recorded in *A Historical Biography of Su Qin* in *Records of the Historian* that the King of Zhao provided Su Qin with "a hundred chariots, two thousand *liang* (about 750 kg) gold, a hundred pairs of white jade, and a thousand bolts of splendid embroidery for him to bribe the other kings to ally with the Zhao State".

Dyeing technology also made great progress in the Warring States period. Bluegrass is the most widely and commonly used dye product with a long history, which is called "indigo plant" as described in *Xun Zi*: "Indigo-blue is extracted from the indigo plant but is bluer than the plant it comes from". According to the *Book of Diverse Crafts*, there are officials in charge of painting five basic colors, namely, green, red, white, black, and yellow; officials named "Zhong Shi" dye the fabric light red, green, red, and black successively with cinnabar and red millet while those called "Mang Shi" remove the sericin by soaking silk in the water mixed with plant ash and clam shell ash. The accumulation and maturity of textile skills were witnessed by the people of this era and recorded in the literature, and they have a long-term influence on later generations.

4.1.2 Han-Tang Period

During the Han-Tang period, Chinese civilization rose and flourished at a high speed, and textile industry also got to a new and higher stage.

The tyranny of the emperors of the Qin dynasty made the people live a hard life and did great damage to the industries. According to *Treaties of Foods and Commodities* in *History of Han*, "Men did their utmost to till the land but were still short of food; women spun and wove cloth but were still short of clothes; the wealth of the world was exhausted to pay their taxation, which was still not enough for their desires". The rulers of the early Han dynasty, who learnt the lessons of Qin, implemented the policy

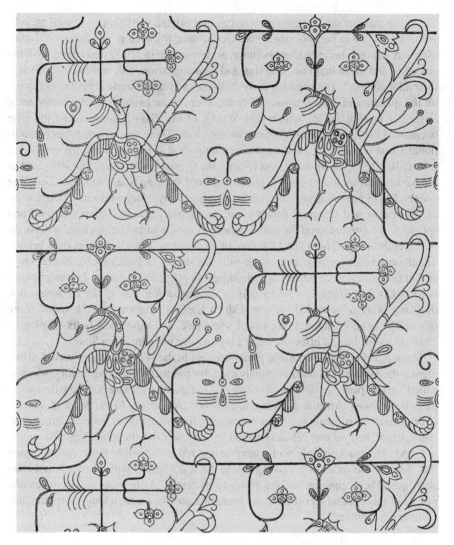

Picture 4.6 A copy of embroidered phoenix and flower patterns unearthed from Chu tombs in Jiangling county, Hubei province

of recuperation, and advocated the recovery and growth of farming and sericulture, resulting in the recovery and growth of textile industry. The center of sericulture and silk weaving was still in the north, but it was spreading to the south. It is recorded in *Treatise of Geography* in *History of Han* that in Zhuya county (nowadays northeast of Hainan Island) "men plowed land, growing rice and hemp, and women reared silkworms, spun and wove silk". *Biography of Wei Sa* in *History of Later Han* also records that Ci Chong, the satrap of Guiyang (nowadays Chenzhou, Hunan province), "taught people to rear silkworms, and they gained benefits from it".

In the Western Han dynasty, the east and west weaving offices were set up in Chang'an, the capital, and the "Sanfu Palace", with thousands of weavers, was set up in Linzi, Shangdong province, which was a special organization to produce high-grade silk fabrics such as white fine silk and colorful silk cloth for the royal family. The popularization of the textile industry led to the improvement of skills. It is recorded in *Weighing the Talents* in *A Comparative Study of Different Schools of Learning* by Wang Chong of the Eastern Han dynasty that "women of many generations in the Qi State are used to doing embroidery, with the unhandy no exception; women in Xiangyi (nowadays Suixian county, Henan province) are good at weaving brocade, the clumsy included. When you see it and do it every day, you'll be skillful at it". In the Western Han dynasty, reeling machines, spinning wheels, winding machines, warping machines, and pedal looms were widely used, and multi-heddle-treadle looms and draw looms were gradually popularized. *Miscellaneous Records of the Western Capital* records that the wife of General Huo Guang "left Chunyu Yan" 24 bolts of brocade and 25 bolts of thin silk with small flower patterns during the reign of Emperor Xuandi (91–48 B.C.) in the Western Han dynasty. The thin silk was made by the wife of Chen Baoguang in Julu (nowadays Pingxiang county, Hebei province), who wove one bolt of such silk on the 120-treadle draw loom in 2 months. *Biography of Zhang Anshi* in the *History of Han* also records that Zhang had 700 servants, all of whom had different manual work to do.

The fabrics of the Han dynasty were rich and varied, and the unearthed objects are represented by more than 200 kinds of silk cloth from the Mawangdui tombs of the Han dynasty in Changsha city, Hunan province, including plain *juan* (silk tabby), *jian* (fine silk tabby), *sha* (plain gauze), and *luo* (gauze), as well as twill *ling* (thin silk), cotton, velvet brocade, sash, and printed gauze. The most amazing among them is the garment of plain silk gauze, which looks as thin as cicada wings. It is 128 cm long, 190 cm wide, and weighs only 49 g (as shown in Picture 4.7). Hemp and kudzu fabrics include coarse and fine cloth such as *xi* (絺), *chi*, *zhu*, *zhou,* and *xi* (绤). There are certain specifications for cloth, for example, each *zong* contains 80 yarns, and the cloth with ten *zong* is the finest. The felt mattress, which is called "*qushu*" in ancient China, is a kind of wool fabric. The cloth and felt products woven by ethnic minorities include *ping* of the Di people, *ji* of the minority groups in the West Regions, cotton cloth, and asbestos cloth widely used in Yizhou and Yongchang in southern regions (nowadays Jinning and Baoshan, Yunnan province). Fabrics in this period were often woven with auspicious expressions such as "all the best" (万世如意) and "prolonging life" (延年益寿), which were embedded among cloud patterns and bird and animal patterns, forming a unique style with patterns and characters (as shown in Picture 4.8). The techniques of over-dyeing and mordant dyeing were further improved, and the mordants used were alum, melanterite, lime, and so on. *Discussing Writing and Explaining Characters* by Xu Shen (about 58–147) contains words of 10 hues and 39 colors, which shows that the chromatography was relatively

Picture 4.7 Garment of plain silk gauze, unearthed from Mawangdui tomb No. 1 of the Han dynasty

complete at that time. It is recorded in *Biography of Merchants* in *Records of the Historian* that "thousands of acres of gardenia star jasmine, and madder; thousands of plots of ginger and leeks; the owner is the equal of a marquis enfeoffed with 1,000 households". Gardenia and madder are commonly used dyes and those with large business equal to a marquis in wealth.

As early as the Warring States period, Chinese silk was transported from the northern grasslands to Europe. In ancient times, the West called China "Seres", which means the country of silk. The word comes from the Greek "Ser". During the reign of Emperor Wudi (141–87 B.C.) of the Western Han dynasty, Zhang Qian was sent to *Xiyu*, or the Western Regions twice, and opened the famous Silk Road. The Silk Road on land started in Chang'an in north-central China, the capital of the Western Han dynasty. The southern route bypassed Hexi Corridor to Loulan, and went westbound to Afghanistan, Iran, and India along the Kunlun Mountains; the northern route ran from Dunhuang to Samarkand, Iraq, and Daqin (Rome) via Turpan and Kuqa. The Maritime Silk Road originally started from Bohai Bay to Korea and Japan. In the Han dynasty, it started from Yunnan to Rome via Myanmar and India, or from Guangzhou to Vietnam, Thailand, and Syria, and reached its peak in the Song and Yuan dynasties. The Silk Road linked the ancient civilizations of the East and the West and greatly promoted trade and cultural exchanges among various regions. China's silk, porcelain, tea, iron smelting, and four great inventions were spread to the West via the Silk Road. Gaius Plinius Secundus (23–79), a Roman historian, praises Chinese silk fabrics in his book *Naturalis Historia* as "dazzling and skillful". At the beginning of the fourth century, silk became a dream of Romans, and its price

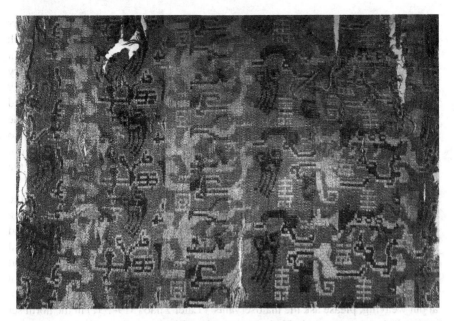

Picture 4.8 Longevity suitable for everybody (延年益寿大宜子孙) brocade of the Eastern Han dynasty

was equal to that of gold. Nobles and rich people were all proud of wearing silk clothes. Meanwhile, musical instruments, horses, gold and silver wares, cotton, and tobacco from Central Asia, West Asia, and Europe were also introduced into China one after another via the Silk Road. The spread of the philosophical ideas of China and Greece during the Axis Time of early world culture as well as the three major religions also benefited from the opening and long-term continuation of the Silk Road. All these are of great significance to the development of material and spiritual civilization around the world, and tracing back to the source, it is undoubtedly owing to the invention of sericulture and silk weaving and the hard work of weavers from generation to generation in China.

During the Three Kingdoms period, the three kingdoms of Wei, Shu, and Wu all implemented the "*tuntian*" system and encouraged farming and sericulture. There were thousands of weaver girls in the harem of Wu, while Shu relied on weaving brocade to maintain its armament.

The place of Shu is famous for its silkworm rearing all along, and its first king was named "Can Cong", meaning the followers of silkworms. The Shu kingdom set up a position in Chengdu in charge of brocade production, hence the nickname of the city "Jinguan City", and the tributary of Minjiang River passing through Chengdu is called Jinjiang River because the weavers washed brocade in it. Zhang Heng, an astronomer in the Eastern Han dynasty writes about brocade in his poem, "My love gives me beautiful brocade as a gift, and I want to give her a sapphire plate in return". According to *Biography of Jiang Qin* in *Records of the Three Kingdoms*, "Jiang's wife and concubines all wore splendid brocade clothes". The Chinese characters "*jin* (brocade)" and "*xiu* (embroidery)" symbolize the acme of beautiful things, such as in idioms "*Jin Xiu Shan He* (land of splendors)", "*Jin Xiu Qian Cheng* (bright prospects)" and "*Jin Kou Xiu Xin* (elegant thought and flowery speech—a fine literary style)". According to *Biography of Medicine and Divination Masters* in *Records of the Three Kingdoms*, "Someone in Fufeng (nowadays in Shannxi province) named Ma Jun was quite creative...He was appointed the *boshi*, i.e. a consultant of the emperor and an official in charge of the imperial library, but was rather poor, so he thought of making improvement to the loom". He made a 12-treadle loom out of the original 60-heddle and 60-treadle one, which is a great improvement of the heddle-treadle loom.

During the Jin dynasty, a large number of people moved southward, and the sericulture and weaving industry moved accordingly. It is recorded in *Book of Song* that the Song (420–479) during the Southern dynasties "was rich in silk, cotton and cloth, which could be made into clothes for the whole country". At that time, a special position named "*pingzhun*" was set up to take charge of weaving and dyeing, and most of the gentry had servants engaged in farming and weaving, so there was a saying that "if you have any questions about farming, please ask the manservants; about weaving, please ask the maidservants". Later Zhao (319–351) in the north set up a weaving department in Yecheng (nowadays Linzhang county, Hebei province), with about a hundred skillful workers, and the Ye brocade they made is as famous as the Shu brocade. During the Northern Wei dynasty, more than 100,000 craftsmen with various skills were moved to the capital, where silk department and weaving department were set up. Dingzhou in Hebei became the silk weaving center in the north.

It was during the Three Kingdoms period that the Wei (220–266) government began to force farmers to pay for textile materials and fabrics in addition to grain. In the Han dynasty, household tax was paid with money but was changed into two bolts of silk tabby and one kilogram of cotton for each household during the reign of Cao Cao (155–220). Therefore, mulberry planting, silkworm rearing, and cloth weaving were necessary for farmers to maintain their livelihood. Afterwards, this kind of heavy burden for farmers grew heavier and heavier in successive dynasties. In the Southern Dynasties period, each grown-up man should pay the government three *zhang* (10 m) cloth, three *zhang* silks, three *liang* (150 g) silk, eight *liang* (400 g) cotton, plus eight *chi* (about 3 m) silk tabby and three *liang* and eight *fen* (about 400 g) cotton as the local officials' salary every year. In the Northern Wei dynasty, each household should pay two bolts of silks, two *jin* (1 kg) cotton fiber, and one *jin* silk every year. Sui and Tang governments implemented the *juntian* system (equal-land system) with land allocated to every grown-up man, who had to pay two *zhang* (about 6.7 m) thin silk and silk tabby, three *liang* cotton, four *liang* (200 g) cloth, and three *jin* (1,500 g) hemp every year as taxes in return. During the Tianbao period (742–756) in the Tang dynasty, 7.4 million bolts of silk tabby, 700,000 *jin* (one *jin* is 0.5 kg) silk, and more than 16 million *duan* (one *duan* is about 6.67 m) hemp cloth were collected, which was unprecedented in quantity. According to the *Biography of Shen Huaiwen* in the *Book of Sui*, "Thousands of bolts of silk tabby and cotton are collected for the use of sacrifice every year within a limited time. So many common people have to buy them to pay the taxes. One bolt of silk tabby costs two or three thousand *qian*, and one *liang* cotton costs three or four hundred *qian*. The poor have to sell their wives and children for the money or are forced to commit suicide".

At the beginning of the Tang dynasty, the Weaving and Dyeing Department was set up under the supervision of the Minor Treasury, with 25 workshops, where artisans took turns to serve or pay for employment. The government-run workshops were quite large-scale. For example, there were more than 360 skillful hands in Ling and Jin workshop alone during the reign of Wu Zetian (690–705). The most important change in silk weaving technology at the time was from warp brocade to weft brocade. In the early Tang dynasty, many patterns of brocade in the Imperial Storehouse were made by Dou Shilun, Duke of Lingyang, and his patterns were called "sample patterns of Duke of Lingyang". After the Mid-Tang, with the expansion of commodity circulation in urban and rural areas, folk workshops greatly increased and had a clear division of labor, such as weaving workshops, dyeing workshops, felt workshops, and carpet workshops. He Mingyuan, a wealthy businessman in Dingzhou (nowadays in Hebei province), owned 500 silk machines, which were far more than those owned by any single person before. As a consequence, the competition between handicrafts became intensified. It is described in *Lyrics to the Weaver Girls* by Yuan Zhen of the Tang dynasty that "one of my neighbors have two daughters, who are skillful weaver girls and can't get married". According to *Laoxue'an Notes* by Lu You of the Song dynasty, "Bozhou produces fine gauze, which is as light as feathers, and the clothing made by it looks like rosy clouds. There are only two families there can weave it, and the family members are married to each other, for fear that other people will learn their skills. Such has been the case for more than 300 years since the Tang dynasty". In both the above cases, the only reason for them was to keep the secrets of their own skills, because there exist some bad rules in the handicraft industry such as "passing on the skills to sons instead of daughters or daughters-in-law".

The frequent and active foreign trade and cultural exchanges in the Sui and Tang dynasties helped promote the diversification of fabric patterns. According to the *Biography of He Chou* in the *Book of Sui*, "The Persian kingdom once presented a golden brocade robe with extremely beautiful fabrics to the Sui Emperor, who ordered He Chou to make a same one. The robe made by He was even better than the original one". The patterns of the brocade unearthed from the Tang tombs in Astana, Turpan city, Xinjiang Uygur autonomous region were quite diversified including pairs of horses, pairs of deer, magpies holding flowers in their mouths, flowers, and birds, which were obviously influenced by Persian culture (as shown in Picture 4.9).

In the early Tang dynasty, the silk weaving industry was still centered around the middle and lower reaches of the Yellow River. However, after the An-Shi Rebellion (755–763), with the economic center moving southward, Yangzhou and Kuaiji quickly became the silk weaving centers. The *liaoling* (dazzling fine silk) produced in the Wuyue Area (regions south of the Yangtze River) was exclusively used by the imperial court. Bai Juyi, a famous poet of the Tang dynasty describes the making of such silk in his poem, "The delicate patterns of silk are required by the royal palace. It is embroidered with ranks of migrating wild geese flying above the clouds

Picture 4.9 Brocade with the pattern of flowers and birds, unearthed from the Tang tombs in Astana, Turpan city, Xinjiang Uygur autonomous region

in autumn, and the colors of silk are like those of river water in the south". The poet also compares the weaver and the wearer in the poem, "Who is the weaver, who is the wearer? The poor girls of Yue brook and courtesans of the imperial palace. … The dancers of Zhaoyang are truly and deeply blessed. A pair of spring clothes worth a thousand gold". Printing and dyeing techniques were popular all over the country in the Tang dynasty, including clamp resist dyeing, which was quite popular, as well as tie-dyeing, wax dyeing, and woodblock printing, as evidenced by the fact that many tie-dyed objects in the Tang dynasty are still preserved in Todaiji Temple in Japan (as shown in Picture 4.10). Hemp and kudzu fabrics were mostly produced in the south. Cotton fabrics appeared in Xinjiang in the Han dynasty and cotton cloth was produced in the Lingnan area (covering Guangdong, Guangxi, and part of Yunnan and Fujian) in the Tang dynasty, but they were still rare in Central Plains. Embroidery stitches developed greatly in the Tang dynasty, such as encroaching satin stitch, long and short stitch, fastening stitch, couching stitch with golden thread, parallel couching stitch with golden thread, and the embroidery of daily clothes and figures of Buddha was quite exquisite.

According to the *Treatise of Geography* in the *History of Han*, Ji Zi went to Korea "to teach the people there to learn etiquettes, farm, rear silkworms and weave" in about eleventh century B.C. If it was true, it showed that as early as the mid-Shang dynasty, the textile technology of China already spread to the Korean Peninsula and then to Japan. There were frequent exchanges between China and Japan during the Tang dynasty. Japan sent envoys to Chang'an many times, and Chinese brocade, silk, and

Picture 4.10 Pattern of tie-dyed flowers and birds of the Tang dynasty, stored in Todaiji Shosoin, Japan

dyed fabrics as well as weaving skills were transported to Japan with them. Southeast Asia, Indonesia, and some other countries also learned the Chinese silk weaving technology through the Maritime Silk Road and developed sericulture during the Han-Tang period. In the tenth year of the Tianbao period (751) in the Tang dynasty, the army led by Gao Xianzhi, the military governor of Anxi (the Western Regions), was defeated by Arabian troops in the Battle of Talas, and 20,000 soldiers were captured, including Le Huan and Lu Li, weavers from Hedong (nowadays Shanxi province). This was the beginning of the spread of Chinese silk weaving technology to the West.

Ancient West didn't produce silk. To get rid of the dependence of silk weaving industry on Persia, Justinian (482–565), the Byzantine Emperor, sent Nestorian missionaries to China in 552 and brought back silkworm eggs in bamboo sticks. Silkworm rearing and silk reeling were seen in East Rome thereafter. Later, during the Second Crusade in the mid-twelfth century, Ruggero II (1095–1154), king of Sicily, plundered 2,000 weavers from Byzantium, and since then Italy became the main country of silk production and center of silk weaving in Europe. It was not until the fourteenth to the sixteenth centuries that Britain, France, and European countries mastered the skills.

4.1.3   Song and Yuan Dynasties

In the Song dynasty, there were various institutes with special functions related to silk making under the supervision of the Minor Treasury, such as Silk and Brocade Institute, Imperial Dyeing Institute, Crafts Institute, and Embroidery Institute. Each of the three Brocade Departments in Hangzhou, Suzhou, and Chengdu employed thousands of craftsmen, and Chengdu also set up a special organization called Brocade Institute of Tea and Horse-trading Offices to weave brocade needed by ethnic minorities in northwest and southwest China. There were many private textile workshops of considerable scale. For example, Puyuan town, Jiaxing city, Zhejiang province was named after Pufeng, son-in-law of Emperor Gaozong of Song (1107–1187), who made and traded silk fabrics there, and the Pu silk produced was well known at that time. Self-employed workshop owners had to pay taxes to the government, but their status was higher than that of tribute weavers in the Tang dynasty. There were also many craftsmen who were employed in workshops or wandered along the streets with their weaving tools for temporary work. The imperial court rewarded those rearing silkworms and weaving, and therefore some writings and artistic works about them were created. For example, *Pictures of Farming and Weaving* by Lou Shu (1090–1162) of the Southern Song dynasty depict in detail the whole process of silkworm rearing and silk making, including bathing silkworm eggs, laying silkworms, picking mulberry leaves, feeding, molting, placing silkworms on screens, mounting silkworms on and off the nests, sorting cocoons, reeling silk, offering thanks, silk winding, tidying warp and weft, and weaving. The pictures are of high literature value.

Frequent wars in the north led to the southward movement of silk weaving center. After the Song government moved across the Yangtze River, the silk industry in the Wuyue area became more developed. *Compilation of Regulations in the Song Dynasty* records that "all the silk supplies in the country are in the southeast". There were special markets for silk floss, raw silks, pillows and hats, clothing, and silk tabby in Lin'an (nowadays Hangzhou), the capital of the Southern Song dynasty. Hemp was also mainly produced in the south, especially in Guangxi. In the Yuan dynasty, the invention of the hydraulic spinning wheel, which was used for spinning hemp and yarn, was a major breakthrough and the earliest spinning machine driven by hydraulic power in the world (as shown in Picture 4.11). According to the *Book of Agriculture* by Wang Zhen (1271–1368), "In the places where ramie is grown, they (the hydraulic spinning wheel) are placed where there are rivers". The wool weaving industry was most famous in the northern part of China such as Lanzhou and Jingzhou (nowadays north of Jingchuan county, Gansu province). The former boasted of weaving *xihe* goat hair fabrics named "Lan Velvet"; the latter boasted of weaving tapestry with fuzz.

Silk fabrics in the Song dynasty were rich in patterns and came in a wide variety. Silk patterns like "large lions", "a pair of wild geese in the sky", "auspicious grass and flying crane", and "twisting golden silk thread on blue and red clothes" are recorded in literature such as *Records of Carriage and Costumes* in *History of Song*, *Notes of Ceasing Farm Labors* by Tao Zongyi (1329–1412) and *Rustic Talks from the East*

Picture 4.11 The restoration drawing of the hydraulic spinning wheel of the Yuan dynasty, cited from *the Book of Agriculture* by Wang Zhen

of Qi by Zhou Mi (1232–1298). The patterns of "vultures and double goats" on the brocade robe of the Northern Song dynasty unearthed in Aral Shehri, Xinjiang are obviously of the Persian style (as shown in Picture 4.12). Great progress was made in *Luo* (gauze) weaving during this period, and the improved gauze draw looms could weave complex ribs. Some famous varieties include gauze with peacock patterns and gauze with treasure and fairy flower patterns. Every year, the government collected more than 100,000 bolts of gauze as tributes.

Picture 4.12 Brocade robe with vultures and double goats' patterns of the Northern Song dynasty

To meet the needs of costume making at the imperial court, and painting and calligraphic work mounting, Song brocade began to be produced in the Southern Song dynasty, with more than 40 kinds such as *qingloutai*. *Kesi* (Chinese tapestry) gradually flourished from the Tang to the Song dynasty, with Dingzhou (in Hebei province) as the main production center. Zhuang Chuo of Song describes the making techniques of *kesi* in *Chicken Rib Compilation*, "When weaving the warps, leave the wefts out first, and then weave the wefts with variegated threads on the warps with small shuttles, thus forming different patterns, which look as if discontinuous. If looked at without any background, it seems that the patterns are hollowed out by knives". Masterpieces of such crafts include "Purple *Luan* and Magpie" and "Purple Heaven Deer", and famous artisans were Zhu Kerou and Shen Zifan (as shown in Picture 4.13).

Printing and dyeing spectrum were quite complete in the Song dynasty, and printed fabrics became the main material of folk clothing, with artists specializing in carving flower plates and printing and dyeing. A craftsman in Jiading (nowadays in Shanghai) with the surname of Gui made blue cloth with designs in white, also called *jiaohuabu* cloth, with the technique of resist dyeing with an alkaline agent. Another dyer dyed the cloth dark purple or *youzi* with ashes of symplocos sumuntia leaves. Both of them were quite popular for a while. In the Southern Song dynasty, Tang Zhongyou,

Picture 4.13 *Picture of Little Ducks in the Lotus Pond* by Zhu Kerou in the Southern Song dynasty (partial)

a wealthy businessman in Wuzhou (nowadays Jinhua, Zhejiang province), dyed thousands of bolts of violet silks, and the Yao people in Guangxi made batik cloth with flower plates, which was called spotted cloth of Yao.

The main achievement of embroidery during that period was "painting embroidery". It meant that famous paintings and calligraphic works were used as drafts and embroidered on cloth, which were characterized by fine needlework, exquisite colors, and vivid embroidery (as shown in Picture 4.14). Representative works include the silk scroll "White Eagle".

During the Tang and Song dynasties, the tax system of summer and autumn was implemented with cloth and silk as the main part of taxes; the system of "fair trade" was also implemented, which meant that the government paid money to the people in advance at the beginning of the spring, and the people should pay taxes with silk tabby to the government in summer and autumn. In order to make peace with the Jin kingdom in the north, it was necessary for the Southern Song dynasty to offer cloth and silks to it, and therefore the deprivation of farmers was even more aggravated. The so-called "fair trade" was actually the plunder from the government. In Chuzhou (nowadays Lishui, Zhejiang province) and other places, some people even killed baby boys to avoid the head tax. Sima Guang, a statesman in the Northern Song dynasty, is quoted in the *Treatise of Food and Loan* in *History of Song*, "The public and private

Picture 4.14 *Flying on a Crane at Yaotai*, embroidery of the Southern Song dynasty, collected by Liaoning Museum

creditors compete with each other... Silks don't belong to the makers even before they are made". It reflects the profound contradiction between the government and the common people.

The rule of the Yuan dynasty was barbaric and cruel. The government wantonly forced craftsmen to do handicraft work, which can be proven by written records. For example, 720,000 civilian craftsmen in the Central Plains were officially drafted in the eighth year of Ogodei Khan (1236) and more than 100,000 in the south of the Yangtze River were drafted in the 21st year of the Zhizheng period (1361). The felt products for Mongolian aristocrats were made by the felt bureaus in Dadu and Shangdu. With the rule that "craftsmen are forbidden to leave the bureau", the workers were actually like slaves, which was a serious retrogression compared with the production relations dominated by independent handicrafts and employing craftsmen in the Song dynasty. After the mid-Yuan dynasty, the government relaxed the law slightly in order to obtain more silks, and therefore the folk textile industry recovered to some extent.

Cotton entered the Shanhaiguan Pass from the northwest in the early Yuan dynasty and spread from Hainan in the south to the Yangtze-Huaihe Region at the same time. Songjiang prefecture (nowadays Shanghai) in Jiangsu province, under the guidance of Huang Daopo, became the center of the cotton textile industry in a short time. Huang, a native of Wunijing, Songjiang, moved to Yazhou in Hainan, where she learned a complete set of techniques of cotton planting, cotton making, and Li brocade weaving from the indigenous Li people. During the Yuanzhen period (1295–1297), she returned to her hometown, led the villagers to "make tools for ginning, fluffing, spinning and weaving cotton", and taught women "to pick-weave and brocade", and "to weave cotton into quilts, mattresses, belts and hats with branches, phoenixes, chess games, and characters, which were bright and vivid". "Many people willingly began to weave after they learned the skills". Since then, the cotton textile industry in the Jiangnan area, the south of the Yangtze River, continued to flourish, making great contributions to the economic and cultural prosperity of this region. Huang Daopo's name on the scroll of fame is admired by the public. People set up temples for her and offer sacrifice to her regularly (as shown in Picture 4.15).

Aristocrats of the Yuan dynasty were fond of costumes made of gold-wefted brocade (called *nashishi* in Mongolian), which was the typical representative of the weaving industry during that period (as shown in Picture 4.16). According to *Biography of Zhenhai* in *History of Yuan*, when Zhenhai oversaw the Weaving Bureau in Hongzhou (nowadays in Gansu province), "there were more than 300 weavers from the Western Regions", who wove the gold-wefted brocade. *The Travels of Marco Polo* records that raw silk on carriages was sold in Beijing in the thirteenth century, and the number of carriages was as many as a thousand every day. Taiyuan, Chengdu, Hangzhou, and Suzhou were also famous for weaving gold-wefted brocade, silk, and gauze.

Picture 4.15 Statue of Huang Daopo

4.1.4 Ming and Qing Dynasties

In the Ming dynasty, the Bureau of Irrigation and Transportation in the Ministry of Works was in charge of weaving and dyeing, and the Imperial Miscellaneous Weaving and Dyeing Bureau was also set up in Nanjing, referred to as the South Bureau for short. There were also Miscellaneous Weaving and Dyeing Bureaus in various regions, and the Suzhou bureau was the largest one with 173 looms (as shown in Picture 4.17) and 667 craftsmen. The quota of fabrics woven in Suzhou, Hangzhou, Songjiang, Jiaxing, and Huzhou was as many as 150,000 bolts during the Wanli period (1573–1620) after being added many times.

Artisans from all over the country were required to take turns to serve in the government. There were 1,343 craftsmen in the Imperial Miscellaneous Weaving and Dyeing Bureau alone, including those of sash, soul cloth (cloth for summoning the soul of the dead in ancient China), embroidery, felt, backstrap loom, thread pulling, pick-weaving, and *kesi*.

Craftsmen in the Ming dynasty were haft free and could produce and do business by themselves, so the folk textile industry developed greatly during that period. *Dream Talk by the Pine Window* by Zhang Han of the Ming dynasty records that a

Picture 4.16 The *nashishi*
shawl for Buddhists in the
Yuan dynasty

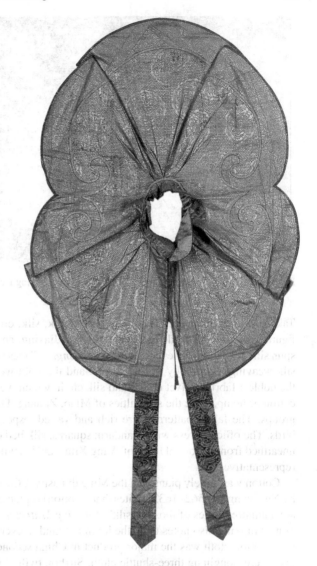

large number of people in the Sanwu area (referring to places in the lower reaches of
the Yangtze River) became rich by weaving. Large workshops usually hired many
craftsmen, and those to be hired often stood at the ends of bridges or on the sides of
streets in the early morning, waiting for the potential employers. Reeling, weaving,
and dyeing in the silk industry and ginning, fluffing, spinning, weaving, and treading
in the cotton industry gradually became independent handicraft sectors, and capitalist
elements developed accordingly.

The differentiation and increase of silk fabric varieties manifested the progress
of textile technology and the increasingly fine division of labor in the Ming dynasty.

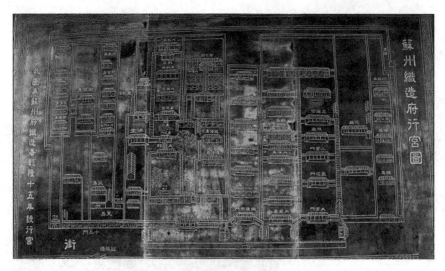

Picture 4.17   Map of the temporary Palace in Suzhou Weaving Department

Take silk as an example. There was thread silk, silk, crepe, and moiré in Suzhou, figured silk, *pu* silk made in Puyuan town in Jiaxing, and water-reeled silk thread-spun silk in Huzhou. The technique of *"zhuanghua"* was representative of the superb silk weaving technology in this period, and the gold-wefted *zhuanghua* satin was the noblest fabric. Velvet cotton and silk cloth, which were made of silk thread and cotton or hemp, were the specialties of Miao, Zhuang, Dong, Yao, and Tujia ethnic groups. The fabric patterns were rich and varied, especially those of flowers and birds. The official dress with Mandarin squares, silk bed curtain, and brocade quilts unearthed from the burial tomb of Wang Xijue and his wife in Suzhou are the typical representatives.

Cotton was widely planted in the Ming dynasty. *A Complete Book on Agriculture* by Xu Guangqi (1562–1633) writes that "(cotton) is planted everywhere…It's merits are a hundred times of those of silk". *Clothing Materials* in *Heavenly Creations* by Song Yingxing also notes that "the loom is found in every ten households". At that time, cotton cloth was the major product in China second to grain, and the famous ones were Songjiang three-shuttle cloth, Suzhou twill cloth, and Jiading spot cloth. Hemp, kudzu, and ramie cloth were produced in Taicang of Jiangsu province, Putian and Hui'an of Fujian, Xinhui of Guangdong, Yulin of Guangxi, and Huangpi of Hubei. Wool fabrics made in Lanzhou and Xi'an in the northwest and felt woven by the Yi people in Shuixi, Guizhou province in the southwest were a tribute to the government.

Weaving in *Code of Great Ming Dynasty* records that the plants used as dye included Brazil wood, *huangdan* (alseodaphne hainanensis merr), gardenia, dark plum, and madder, and the government even set up the Pigment Bureau for this purpose. Resist dyeing technology was more advanced, which included Shu dyeing, brocade dyeing, deer's embryo dyeing, etc.

The major progress of embroidery skills in the Ming dynasty was the invention and development of Gu embroidery. The daughter-in-law, granddaughter-in-law, and great-granddaughter of Gu Mingshi, a scholar during the Jiajing period (1522–1566), were all famous for painting embroidery. Their works were noble and elegant with delicate skills, which had great influence. Many of the embroidery workshops in various places were named after Gu embroidery, which was a general name for embroidery at that time (as shown in Picture 4.18).

Suzhou, Hangzhou, Huzhou, and Jiaxing in Taihu Lake Basin had been the centers of silk weaving industry since the Southern Song dynasty. During the Ming and Qing dynasties, most Suzhou residents were engaged in craft skills. Especially in the east part of the city, all of them learned weaving skills, and every family was engaged in weaving. In Puyuan town, Jiaxing, thousands of families wove silk for a living.

Picture 4.18 *Picture of Washing Horse* in *Book of Famous Paintings in the Song and Yuan Dynasties* by Han Ximeng, representative of Gu embroidery in the Ming dynasty

On this basis, Shengze in Wujiang of Jiangsu, Fengjing in Songjiang, Wangjiangjing in Jiaxing of Zhejiang, and Shuanglin in Huzhou all developed from small villages with dozens of households into new towns with tens of thousands of residents in just a few decades. The weaving workshops were clustered, the traffic and commerce were busy, and people all over the country gathered in those towns. The largest weaving center in the north was Lu'an in Shanxi. According to the *Annals of Lu'an Prefecture*, there were thousands of weavers in this area in the Ming dynasty, which was praised as "Lu silk is all over the world". What's more, Shu brocade of Sichuan, Yue satin, and Guang gauze of Guangdong, as well as the silk and velvet of Fuzhou and Quanzhou were also famous products.

In the Qing dynasty, the folk silk weaving industry was concentrated in Jiangsu, Zhejiang, Guangdong, Sichuan, Anhui provinces, and there were more than 30,000 looms in Jinling (nowadays Nanjing) alone. With such a strong basis, the Qing government set up silk industry centers directly under the royal family in Jiangning, Suzhou, and Hangzhou, which were known as the three major weaving centers in Jiangnan area. According to *Records of Laws and Systems of the Qing Dynasty*, during the Qianlong period (1736–1796), Jiangning Weaving Bureau had 600 looms and 2,547 craftsmen; Hangzhou Weaving Bureau has 600 looms and 2,330 craftsmen; Suzhou Weaving Bureau has 663 looms and 2,175 craftsmen. The numbers of looms and craftsmen were nearly four times those of the Ming dynasty. Ancient China was a patriarchal society with autocratic regime, and the hierarchy was distinguished by etiquette. As early as the Xia dynasty, the clothing system was already in existence. For example, emperors wore clothes with patterns of the sun, moon, stars, rice, and axes (黼黻, *fu fu*); the upper outer garment with paintings and petticoats with embroidery were called "twelve ornaments" (as shown in Picture 4.19). All the other dynasties have followed the system since then. The civilians wore clothes made of hemp while the nobles wore those made of silk. In the Tang dynasty, officials with the ranks above the fifth were dressed in purple and those with lower ranks were dressed in green. The emperor robes of the Ming and Qing were designed with five-claw dragon patterns (as shown in Picture 4.20), and the official robes were designed with four-claw python patterns. The Mandarin square of officials was decided by their ranks, with poultry patterns for civil servants (as shown in Picture 4.21) and beast patterns for military officials. The Qing government devoted huge manpower and financial resources to maintaining the Jiangnan weaving production regardless of cost, which pushed the traditional silk weaving technology to its peak, and at the same time gave full play to the royal family's tyranny and extravagance. To restrain the development of folk textile industry, the government often took various measures such as reducing the numbers of weaving machines, imposing heavy taxes, allocating official weaving work at low prices, etc., while the government-run textile bureau made huge profits by reducing the workers' salary and rations, and resorting to fraud when purchasing raw materials and transporting finished products. Therefore, officials of the weaving bureau in the Qing dynasty had lucrative posts, and most of the officials appointed were banner people to whom the government intended to give benefits. In this case, folk workshops managed to survive and develop only by trying to break through the official shackles. For example, during Emperor Kangxi's reign

Picture 4.19 A duplicate of the dragon robe

(1662–1722), many owners appealed to Cao Yin, who was in charge of the Jiangnan Weaving Bureau, for tax exemption and were satisfied. After the Daoguang period (1821–1850), large workshops with thousands of looms, huge capital, and a large number of hired workers appeared in Suzhou and Hangzhou.

As early as the Song dynasty, some guilds appeared in the textile industry. During the Ming and Qing dynasties, guild halls and industry associations were set up. Stricter rules and regulations were established to stipulate the rights and obligations of the workshop owners, the temporary or long-term helping hands, and the ordinary employees, and decide the wages, the apprenticeship system, the distribution of raw materials in the guild, product specifications and prices, location of workshops, and prohibition of outsiders' operation. The head of the guild, who was elected by all the members or changed through rotation, would oversee regular affairs, offering sacrifices and public welfare relief. As for the household spinning and weaving activities, although they were quite common in the vast rural areas and occupied a very important position in farmers' economic life, they were always in a scattered and unorganized state, and thus their situation was the most miserable.

Compared with the Ming dynasty, the types of silk fabrics of Qing were more varied. For example, there were different types of satin in Jinling (nowadays Nanjing city) such as the number one, number two, number three, *basi,* and *changtou.* Even in Zunyi, Guizhou, which is located in the backland of China, its silk products could

Picture 4.20 Emperor Qianlong's dragon robe (ceremonial robe)

compete with Wu Ling and Shu Jin because of their low price and good quality. Cotton cloth was also finer, and the blue cloth in Shanghai, the purple cloth in Nanjing, the kapok cloth in Wucheng, Zhejiang, and the small cloth in Baling, Hunan were all very popular among the people. The hemp cloth and kudzu cloth made in Chaoyang, Xinhui, Leizhou, and Zengcheng of Guangdong province were of the best quality, and they were also produced in Jiangxi, Sichuan, and Fujian provinces. The progress of embroidery was no more than the maturity and finalization of various embroidery types. Suzhou embroidery, Shu embroidery, Xiang embroidery, Yue embroidery, Bian embroidery, and Lu embroidery all established their status during this period. Many textile centers made their cities more prosperous. For example, there are 14 silk workshops and 16 cotton workshops in the scroll painting *Prosperous Suzhou* (also named *Flourishing City of Gusu*) by Xu Yang, a painter during the Qianlong

Picture 4.21 Mandarin square patterns on civil servants' official clothes in the Ming dynasty

period (1736–1796) (as shown in Picture 4.22). Nevertheless, under the exploitation of the government and businessmen, the life of farmers and craftsmen engaged in weaving was still extremely hard, which often caused fierce resistance. In the late Ming dynasty, Lu Bao, a eunuch, was sent to Hangzhou to urge the satin taxes, and many workshop owners were forced to flee. Contradictions in the Qing dynasty were even more acute. For example, in the ninth year of the Kangxi period (1670), the treading craftsmen in Suzhou, headed by Dou Guifu, stopped working and gathered by means of leaflets, demanding a pay rise. In the 32nd year of the Kangxi period (1693), Luo Gui and Zhang Erhui gathered people at the treading guild, threatening with suicide, and in the 40th year (1701), thousands of treading craftsmen gathered in

Picture 4.22 *Prosperous Suzhou* (partial)

groups and kicked up a row almost every day. In the 12th year of the Yongzheng period (1734), weaving machine owners went on strike against the dismissal and for the increase in salary. Under such circumstances, technical improvement was impossible, let alone breakthrough. To sum up, the Qing dynasty made little achievements in textile machines as well as weaving and dyeing skills, and new changes could not come until the new ages.

In the late Qing dynasty, the steam silk reeling machine and electric jacquard machine were introduced from the West to China. The large amount of dumping of European textiles and factories established by foreign businessmen in China dealt a devastating blow to the traditional Chinese textile industry. In 1880, Zuo Zongtang, a statesman of the late Qing dynasty, founded Lanzhou Textile Bureau, and its products were only for military supplies. Yarn mills, cloth mills, hemp mills, and silk weaving mills sprung up in many places and they began to take shape in 1911 when there were 32 cotton spinning mills with 830,000 spindles. However, it was still difficult for the Chinese mills to compete with foreign businesses because of their meager strength. After the First World War, the number of Chinese cotton spinning mills increased to 54, with 1.54 million spindles, but they were still struggling to survive. After the founding of the People's Republic of China in 1949, the textile industry was vigorously developed by introducing advanced technology, manufacturing new equipment, and building textile machinery plants, chemical fiber factories, and textile mills in a planned way, which changed the original backward situation and basically met the needs of the people and various industries for textiles and even exported them in large quantities. Since the beginning of the twenty-first century, traditional

Chinese weaving, dyeing, and embroidery skills have attracted the attention of the public and the government. Now, many precious skills have been listed as national intangible cultural heritage, which are expected to be protected and inherited.

4.2 Raw Textile Materials and Technology

Raw materials are the first element of textile production and are crucial to the design of production processes and equipment. In ancient times, what was used for textile were natural fibers, generally the staple fibers including wool, hemp, and cotton, or one of them. But in ancient China, a fiber besides those three was also creatively used, namely the long-fibered silk. The utilization of silk fiber greatly promoted the progress of textile processing and textile machinery and greatly influenced the textile printing and dyeing technology of hemp, cotton, and wool. What is particularly important is that the Silk Road, which only came into existence because of silk trade, greatly promoted the exchange and assimilation of economy, technology, culture, and religion between the East and the West.

4.2.1 Silk-Weaving Technology

A bolt of gorgeous, elegant, and soft silk (as shown in Picture 4.23) must go through many processes such as silkworm rearing, silk reeling, and weaving. The traditional silk weaving processes are as follows:

Silkworm rearing—silk reeling—silk winding, doubling and throwing—weft winding, warping, drawing-in, and warp sizing—weaving—scouring—printing and dyeing.

4.2.1.1 Silkworm Rearing

To rear silkworms well, we must first understand the several steps in their growth process (silkworm eggs, hatching, pupation, cocooning, and eclosion) (as shown in Picture 4.24).

It is recorded in *Typical Weather in a Given Season* in *Classic of Rites* that "it is forbidden to rear silkworms twice in a year". *Instructions for a Grand Nation* in *Huai Nan Zi* also records that "it is not unfavorable to rear silkworms twice in a year, and the reason for forbidding it is that it will do harm to the mulberries". The citations prove that the Chinese mastered the basic knowledge of silkworm rearing as early as the Zhou dynasty. The premise of implementing the measures to ban the rearing of silkworms twice a year was to control eclosion since it is impossible to realize the same result in the natural environment. During the Spring and Autumn period and

Picture 4.23 Silk of various colors

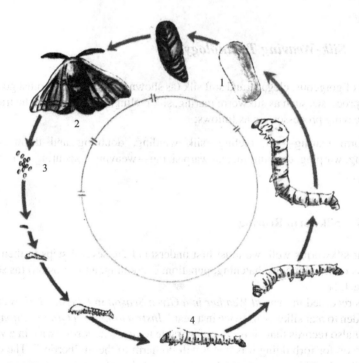

Picture 4.24 Several forms of silkworm during growth. 1. silkworm pupa, 2. adult silkworm (moth), 3. eggs, 4. larva

the Warring States period, sericulture techniques and tools were gradually perfected, and have now been applied for more than two thousand years.

Sericulturists have paid special attention to the factors affecting the growth of silkworms, such as silkworm rearing rooms, egg sorting, hatching, breeding, mounting silkworms on the nest, and silkworm diseases, and summed up many effective methods.

Silkworm rearing rooms. To ensure a suitable growth environment for silkworms, a silkworm rearing room should be built with special care. It should be near rivers to ensure fresh air; thorns should be placed on the walls and the exit should not lead directly to the outside to ensure that the temperature and humidity in the room remain as stable as possible. The best choice for the room is that it faces north, the second best is facing south or west, while it is taboo for it to face east. The room should be high and wide because lower ones are usually too hot.

Sorting eggs. Sorting silkworm eggs is the first step to rearing silkworms, and the quality of eggs will directly affect the silk yield and egg yield. Those cocoons which are directly exposed to the sun or on the top of the nest should be selected for future eggs as they are the best. If there are sick or weak moths or moths that come out first or last, they should be removed, leaving only those that are completely mature and healthy.

Hatching. At normal temperatures, silkworm eggs hatch in seven or eight days. If they are not controlled artificially, it is often inconvenient and irregular to try your hand at sericulture. In order to control the hatching time of silkworm eggs, sericulturists often adopt the method of low-temperature incubation. The specific method is to place an urn containing silkworm eggs in cold spring water to cool it down, and thus control the hatching time in 21 days.

Rearing silkworms. The external factors that may influence silkworm growth should also be noted, such as the mulberry leaves, fire, cold, summer heat, dryness, and dampness. These all contribute to carefully creating an environment suitable for silkworm growth.

Putting silkworms on the screen. When the mature silkworms are put on the screen, water accounts for about half of their body weight, and it will be distributed to the cocoon room when silkworms are spinning. The moisture will not only lead to poor cocoon quality but also reduce the reelability percentage during silk reeling if not discharged in time. Heating and ventilation are generally used to discharge moisture in the room. People in Jiaxing and Huzhou use high shed and umbrella-shaped cocoon rooms, in which basins with charcoal fire are placed. There are about 20 basins in some large rooms and 18 in small ones. A high shed is convenient for ventilation, and the charcoal fire makes the room dry as well as at the most suitable temperature for cocooning.

Silkworm diseases. Silkworms are often infected by bacteria or viruses during their growth process. Different treatment methods should be adopted for silkworm diseases with different symptoms. For example, flacherie of silkworms is infectious, and the infected ones should be removed immediately for fear of infecting the others; for viral diseases, preventive measures such as sprinkling fresh weathered lime on silkworm tools can be taken.

Traditional sericulture tools include silkworm screens, silkworm baskets, silkworm shelves, silkworm nests, silkworm trays, and silkworm nets. The silkworm screen, one of the utensils for placing young silkworms, is a reed-woven screen, which can also be made of bamboo strips or grass stems. The silkworm basket, also known as *"fei"*, is the residence of silkworms, and they spend most of their time in it before cocooning. It is mostly round or oval and is placed on the silkworm shelf layer by layer (as shown in Picture 4.25). The silkworm shelf, also known as "silkworm *chui"*, is mostly rectangular (as shown in Picture 4.26). The silkworm nest, also known as *"ru"*, is a tool for silkworms to cocoon when they are mature. In the early days, it was usually a bundle of thatch or wormwood, but later, it was made of rice straw or wheat straw. The nest is an essential place for cocooning because there are many small spaces in the nest for silkworms to choose from. Those small spaces should be properly packed to ensure air circulation for discharging moisture, and there should also be uniform light for convenient cocoon gathering (as shown in Picture 4.27). The silkworm net is a net woven with ropes, with handles at four corners for lifting young silkworms. According to the *Book of Agriculture* by Wang Zhen (1271–1368), when moving silkworms, first cover the silkworms with the net, then put some mulberry leaves on it, and the silkworms will climb through the holes on the net to eat them. When all the silkworms are on the net, lift it and move them elsewhere. It is an efficient and labor-saving method of moving silkworms.

4.2.1.2 Silk Reeling

The main components of silk are fibroin and sericin. The former, a nearly transparent fiber, is the main part of cocoon silk, while the latter is a viscous substance wrapping the fibroin. Fibroin is insoluble in water, while sericin is soluble in water, and the higher the temperature, the greater the solubility. The difference between fibroin and sericin makes it easy to decompose cocoons. The process of drawing silk fiber is called silk reeling, which is a seemingly simple but in fact a complicated technological process. It basically includes three steps: (1) Cocoon sorting and cocoon stripping; (2) Cocoon cooking; and (3) Reeling.

Cocoon sorting means eliminating the rotten, moldy, and misshapen cocoons, and sorting the good cocoons according to their shape and color. It was in the Zhou dynasty at the latest that people purposefully started sorting cocoons. In *Meaning of Rituals* in *Classic of Rites* is recorded that "women who finished their work of rearing silkworms at the end of spring showed the cocoons to the King and offered them to the Lady". It is also recorded in *Typical Weather in a Given Season* in *Rites of Zhou* that "women are allocated with cocoons to reel them, and their performance is assessed by the weight of their silk, which will be used to make sacrificial garments for sacrifice ceremonies". If you offer cocoons as tribute, you must choose the best ones; When cocoons are reeled into silk, they must reach a certain standard. The above two citations prove that the relationship between the cocoon's quality and the silk's quality as well as the necessity of cocoon sorting were understood at that time. Cocoon stripping means removing the loose cocoon husks that are not suitable for

Picture 4.25 Silkworm
basket

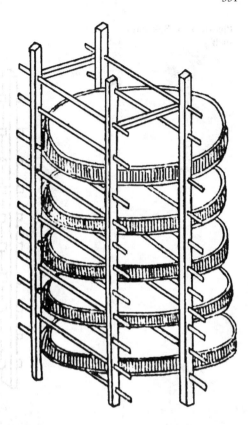

weaving on the surface. The cocoon husks removed can also be used to make silk
floss, as became apparent when silk floss robes were unearthed from the tombs of
the Western Zhou dynasty in Chaoyang county, Liaoning province.

The function of cocoon cooking is to soften sericin to reel off the silk. There are
the consistent warp and weft lines on the silk pieces of 4,000 years ago, which were
unearthed at the Qianshanyang site, Huzhou city, Zhejiang province, and it shows
that silk reeling techniques had already reached a comparatively high level in the
late Neolithic Age. During the Shang and Zhou dynasties, the method of reeling
silk in hot water was popularized. Later, in the Qin and Han dynasties, there were
many literary descriptions about cooking cocoons in "boiling water" as well as the
influence of water temperature on silk reeling, especially on the quality of the silk
thread. This indicates that systematic and complete experience had been accumulated
in this field at that time.

There are roughly two methods of silk reeling through cocoon cooking. One is
called the "hot water pot" reeling method, which means to put the pot with cocoons
in water directly on the stove and reel silk as it is boiled; the other is called the
"cold water basin" method, which means boiling the cocoons in a hot water pot for
a few minutes, then move them into a "water basin" with cold water next to it, and

Picture 4.26 Silkworm
shelf

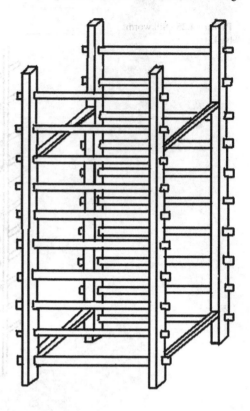

then reel the cocoons in this "cold water basin" (as shown in Picture 4.28). With the
first method, it is more efficient to reel silk and the silk thus reeled is called "fire
silk"; however, the silk reeled with the second method is of good quality, which is
called "water silk". Generally speaking, superior cocoons produce water silk while
secondary cocoons produce fire silk. According to the *Book of Agriculture*, water silk
is "glossy and strong, which is of top grade and is made into brocade, embroidery,
satin and gauze".

In most cases, the water used to cook cocoons should be clear, spring water being
the most preferred, river water second, and well water third. Qili silk, which is a
famous kind of silk, is white and strong. This quality is closely related to the quality
of the local water. According to *Essays on the Emergence of Pillars with Buddhist
Inscriptions* by Zhu Guozhen, "Of the Huzhou silk, Qili silk is the best, which sells
at a higher price than the others". "Qili" is a small village seven miles away from
Nanxun town, Huzhou. In the 20th year of the Daoguang period (1840) in the Qing
dynasty, the *Annals of Nanxun Town* records that "the Xuedang River goes through
the Jili village, south of the town, at the Chuanzhu bay, and the silk reeled in the clear
river water is of lovely luster". The "Jili" mentioned in the annals is "Qili", which
is famous for its silk production. In order to promote their silk, the silk merchants
in Nanxun town refined "Qili (which means seven *li*)" into "Jili", probably because

Picture 4.27 Silkworm nest

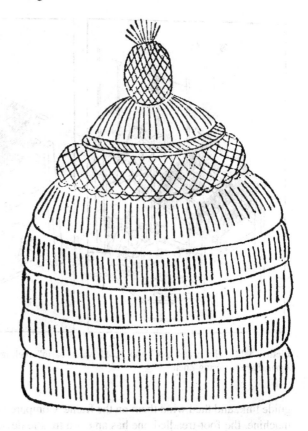

the pronunciation of "seven" in Chinese is similar to that of "ji", and "ji" also means "silk reeling".

The first step of reeling is to pick ends, which means to stir the water in the cocoon basin to make the silk ends float, and to pick them up with wooden chopsticks or small stems of hairy plants to draw out the long filaments; the second step is to clear up the silk ends by removing the thick silk heads and combining several filaments into one thread; the final step is to wind the arranged thread on the reeling machine by passing it through the guide eyes and over the guide link (as shown in Pictures 4.29 and 4.30).

There are two kinds of silk-reeling machines, namely hand-cranked and foot-treadled. The former was popularized in the Tang dynasty at the latest, while the latter, improved on the basis of the former, appeared in the Song dynasty. They are similar in structure, which is mainly comprised of a stove, pot, guide eyes, pulley, rack, guide link, frame, and crank. Among all the parts, the role of the guide eyes is to combine filaments into threads; the pulley is a wire guide, used to remove the nodes on the threads; the rack is a traverse for the silk threads to wind onto the frame in layers. When reeling silk, it is necessary to pass the silk ends of the cocoons in the pot through the guide eyes, place the threads over the pulley, pull them over the rack and

Picture 4.28 "Hot water pot" and "cold water basin" reeling methods, cited from *Book of Agriculture* by Wang Zhen

guide link, and then wind them on the frame. Compared to the hand-cranked reeling machine, the foot-treadled one has an extra treadle device. By using feet instead of hands, silk reeling workers can use both hands to pick up the ends and add threads, thus greatly increasing efficiency.

4.2.1.3 Silk Winding, Doubling, and Throwing

Silk winding. It refers to the technological process of winding silk skeins onto the silk spool. When winding, the silk thread needs the proper tension, so that it is well-formed, easy to unwind, and various defects on the thread can also be removed during the process. Many stone reliefs of the Han dynasty, unearthed in Longyang town, Tengxian county, Shandong province, in 1952, and later in other places, are engraved with characters who are operating silk winding machines, which shows that such devices were a widely used textile tool as early as the Han dynasty.

The winding machine is composed of the *ni*, which is used to hold the skein and the spool. The *ni* consists of four wooden sticks erected on the ground or a wooden frame with a base. When operating it, the wooden frame is placed on the ground, four bamboo sticks are inserted, and the silk is stretched around them. Another small bamboo stick is hung at a higher place beside the bamboo sticks at an angle, with a

Picture 4.29   Silk reeling, cited from *Heavenly Creations*

crescent hook on one end of it, through which the silk thread is passed. The operator rotates the spool in one hand to wind the silk thread, which will be ready to be converted into warp and weft yarns. If some thread breaks off during the process, the hook can be lowered by means of a lever that consists of the small bamboo stick with a suspended stone at the other end to function as a weight (as shown in Picture 4.31).

Silk doubling and throwing. The former refers to the process of drawing more than two threads together according to weaving requirements, and the latter means that the silk receives a twist. Twisting can improve the strength and wear resistance of silk thread, reduce fuzzing or broken ends, increase the elasticity of the fabric, and make the gloss of fabric soft. Usually, doubling and throwing are carried out at the same time, and the machines used are the spindle and the spinning wheel.

Picture 4.30 Contemporary folk silk reeling

The former consists of a disc-shaped whorl and a shaft. It skillfully makes use of the weight of the object itself and the force generated when rotated, so that the messy fibers are drawn, twisted, and thrown into yarn. The outer diameter and weight of the whorl are the keys to determine the fineness of the yarn. If they are larger, the moment of inertia is larger and the spun yarn is thicker; if they are smaller, the spun yarn is finer and more uniform (as shown in Picture 4.32). The appearance of the spindle can be traced back to the late Neolithic Age, and it has a far-reaching influence on spinning tools in later generations. To this day, it is still used in many areas (as shown in Picture 4.33).

There are three types of spinning wheels, namely, hand-cranked, foot-treadled, and large silk-spinning wheels. The first two kinds are usually used in household spinning, and the large kind is used in silk workshops of a certain scale.

It is speculated that the hand-cranked spinning wheel appeared during the Warring States period, and was widely used in the Han dynasty. It generally consists of a wooden frame, a spindle, a string wheel, and a handle. The spindle is made of bamboo or wood, with one end interspersed between the two bars on a small wooden frame on the left, and the other end extending beyond the frame. The top end of one bar is covered with the string cord, and the top end of the other near the spinner is covered with a bamboo or reed tube, which is called the top with the spun silk thread (as shown in Picture 4.34).

Picture 4.31 Winding silk, cited from *Heavenly Creations*

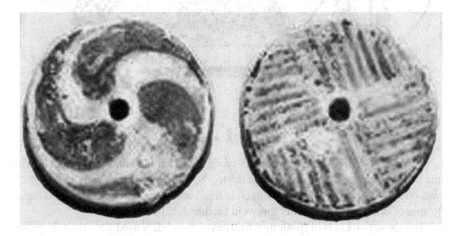

Picture 4.32 Pottery spindles with patterns of Taiji of the Qujialing culture (3300–2600 B.C.)

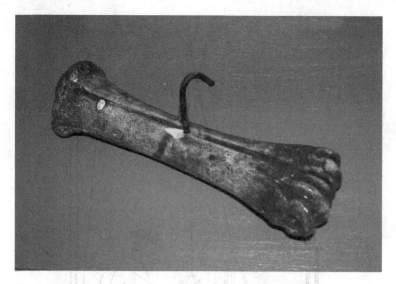

Picture 4.33 Modern folk spindle

Picture 4.34 Hand-cranked spinning wheel

The foot-treadled spinning wheel is made based on the hand-cranked one, with the main difference of an extra treadle device, so that the spinner's right hand, which was originally used to rotate the wheel, is freed and the spinner can use both hands to spin or double the thread (as shown in Picture 4.35).

The above two kinds of spinning wheel have been widely used since their invention because of their simple structure and easy operation. They can still be seen in the countryside in China today (as shown in Pictures 4.36 and 4.37).

Picture 4.35 Three-spindle foot-treadled spinning wheel, cited from *Tao Ying, the Widow of Lu State* in *Biographies of Exemplary Women*

The large silk-spinning wheel is a kind of spinning wheel with dozens of spindles, which was invented in the Yuan dynasty and perfected in the Qing dynasty (as shown in Picture 4.38). It is composed of four parts, namely, the frame, the yarn-output mechanism, the yarn-winding mechanism, and the transmission mechanism. Among them, the yarn-output mechanism consists of bamboo boards, thread-passing bamboo sticks, humidifying grooves, and 50 or 56 spindles. The yarn-winding mechanism consists of a frame, a yarn-guide rod, and yarn-winding rods. The yarn-guide rod is of the same length as the frame, placed under the winding rods, with the role of guiding and positioning the yarn. The transmission mechanism is composed of a driving wheel, a passive wheel, a pulley, and a transmission belt. The driving wheel is located on one side of the frame, and the passive one is on the other side. Both of them rotate counterclockwise, and the frame rotates clockwise. The driving wheel and the frame basically move synchronously, so that the yarn is wound on the frame going through various parts of the yarn-winding mechanism, and the yarn is thrown due to the speed difference between the spindles and the frame.

Picture 4.36 Hand-cranked spinning wheel still used in minority areas in China

4.2.1.4 Weft Winding, Warping, Warp Drawing-In and Sizing

Weft winding. It is also known as top winding, which is carried out on a spinning wheel by winding the silk thread on the spool onto the top tube. When weaving fabric, the top is put into the shuttle.

Warping. The function of warping is to draw warp yarns of specified length and number in parallel on the weaving shaft according to a certain rule, which was called "silk pulling" (*zhensi*) in ancient China, and the tools used were called warp creel, warp frame or pulling frame (*zhenchuang*). There are two types of traditional warping: the warp rack type and the warp creel type. The warp rack consists of two vertical wooden columns and two rungs, each of the columns nailed with 30 small wooden stakes. The spools with silk threads are arranged according to a certain rule, and the threads released from the spools are bundled together with the help of a wooden pole and wound around the small wooden stakes (as shown in Picture 4.39). The warp creel type, first seen in the Southern Song dynasty, consists of the spools, the warp creel, and the round frame. One operator arranges the silk threads on the spools in order, the operator in the middle holds the "combing harness" to smooth the kinks of the warp threads, and the third one rotates the round frame to wind the warp threads (as shown in Picture 4.40).

Picture 4.37 Foot-treadled spinning wheel still used in minority areas in China

Warp drawing-in. The task of this process is to draw the warp yarns through the heddle and harness according to the weaving requirements of the fabric, so as to form the shed during weaving.

Sizing. It means applying glue to the warp yarn, which improves its ability to bear repeated stretching and bending of the heddle and harness after being put on the weaving machine and avoid a large amount of fluffing caused by repeated friction between the yarn and the heddle and harness. In ancient China, sizing warps were called "*guohu*" or "*jiangsha*". For undyed yarn, wheat or japonica flour is used as sizing liquid, and ox glue is used for dyed yarn. The method of sizing is as follows: First, the sizing liquid is evenly coated on the warp silk with a brush; then it is also brushed on the combing harness, and the harness is moved back and forth through the warp silk to size it evenly (as shown in Picture 4.41).

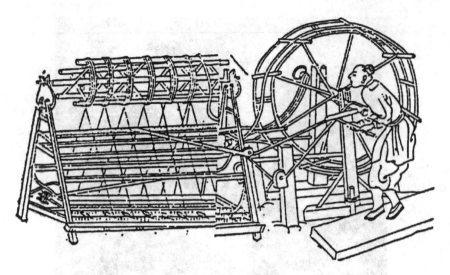

Picture 4.38 Large silk spinning wheel in the Qing dynasty, cited from *Collection of Important Essays on Sericulture*

Picture 4.39 Warping with the warp rack, cited from *Heavenly Creations*

Picture 4.40 Warping with the warp creel, cited from *A Complete Book on Agriculture*

Picture 4.41 The warp rack
for separating and sizing
warp threads, cited from
Heavenly Creations

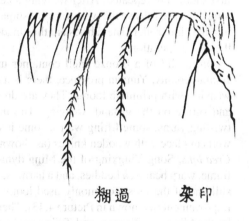

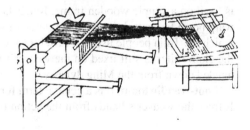

4.2.1.5 Weaving

After the warp beam is placed on the loom, the warp shed on the loom can be controlled and changed according to the texture of the fabrics, and the warp and weft threads are interwoven to make silk fabrics. There are many types of traditional looms, and several representative ones are described in this part.

The backstrap loom. This type of loom can be traced back to the late Neolithic Age. The main components of the primitive one include the front and back crossbars, which are equivalent to the cloth roller and warp beam of modern looms, the beating-up knife, the weft top, the warp-separating rod, and a thinner heddle rod. The warp-separating rod separates the warp yarns into upper and lower layers according to odd and even numbers to form a natural shed. The heddle rod is connected to all the odd or even warp yarns so that they can be lifted or lowered to form another shed. When weaving, the weaver sits on the floor and ties both ends of the warp yarn to the two crossbars, one of which (the cloth roller) is tied around his/her waist and the other is held by his/her feet. The tension of the warp yarn is controlled by the weaver's waist and back, and the weaver then inserts the weft with the top through the shed formed by the warp separating rod and beats up the weft with the knife. When weaving the second shuttle, the heddle rod is lifted to move the lower warp yarn to the upper layer to form the second shed, the knife is set up to fix it, and the actions of weaving the first shuttle are repeated. After weaving a certain length of cloth, the warp beam is turned over to release the warp yarns of some length, and the finished cloth is wound into the cloth roller. The fabrics are thus made inch by inch through the repetition of the above operation.

On the lid of a bronze shell container unearthed from the Shizhaishan site in Jinning county, Yunnan province, there is a cast image of a group of local women weaving with primitive looms. They are dressed in coarse double-breasted clothes and sitting on the ground, weaving. The images are lifelike, with some of them twisting yarns, some lifting warps, some inserting wefts, and some beating up the wefts in place with wooden knives (as shown in Picture 4.42). In his book *Heavenly Creations*, Song Yingxing of the Ming dynasty describes a backstrap loom with a frame, warp beams, a heddles, and a harness, which shows that this type of loom was still one of the most commonly used looms in the Ming dynasty after continuous improvement (as shown in Picture 4.43). There are two types of backstrap loom used by the Miao, Li, Zhuang, and Tujia people. One is the type with the weaver's feet fixing the warp beam and it still maintains the shape of the primitive one (as shown in Picture 4.44); there is also a backstrap loom with a suspension warp beam, which is fixed on a simple wooden frame. Ratchets are often provided at both ends of the warp beam to control the gradual release of warp yarns. The front end of the loom, which the Tujia people use to weave brocade, has evolved into a wooden frame, but the cloth roller is still fixed on the weaver's waist. It is clearly the backstrap loom handed down from the Ming dynasty.

Foot-treadle loom. It is a general term for looms with the treadle-heddle device. It frees the weaver's hands from the action of lifting the heddle, so they are solely

Picture 4.42 A bronze shell container from the Han dynasty unearthed from the Shizhaishan site in Jinning county, Yunnan province

engaged in shuttle throwing and beating up the wefts, thus greatly improving efficiency. This type of loom is definitely a major invention in the history of loom development.

Based on research pertaining to the subject, scholars speculate that the invention of the foot-treadle loom can be traced back to the Warring States period at the latest. For example, according to a story in *Questions of Tang* in *Lie Zi*, Ji Chang learned archery by "lying down under his wife's loom and gazing at the shuttle attentively"; it is also recorded that the cloth given as gifts between dukes and princes during the Warring States period was 100 times more than that in the Spring and Autumn period; what's more, many stone reliefs of the Han dynasty engraved with foot-treadle looms have been unearthed in various places in recent years. In the Qin and Han dynasties, the foot-treadle loom was widely used in vast areas of the Yellow River Basin and the Yangtze River Basin. With its invention, the productivity of plain fabrics increased by 20–60 times compared with the primitive backstrap loom, making it possible for each weaver to weave 0.3–1 m of cloth per hour.

There are three types of foot-treadle loom: double-treadle-single-heddle, single-treadle-double-heddle, and double-treadle-double-heddle. The main components are the frame, the horse head (or *ya'ermu,* meaning crow wood), the treadle, *Fu, Teng,* the heddle, and the harness, among which the heddle, the treadle, and the horse head (or crow wood) are the heddle-lifting device. The harness is to control warp density, cloth width, and beating-up of the wefts. There are two ways to install the harness on the loom. One is to connect the bamboo harness to the oscillating rod and beat up the weft with the weight of the rod itself; another way is to hang the harness under two bent bamboo sticks with strings and beat up the weft with the help of the elasticity of the sticks.

Picture 4.43 Backstrap
loom, cited from *Heavenly
Creations*

There are two treadles to control the lifting of one heddle on a double-treadle-single-heddle loom, and the heddle only plays the role of lifting warp. The slant looms on the stone reliefs in the Han dynasty belong to this type (as shown in Picture 4.45). The upright posts on the left and right sides of the frame are respectively equipped with a horse head for lifting the heddle. The front end of the horse head is tied with the heddle, and the middle and rear ends are equipped with two crossbars, with the middle one being the central beam to pressure warp and the rear one to open the shed. One of the treadles under the loom base is connected with the heddle-lifting rod, which is then connected with the horse head; the other is connected with the lower end of the heddle. When the treadle connected with the heddle-lifting rod is stepped down, the horse head leans forward and upwards, the heddle lifts the ground warp, and the middle beam presses the surface warp downward, thus forming a triangular shed. When the treadle connected with the heddle is stepped down, the ground warp also drops and regains the initial shed shape with the surface warp.

Picture 4.44 Weaving with a backstrap loom

For the single-treadle-double-heddle loom (as shown in Picture 4.46), one-foot treadle controls the lifting of one heddle, and the upright posts on the left and right sides of the frame are respectively equipped with a *ya'ermu*. Its front end is tied with the heddle, and the back end is connected with the treadle. The crossbar connecting the back ends of the two *ya'ermu* is actually the warp-pressing rod. When the treadle is stepped down, the *ya'ermu* tilts up and lifts the ground warp, and the warp-pressing rod lowers and presses down the surface warp, thus forming a shed. When the treadle is released, the *ya'ermu* and the heddle restore their normal positions by their own weight, and the shed also restores its normal shape.

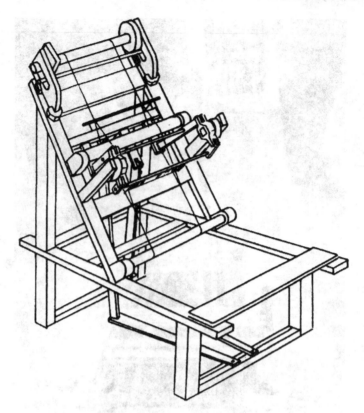

Picture 4.45 Foot-treadle slant loom in the Han dynasty

For the double-treadle-double-heddle loom, two treadles control the lifting of two heddles respectively, and either heddle has the functions of lifting and pressing warps in turn. There are two ways to connect the treadles with the heddles. One is that the two treadles are respectively connected with one end of the two levers on the frame, and the other end of the levers is connected with the upper part of the two heddles, respectively. This type of loom is recorded both in *Painting of Sericulture and Weaving* by Liang Kai in the Southern Song dynasty and *Book of Agriculture* by Wang Zhen of the Yuan dynasty. The other way is that the two treadles are respectively connected with the lower end of the two heddles, the upper end of which are respectively connected with the two ends of a lever above the frame. When one of the two treadles is stepped down, the connected heddle is lowered, while the other one is lifted by the lever, thus forming a shed; and it is the same with the other treadle. The silk-weaving loom recorded in the *Collection of Important Essays on Sericulture* by Wei Jie of the Qing dynasty belongs to this type (as shown in

Picture 4.46 Single-treadle-double-heddle loom (The picture was originally stored in Paris)

Picture 4.47). The traditional wood loom still used by the Bouyei people in China is similar (as shown in Picture 4.48).

The multi-heddle-treadle loom, invented during the Warring States period, is used to weave fabrics with complex geometric patterns. It is characterized by the same number of treadles and heddles, with one heddle controlled by one treadle. The number of heddles and treadles can be increased or decreased as needed.

The *Dingqiao* (bridge piers) loom, which was once used in Shuangliu county, Chengdu city, Sichuan province, is a typical one. It is named this way because the treadles are full of bamboo nails, which makes it look like the piers of a bridge. Its structure, shown in Picture 4.49, includes the frame composed of numbers 1–9, the shedding composed of numbers 10–22, the weaving harness composed of numbers 23–28, the warp beam composed of numbers 29–33, the warp-separating rod number 34, the cloth roller composed of numbers 35–39 and the seat number 40. There are

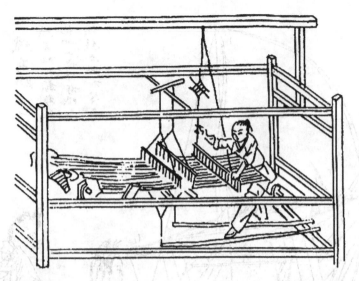

Picture 4.47 Double-treadle-double-heddle loom, cited from *Collection of Important Essays on Sericulture*

Picture 4.48 A Bouyei woman weaving

two kinds of heddles on this type of loom. The 1–8 heddles in front of the frame are called ground heddles or *zhanzi*, which are specially used to control the ground warp. The shedding is formed as shown in Picture 4.50. Step down on the treadle, which pulls the lower rim of the *zhanzi* down through the crossbar, so that the warp threads lower as well; when the treadle is released, the *zhanzi* restores its normal position by the elastic force of the bow on the top of the loom, and the warp threads also return to their normal position. Beside the ground heddles, the rest of the heddles on the loom are figure heddles, also known as *fanzi*, which mainly control the movement of the figure warps. The shedding is formed as shown in Picture 4.51. Step down on the treadle, the *ya'ermu* pulls up the *fanzi*, which lifts the warp threads; when the treadle is released, the heddle restores its normal position by its own weight and the tension of the warps. Since there are many treadles, the positions of nails on adjacent treadles are different so as to avoid stepping mistakenly on the adjacent one and affecting the correct movement of the heddle. Generally, every few treadles are installed with the same pattern of nails. In addition, when weaving *ling* (thin silk) with figures, it is inconvenient to operate if the treadles controlling the figure heddles and those controlling the ground heddles are arranged separately since the width is too large. In this case, the treadles controlling the ground heddles can be placed in the center of those controlling the figure heddles. When the weaver steps down on the left side, the left foot controls the figure heddles, and the right controls the ground ones; when he/she steps down on the right side, the right foot controls the figure heddles, and the left controls the ground ones.

With this type of loom, weavers can weave dozens of kinds of lace patterns such as phoenix eyes, tides, dotted flowers, and ice plums as well as over a dozen *ling* with figures and pattern brocade such as water waves, swastikas (Buddhist symbol), turtles, and sweet-scented osmanthus. The patterns are usually across the whole width of the fabric, with different lengths, but not more than a few centimeters. The number of heddles and treadles depends on the complexity of the fabric patterns. For example, 32 heddles and 32 treadles are used to weave the lace pattern of "five plum blossoms". A skillful weaver can pick the shuttle 110 times per minute on average when inserting wefts, and an average weaver 80–100 times per minute.

Hualou drawloom (figure tower drawloom) is a kind of loom for weaving complex patterns and large repeated patterns, and its most important feature is that the figure drawing information is stored by figure designs. The drawing warp is controlled by thread heddles instead of frame heddles. The number of thread heddles equals that of the drawing warps, and the thread heddles moving in the same direction are tied together and hung on *hualou*. It represents the highest level of drawing technology in ancient China.

According to the unearthed cultural relics and literature records, the *hualou* drawloom took shape in the Han dynasty and was quite mature in the Song and Yuan dynasties after continuous improvement in the Six dynasties, Sui and Tang dynasties. There are mainly two types, namely, the small and the large *hualou* drawloom. Their main differences are that: Weavers can weave various large and complex patterns on the large type, but only relatively simple patterns on the small type; the former has more than 2,000 drawing threads, while the latter has only about 1,000; the figure

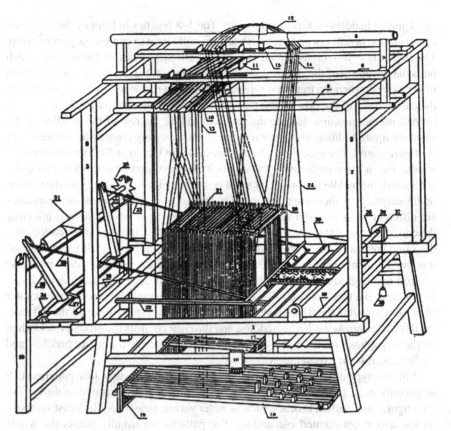

Picture 4.49 Schematic diagram of a *Dingqiao* loom

designs of the large one can only hang around the loom because they are quite large, while those of the latter one only need to hang upright separately.

The structure and operation method of the small *hualou* drawloom, as described in *Heavenly Creations* (as shown in Picture 4.52), is that: The loom is made up of two parts, with a total length of 1.6 *zhang*, which is about 5.3 m. The high part of it is called the *hualou* or figure tower. In the middle is the *qupan* (drawer board) and the *qujiao* (rigid rods) hang below. There are altogether 1,800 *qujiao*, which are bamboo rods polished with water. Dig a pit about two *chi* (67 cm) deep under the *hualou* to hold the *qujiao*. If the ground is damp, a two-*chi*-high frame is built to replace the pit. The thread puller sits on the wooden frame of the *hualou*. At the back end of the drawloom, the warp beam winds the silk, and two wooden poles, called driving shafts, in the center of the loom vertically connect two beams, about four *chi* long, with the ends plugging into the two ends of the harness. To increase the force of the driving shafts, one part of the loom lies level, while the part from the *hualou* to the weaver inclines down about one *chi*. The driving shafts for weaving the gauze should be more than 10 *jin* (5 kg) lighter than the one used for weaving thin silk and tabby

Picture 4.50 Schematic
diagram of shedding formed
by *zhanzi*. 1. bow, 2. *zhanzi*,
3. ground warp, 4. crossbar,
5. *dingqiao*

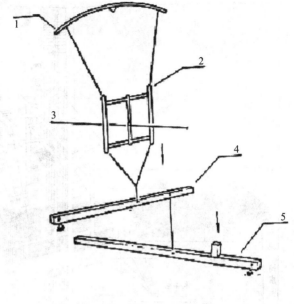

Picture 4.51 Schematic
diagram of shedding formed
by *fanzi*. 1. *ya'ermu*, 2. *fanzi*,
3. figure warp, 4. *dingqiao*

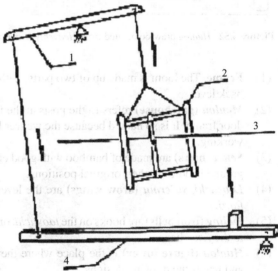

silk. For weaving soft thin gauze or thin silk and tabby silk showing small designs
such as water waves and plum blossoms, only two more heddles need to be added
to the lighter loom used for weaving plain gauze. In this case, one weaver operating
the loom with treadles is enough to complete the task, and no thread puller is needed
to man the *hualou*, nor *qupan* and *qujiao* need to be installed.

The shape and function of the main components of a small *hualou* drawloom are
as follows:

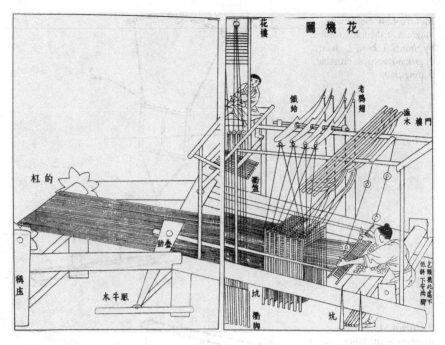

Picture 4.52 *Hualou* drawloom, cited from *Heavenly Creations*

(1) **Frame**. The loom is made up of two parts, with the front part inclined and the back level.

(2) ***Menlou* (gatehouse)** refers to the posts at the front of the loom shaped like a door frame. It is so named because the weaver has to enter the *menlou* to start working.

(3) ***Semu* (bows)** are made of bamboo with good elasticity and used to restore the ground heddles to their original positions.

(4) ***Laoyachi, ya'ermu* (crow wings)** are the levers for lifting and lowering the *fanzi*.

(5) ***Tieling* (iron bells)** are hooks on the *laoyachi*, onto which the strings connected to the *fanzi* hang.

(6) ***Hualou* (figure tower)** is the place where the thread puller lifts the threads and hangs the drawing designs.

(7) ***Tongsi* (thread beams)**, also known as fiber beams, can lift or lower the warp thread separately. Its number depends on the number of pattern cycles, and under each beam are hung 2–7 *quxian*.

(8) ***Quxian* (drawing threads)** are hanging under the *tongsi*, with a heddle eye on each of it. All the lifted warp threads pass through the heddle eyes, with only one thread passing through one eye.

(9) *Qupan* (**drawer board**) is made of arranged bamboo poles. It is placed at the upper part of the *quxian* with functions to prevent them from entangling with each other.

(10) *Qujiao* (**rigid rods**), with their fixed length and weight, mainly help the *tongsi* hang vertically and remain stable.

(11) **Pit** is where the *qujiao* and bamboo ends are hidden.

(12) **Driving shafts**, also known as the stands, are part of the beating-up mechanism of the loom. They push the striking bar and the harness frame to beat up the wefts by relying on the inclined gravity when they are swinging.

(13) *Mianniumu* (**sleeping ox rods**) are the bases for the driving shafts, with several grooves on either of them to adjust the standing point of the shaft according to the force needed for beating-up of different fabrics.

(14) The teeth at both ends of the *Digang* (**warp beam**) are the control mechanism for letting off warps.

(15) *Chengzhuang* (**supporting posts**) are where the warp beam is placed.

Below are the structure and operation methods of the large *hualou* drawloom, with Nanjing *zhuanghua* brocade drawloom as the representative model (as shown in Picture 4.53): There are five parts for this type of drawloom, namely, the frame or the supporting part, the *hualou* or the thread-drawing mechanism, the shedding mechanism, the weft-beating-up mechanism and the warp-letting-off mechanism. The overall structure and names of each part are shown in Picture 4.54.

Numbers 1–6 are parts of the warp-letting-off mechanism, and their corresponding names are *paiyan* (parallel swallows), *tuoni* (base stretchers), *panguan* (judges), *digang* (warp beam), *yangjiao* (cleats) and *dajiaofang* (angle butts).

Numbers 7–18 are parts of the supporting frame, and the corresponding names are *jishen* (frame bars), *jitui* (frame legs), *gounao* (dog brain), *jutou* (bar ends), *menlouzhu* (posts of the gatehouse), *toudao louzhu* (the first house posts), *erdao louzhu* (the second house posts), *menlouliang* (arch over the gatehouse), *toudao huolunquan* (the first fire ring), *erdao huolunquan* (the second fire ring), *yanchi* (swallow wings) and *bazicheng* (splayed stay bars).

Numbers 19–24 are parts of the thread drawing mechanism, and the corresponding names are *chongtianzhu* (towering posts), *chongtaingai* (towering beam), *huaji* (figure bar), *huaben* (figure adze), *qianjintong* (jack tube) and *jiahuazhu* (figure hanging bamboo poles).

Numbers 25–34 are parts of shedding mechanism, and the corresponding names are *liren* (stands), *zhuanggan* (striking bar), *lirenxiao* (stand bolts), *guilian* (ghost face), *zhuangjishi* (bump stone), *lirenpan* (stand bases), *koukuang* (harness frame), *niuyanjing* (ox eyes), *jiangjunzhu* (general posts) and *gaoyaban* (high pressure board).

Numbers 35–44 are parts of weft-beating-up mechanism, and the corresponding names are *sanjialiang* (three beams), *yazizui* (duck mouth), *gongpeng* (bow), *zhangzi* or *zhanzi, fanzi, yinggejia* (parrot rack), *chengqiangduo* (battlements), *yingge* (parrot), treadles, and crossbars.

Picture 4.53 Large *hualou zhuanghua* brocade drawloom

Picture 4.55 is the schematic diagram of heddle lifting and thread drawing, the process of which is as follows: After the installation of silk threads, the weaver and the thread puller take their places on the loom. The latter first lifts an *erzi* thread (ear thread) at the conjunction of the *jiaozi* threads (foot threads) and the *xian* threads (fiber threads) according to the arrangement order of the figure design and separates the *jiaozi* threads in the upper part with one hand holding the separated *jiaozi* threads and *xian* threads, and the other hand tidying the remaining *jiaozi* and *xian* threads. The puller raises his/her hand holding the *jiaozi* and *xian* threads, which pull up the warp threads, forming a shed. After the weaver inserts the weft, the thread puller puts down the remaining threads and pulls the ear thread out from the upper *mokou*, and passes it to the lower *mokou*, thus crossing the conjunction of the *jiaozi* threads and the *xian* threads. The same process being repeated, the whole figure design is thus shifted from the upper *mokou* to the lower one. The *jiaozi* threads butt into rings, so the figure design can return to the upper *mokou* for another round of drawing

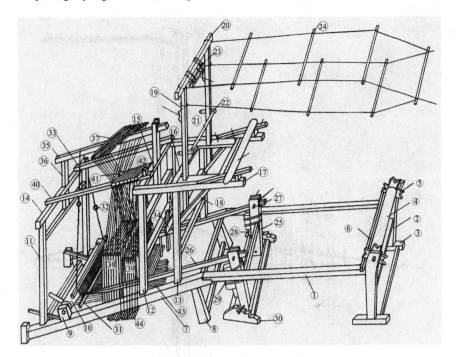

Picture 4.54 Structural diagram of the *zhuanghua* brocade drawloom

by turning the *jiaozi* thread rings. If it is to weave symmetrical patterns from top to bottom, the two ends of the *jiaozi* threads can be bundled separately instead of being butted. After the threads on the figure design are all drawn, reverse the order of thread drawing and turn the figure design back from the lower *mokou* to the upper one.

Making figure designs for weaving. Weaving patterned fabrics with the drawloom is indispensable to the figure design, which is a bridge connecting patterns on drawings to those on fabrics, as well as the basis for lifting and lowering figure heddles. It is like a memory, in which all the programs for warp movements are stored. The weaver can weave the expected patterns as long as the order of the ear threads on the figure design is lifted in turn. Making figure designs is a special technique, which requires very high skills. It became an independent profession at an early age in China. According to *Heavenly Creations*, "The artisan who makes figure designs for weaving has to be the most ingenious. A painter first paints the patterns and colors of a fabric design on a piece of paper. The artisan exams the fabric design, and then figures out how to make it without any mistake. The design is hung on the *hualou* of the drawloom. Even if the artisan doesn't know what the figure will look like, it will appear at last as long as the design is followed. The appearance of the figure on thin silk and tabby silk depends on the warps, while the appearance of the figure on yarn and gauze depends on the wefts. In weaving the former two kinds of fabrics, the warp is lifted at every pick of the shuttle, but in the case of the latter two kinds,

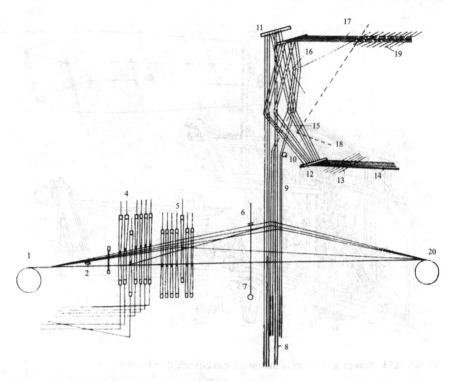

Picture 4.55 Schematic diagram of heddle lifting and thread drawing. 1. cloth roller 2. shuttle 3. harness 4. *zhangzi* (connected with the bow) 5. *fanzi* (connected with the *yingge*) 6. thread pressing board 7. weight stone 8. *yajiao* 9. *xian threads* (fiber threads) 10. *huaji* (figure bar) 11. *qianjintong* (jack tube) 12. figure design hanging tube 13. figure design 14. *jiaozi* threads (foot threads) 15. lower *mokou* 16. upper *mokou* 17. *erzi* thread (ear thread) 18. the *erzi* thread passed to the lower *mokou* after drawing 19. figure design 20. warp beam

it is lifted only every other pick. The weaving skills of Vega (the legendary weaver girl in ancient China) in heaven have been mastered and embodied by the skillful artisans here in China".

To make a figure design, it's necessary to make the notation first. The process, which is called "making figures on profile-paper", is detailed in the *Collection of Important Essays on Sericulture* by Wei Jie in the Qing dynasty. "To draw out the figure, five pieces of paper should be used. Firstly, with a new figure in mind, draw it on a piece of paper; secondly, refine the figure by drawing it on another; thirdly, ask a good artist to copy the figure; fourthly, stick the figure on a piece of well-fitted paper; fifthly, make a rubbing of the figure on a thin, bright and fine piece of paper, and draw horizontal and vertical lines to form grids on it. Brush it with water mixed with lead powder to prevent the paper from hurting the eyes. Mark different numbers on the paper with pigments such as red and green after the powder dries, for future use. One grid on the paper is called one piece of empty space, the number and size of which vary. The number of horizontal grids is that of the shuttle insertions, which decides the making of figure design". The horizontal grids mentioned in the passage represent wefts, and the vertical ones represent warps, that is, the grids represent the arrangement order of interwoven warps and wefts.

There are three methods of making figure designs, namely, picking the figure design, duplicating the figure design, and joining the figure designs, which can be used alone or together depending on the figure.

Picking the figure design means picking the patterns with silk or cotton threads according to the arrangement order of interwoven warps and wefts, i.e. picking warps when the figure is woven on the warp and picking wefts when the figure is woven on the weft. The threads for picking warps are called the *jiaozi* threads (foot threads), and those for picking wefts are called the *erzi* thread (ear threads). The number of foot threads is the same as that of warp threads in the same figure, but the length varies with the colors and length of the figure, which is generally about 10 times that of the figure. If there are quite a few colors, the length should be properly added. The number of the ear threads is based on the weft density multiplied by the length of one figure alone and the numbers of colors, that is, the number of wefts in one single figure is the same as that of the ear threads, the length of which is generally over six *chi* (2 m).

Duplicating the figure design is a labor-saving method when the figure repeats symmetrically. It means duplicating a symmetrical and identical figure design of the partially finished one. For example, when picking the opposite or quadruple patterns, the artisan only needs to pick one unitary figure design, duplicate one or several more, and then combine them into a complete figure design.

The method of joining the figure designs is used when there are too many warp threads to be hung on the frame when picking large-area figures. It comprises the following steps: first, the foot threads of the upper and lower figure designs are juxtaposed, and secondly, the ear threads of the lower figure design are replaced by those of the upper one. During this process, attention should be paid to the arrangement order of the foot threads and ear threads, which must be exactly the same. If there is an error, it will cause wrong figures after the threads are drawn.

Figure designs should be made on the figure picking frame, which is also introduced in *Collection of Important Essays on Sericulture*. "The frame is made of four hard fine wood sticks, two long and two short, about 1.5 *cun* (5 cm) wide and 2 *cun* (about 6.7 cm) thick. The two long sticks are chiseled at both ends, connected by the two short sticks, with the frame 3.6 *chi* (1.2 m) long and 1.8 *chi* (0.6 m) wide. Ten round bamboo nails are nailed on the upper horizontal sticks to hang the *xian* threads (fiber threads), and ten round eyes are drilled on the lower horizontal stick for the fiber threads to pass. A horizontal crossbar is made next to the left long stick, and a bamboo pole is used to hang the threads. Two harness frames are put near it to separate them".

The figure design is used to store the thread-drawing information and control the thread heddles, and it pioneers the control of warp movement with arranged programs, which has had a far-reaching impact on the textile industry worldwide. In 1725, inspired by the Chinese drawloom, French engineer Bouchon replaced the figure design with the "perforated paper tape" to control the weaving movement of the

Picture 4.56 The jacquard device on a western loom in the eighteenth century

drawloom. Around 1801, Joseph Marie Jacquard (1752–1834), a French engineer, designed and replaced the perforated paper tape with punched tape, and completed the design and manufacturing of the "automatic jacquard loom", named after its inventor. The invention of the jacquard loom preluded the automation of the drawloom. In 1888, H. Hollerrith (1860–1929), a German-American, invented the automatic data-processing machine inspired by the method of storing data on the punched tape. His invention made it possible to analyze a large amount of data unimaginable before, and was considered the beginning of modern computer-data processing technology. Therefore, it can be said that "making figure designs" is a typical example of the internal connection between traditional skills and modern science and technology (as shown in Picture 4.56).

4.2.1.6　Silk Scouring

Although some symbionts and impurities can be removed from the silk fiber during silk reeling, it is not very thorough. If such fiber is directly used for weaving and dyeing, its good textile characteristics can't be displayed in many cases, and the color fastness and brightness are not very satisfactory, so further scouring is needed. If not scoured, the reeled silk fiber feels rough and is called raw silk. The silk made with it is called raw silk fabric. Raw silk and raw silk fabric are unlike the soft and smooth silk we associate with "silk". Only after the raw silk fiber is scoured can the sericin and impurities be removed from it. Then after dyeing, the scoured silk fiber is called "boiled-off" and will demonstrate its attracting qualities such as lightness, softness, smoothness, elegance and colorfulness.

Traditional silk-scouring methods include water scouring, ash scouring, enzyme scouring and pounding scouring.

The technological principle of water scouring is as follows: Firstly, the ultraviolet radiation of sunlight will melt the sericin and degrade the pigment, thus degumming and bleaching it; secondly, the residual pigment, sericin and other impurities in silk fiber are precipitated and dissolve in water with the help of the temperature difference between day and night, more specifically, the effect of expansion with heat and contraction with cold caused by repeated alternation of sunlight and water washing. The reason for washing the fiber with well water is that the minerals and microbial activities in it are relatively stable, which is conducive to the decomposition of impurities such as sericin and pigment, and is also conducive to improving its whiteness and purity.

Ash scouring refers to adding alkaline ash burnt from shells such as oysters and clams to the scouring solution. The whole process includes three steps, namely, scouring it twice with ash solution and then once with water. To be more specific, the silk fiber is first scoured in the neem ash water with higher concentration and stronger alkalinity to rapidly expand and dissolve sericin, and then soak it in oyster ash water with a lower concentration and weaker alkalinity. The second step is to wash the sericin and make the alkali agent penetrate into the silk fiber evenly for degumming. Finally, water scouring is used to reduce the effect of the alkali agent so as to degum it gradually (this step has the dual functions of refining and washing). The whole process is like going from violent to medium strong and then to mild states, thus avoiding the phenomenon that some parts of the silk fiber are boiled-off while some other parts of it are still raw due to uneven degumming.

Enzyme scouring, invented in the Tang dynasty, uses biological enzymes as scouring agents, most of which are synthesized from porcine pancreas and plant ash. Porcine pancreas contains a lot of protease, which has a low intensification ability and strong specificity after hydrolysis. Sericin is unstable to protease and easily decomposes. Generally speaking, a high degumming rate at room temperature can be achieved with this method without damaging the fibers. The degumming and degreasing effect of plant ash comes from the large amount of potassium carbonate it contains, which can expand gum and saponify oil, thus dissolving them.

Pounding scouring, which first appeared in the Han dynasty, has become one of the main ways to scour silks since the Wei and Jin dynasties. It is vividly described by Cao Pi of the Eastern Jin dynasty in his poem named *Listening to Someone Pounding Cloth at Night*, "Slender hands fold light silks; ringing mallets rap resonant fulling stones". During the process of pounding, special attention should always be paid to the silk fiber to examine if it is boiled off; otherwise, the folds will break easily. There is a painting named *Court Ladies Preparing Newly-woven Silk* in the Museum of Fine Arts in Boston, USA, which is a facsimile of Zhang Xuan's original work by Zhao Ji (1082–1135), dating back to Emperor Huizong of the Northern Song dynasty. It vividly reproduces the scene of women pounding scouring silks and the

Picture 4.57 *Court Ladies Preparing Newly-woven Silk*, painted by Zhuang Xuan

pounding tools used in the Tang dynasty. In the painting, there is a rectangular stone anvil block, on which some raw silk fiber tied with strings is placed, with four women beside it. Two of them are holding wooden pestles for pounding, and the other two are helping them (as shown in Picture 4.57).

4.2.2 Primary Processing of Cotton, Hemp and Wool Fibers

For a long time, cotton, hemp, wool and silk have been the most used textile fibers in China. Even nowadays, with the great development of chemical synthetic fibers, these four fibers still occupy a considerable proportion in textile production. As far as the whole textile technology is concerned, no matter what kind of fiber it is, the spinning, weaving, printing and dyeing processes are roughly similar, with the only difference lying in how the raw materials are processed. Taking cotton weaving as an example, its technological process is as follows:

Primary processing of cotton fiber—cotton reeling—thread winding, doubling and throwing—weft winding, warping, drawing-in and warp sizing—weaving—printing and dyeing.

The primary processing of cotton, hemp and wool fibers will be briefly introduced in the following part.

4.2.2.1 Primary Processing of Cotton Fibers

Although it was widely used much later than silk and hemp in China, cotton quickly replaced kudzu and hemp, and became a bulk textile raw material as important as silk after the Song and Yuan dynasties due to its excellent textile properties.

The primary processing of cotton fiber includes three processes, namely, cotton ginning, cotton fluffing and rolling, with the aim of separating cottonseed and loosening cotton fiber. Cotton containing cottonseeds, called seed cotton, cannot be directly used for weaving. Only after the seeds are removed can it be spun and woven.

Cotton ginning was called "*gan*" in ancient China. Cotton was peeled by hand initially and other methods began to be used in the Song dynasty, such as "grinding off seeds with iron bars" and "driving out seeds with iron sticks". "*Jiaoche*", a cotton ginning cart also called "*tache*" or "*yache*", was invented in the middle of the Yuan dynasty. Its work method is to continuously put cotton between two wooden cylinders rotating in opposite directions, thus squeezing out the seeds (as shown in Picture 4.58).

The purpose of **cotton fluffing** or bowing is to loosen cotton fibers. Before the Yuan dynasty, it was inefficient to fluff cotton with a small string bamboo bow, and the fiber produced was poor in quality. The wooden cotton bow was invented in the middle of the Yuan dynasty, with the thin string cord replaced by a thick rope cord, which increased its elasticity and efficiency. At the end of the Yuan dynasty, the "hammer bow" was invented. The hammer is made of rosewood or mahogany, with a small and a big end. When it hits the bowstring, the cotton will bounce up in all directions and the fibers will be loosened (as shown in Picture 4.59). Such kind of cotton bow is still in use now.

Picture 4.58 Cotton grinning, cited from *Heavenly Creations*

Picture 4.59 Cotton
fluffing, cited from *Heavenly
Creations*

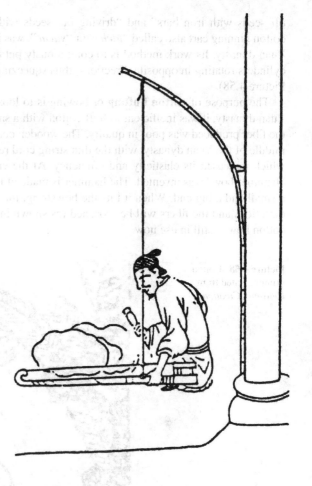

Cotton rolling, also known as cotton fiber straightening and rolling, is equivalent to modern carding into slivers, and this process is to make the loose fibers of the fluffed cotton take on the shape of cylindrical strips, so that they can be continuously and smoothly led out of the slivers during spinning. There's no need to go through this process when spinning with the spindle or cotton twisting shaft, and the loosened cotton can be spun directly on them. But it is necessary to roll the fluffed cotton before spinning on the spinning wheel, because the spindle rotates so fast that there's not much time to handle the cotton, thus making it difficult to keep the yarn even. The main tool used is a small knotless bamboo stick, which rolls the fluffed cotton on the table, and is taken out of the slivers can afterwards (see Picture 4.60).

Picture 4.60 Straightening
and rolling cotton fibers,
cited from *Heavenly
Creations*

4.2.2.2   Primary Processing of Hemp Fibers

Hemp weaving refers to weaving cloth fabrics with phloem fibers of hemp, which includes many kinds such as hemp, ramie, abutilon, kudzu, flax and jute. Among them, hemp, ramie and kudzu have been used for weaving since the Neolithic Age, with hemp and ramie being the most important plant fiber textile materials before the Yuan dynasty. Since then, hemp weaving has still been an important part of textile production even after cotton became the main textile material.

The phloem of hemp plants is composed of cellulose, lignin, gum and other impurities. The process of separating and extracting the spinnable fibers from impurities around them to make them ready for spinning is called "degumming" in modern textile processing. Traditional methods include retting, boiling-off and ash treatment.

The method of retting is to put the stem bark of plants into water for retting, so that the fibers are exposed in bundles. During the process of retting, various carbohydrates are decomposed from the stem bark, which become nutrients for microbial reproduction. Microorganisms secrete a large number of biological enzymes during growth, and gradually decompose some hemicellulose and gum, whose structure is much looser than cellulose, so that fibers are exposed and separated in bundles. This kind of method to make plant fibers in ancient China is still in use today. The process is usually carried out in a retting pond, and it is better to fill it with clear water to maintain the original color of the fiber.

The method of boiling-off means soaking the plant stems and branches or peeled stem bark in water and boil them, so that the gum is gradually removed and dissolved in water. Then fish them out and gently beat them with a wooden stick to obtain the dispersed fibers. Compared with retting, the water temperature and time of boiling are easy to control, which effectively controls the degree of degumming. Phloem plants with short fibers only need "semi-degumming", so it is more suitable to apply this boiling-off method.

Ash treatment method is roughly the same as refinement in modern hemp scouring technology, which is to twist semi-degummed fibers into yarns, and then soak or boil them in alkaline solution to make the gum fall off as much as possible, so that the fibers are finer, softer and whiter, and can be woven into higher-grade fabrics. The "ash" for making alkaline solution includes mulberry wood ash, millet stalk ash, lime and so on.

In addition to degumming, there are more processes needed to obtain spinnable fibers from plant stem bark, such as peeling, scraping and splitting. For the retting method, degumming is the first step, and followed by peeling, scraping and splitting, while for boiling-off, the sequential steps are peeling, scraping, degumming and splitting. The bark of hemp, ramie and kudzu is peeled from tips to roots with bamboo knives or iron knives, and the tendon-like material under the bark is then boiled. To scratch hemp in winter, it is easier when it is wet with warm water. Splitting means splitting the stem bark or degummed fiber bundles by hand into as fine fiber strips as possible for spinning.

4.2.2.3 Primary Processing of Wool Fibers

The purpose of primary processing of wool fibers is to remove impurities, loosen and card the fiber, and make it suitable for spinning or being processed into other products. Taking wool as an example, there are generally three processes including wool taking, wool cleaning and fluffing.

Wool is usually taken directly from sheep or goats with scissors. Sheep can have their wool sheared three times a year and goats once a year. According to Volume 6 of *Essential Techniques for the Peasantry*, "In March every year, white sheep feed on spring grass and they should be sheared as soon as the new wool grows. After that, they should be washed clean in rivers so that the future wool will be white. Again, in May when the sheep wool is about to drop, shear them again and wash them as before. Then at the beginning of August, shear the sheep again before the cockleburs are ripe, and wash them just as before". However, goats are not cold-resistant, so they can only "be sheared at the end of April and the beginning of May. If sheared too early, they will freeze to death". It is preferable to shear the sheep for the third time before the beginning of August, because many plants that easily stick to wool have not yet grown at that time, and the sheared wool is clean and easy to process. If "the sheep are sheared when the cockleburs are ripe", it will be "difficult to process the wool with them"; what's more, "the new wool on the sheep will be too short to protect themselves when they are sheared later than August". In the cold areas of *Mobei* (Gobi Desert) in China, sheep can only be sheared twice a year because "if they are sheared in August, they wouldn't be able to resist the cold".

There are two ways to gather cashmere; one is combing and the other is plucking. Combing cashmere means combing down the coarser cashmere that has been taken off or will be taken off with a bamboo comb, while plucking means plucking cashmere directly from goats by hand. According to the part of "Woolens and Felts" in *Clothing Materials* in *Heavenly Creations*, both the methods came from northwest China and were not introduced to the Central Plains until the Tang dynasty. Woolens made of the cashmere obtained with the method of combing are also called "*hezi*" and "*bazi*", which are coarse, while those made of cashmere obtained by plucking are as soft and smooth as silks. The production efficiency of plucking cashmere is very low. One person works a whole day, only to pluck enough cashmere to make thread weighing one *qian* (five grams), and it takes them half a year to gather enough material for making a bolt of cloth. However, the amount of cashmere obtained by combing is several times that of plucking every day.

The quality of wool is affected by the health of the sheep. The wool taken in fall is worse than that in spring because the sheep lack fodder in winter while they are well fed by abundant grass in spring and summer.

Wool cleaning refers to the removal of grease and impurities attached to the wool, and the method of scrubbing it in hot water is often used. In modern times, ethnic minorities in northwest and southwest China first rub wool with yellow sand to remove grease, and then pick out impurities.

Wool fluffing means fluffing dry wool into single loose fibers. The traditional methods and tools for fluffing wool are basically the same as those of fluffing cotton, but as wool fibers are longer than cotton fibers, and its elasticity and strength are also greater, the size of the wool bow is larger than that of the cotton bow. It is also equipped with a bar and annular shoulder holster, which are connected to each other. When fluffing the wool, hang the back of the bow on a high place, face the wool with the bowstring, and put on the shoulder holster.

People in the autonomous regions of Inner Mongolia, Ningxia and Xinjiang, and in Gansu province still adopt the more primitive methods of wool fluffing in modern times. They build a stone or mud platform on the ground, and spread wool about 5 cm thick on it. Two wooden stakes are erected at each end of the platform, and a leather strip (or rope) is tied to each stake and leads to the opposite side of the platform. When fluffing wool, one person stands at each end of the platform, holding a leather strip in each hand, waving it while moving, fluffing the wool to loosen it.

4.3 Dyeing and Finishing Techniques and Varieties

In ancient China, the process of dyeing fabric was called "*zhangshi*", which was classified as "stone dyeing" (mineral pigment) and "grass dyeing" (plant dyes), according to the different dyeing materials used. The latter was the most commonly used.

Traditional dyeing and finishing techniques are rich in content and variety, which can be summarized as the preparation of pigments and dyes, dyeing, printing, and finishing. Before the invention of synthetic dyes in 1856, China's printing and dyeing technology had always been at the leading level in the world.

Although mineral pigments and plant dyes are coloring pigments, their functions are quite different. The former is applied to the fabric surface by adhesion, which can't be compared with plant dyes; the applied colors can't stand washing although they have specific colors. However, the latter are different. The pigment molecules in plants can be intimately integrated with fabrics due to chemical adsorption during the process of dyeing, thus changing the color of fibers. The color does not or rarely fades, even after sun exposure and water washing, so they are called "dyes" rather than "pigments".

There are many kinds of traditional plant dyes. The red ones include safflower, madder and sapanwood; yellow dyes include hispid arthraxon, gardenia, curcuma aromatica and turmeric; the purple dye is arnebia euchroma; the green dye is rhamnum; the black dye is acorn shells; the blue dye is mainly indigo. There are various ways to use these plant dyes to dye fabrics.

4.3.1 Dyeing

There are three types of dyeing, namely, redyeing, over dyeing and mordant dyeing, but many colors need to be obtained by combining these three types of processes.

Redyeing refers to dyeing textile fibers or fabrics with the same dye solution, so that the color is gradually deepened. Between twice dyeing, the fabric must be dried without being wrung so that it can absorb more pigment during the second dyeing. Indigo dyeing, the most common one, belongs to this type.

The principle of **over dyeing** is basically the same as that of redyeing, which is also multi-dyeing, but the main difference between the two lies in that over dyeing is to dye the fabric for a second or third time in two or more different dyes to get an intermediate color. For example, over dye the red with the blue, and you will get purple. Dye with indigo and then with the yellow dye, and you will get green. With the method of over dyeing, more colors can be produced with only a few dyes.

Mordant dyeing means helping adhere pigments in dyes to fabrics by means of mordants. Most of the dyes, except gardenia and curcuma aromatica, do not have strong dyeing property on fibers and cannot be dyed directly. However, most dyes contain mordant genes and can be dyed through the method of mordant dyeing. This is because mordant dye molecules can react with metal ions to form complexes, and insoluble colored precipitates can be precipitated on fabrics after mordant treatment. Mordant dyes have much better dye uptake, light resistance, acid and alkali resistance and color fastness, but the dyeing process is more complicated than other dyeing methods. If the mordant is used slightly improperly, the dyed color will greatly deviate from the original standard, and it is difficult to change. The traditional aluminum mordants are alum [$K_2SO_4 \cdot Al_2(SO_4)_3 \cdot 24H_2O$] and ash of aluminum-containing plants such as wormwood and symplocos sumuntia. Iron mordants include melanteritum ($FeSO_4 \cdot 7H_2O$) and river mud containing iron ion metal salt.

4.3.2 Xie Dyeing

Xie dyeing, a kind of traditional Chinese pattern dyeing, includes clamp dyeing, wax dyeing and tie dyeing, all of which are resist dyeing, that is, using "*xie*" to resist dyeing in some parts of the fabric.

Clamp dyeing is actually a resist printing with hollow plates engraved with the same patterns. Fold the cloth in half and clamp it tightly between the two plates, and then brush dye or color paste on the hollow part. Remove the hollowed-out plates, and the hollowed-out part on the cloth will show symmetrical patterns since the part covered by the plate is not dyed.

At present, the traditional clamp dyeing process is still preserved in Badai village, Yishan town, Cangnan county, Zhejiang province. The specific procedures are as follows. Firstly, cut the wood board into plates, plane them flat and smooth, and soak them in water for about a week to remove the resin. Stick the drafts of sample patterns firmly on the plates, draw the outline of patterns with an oblique knife, and then make grooves about 0.4 cm deep from left to right with a round knife and a flat knife to chisel out the blanks of the patterns. In addition, it's necessary to dig blind passes between the individual patterns for the dye to flow freely (as shown in Picture 4.61). When applying color, fold the white cotton cloth in half, clamp it in the middle of the pattern plates, and then soak them in the indigo liquid. The dyeing platform is made up of 8 dyeing vats, with a height of 1.25 m, a diameter of about 1 m, a large belly, and a sharp bottom. The big belly is to hold more dye solution and accommodate the clamping plates. After dip dyeing the cloth for about half an hour, lift the plates up out of the dyeing vats, and hold them in the air for a while before the second dip dyeing, after which turn the plates up and down and do the third and fourth dip dyeing in the same way. All the four times of dyeing completed, remove the cloth from the plates, spread it flat in the river and rinse it, and then dry it on a high bamboo frame (as shown in Picture 4.62). Ms. Zhang Qin has made a detailed study of this process and collected a large number of complete sets of engraved plates, making an important contribution to the protection of this precious heritage.

Wax dyeing is also called batik. The traditional batik is to heat and melt the beeswax first, dip a wax knife made of bamboo brush or copper sheet (as shown in Picture 4.63) in the wax liquid and draw patterns on the fabric (as shown in Picture 4.64). Dye the fabric after the wax gets congealed, and then cook it in boiling

Picture 4.61 Plates for clamp dyeing

water to remove the wax. In this way, where there is wax on the fabric, the dye solution is prevented from immersing. The white flower patterns are presented against the background of the surrounding dyed colors. Due to the shrinkage of wax after condensation and wrinkles of the fabric, there are often many cracks on the wax. After dyeing, the pigment will penetrate the cracks after dyeing and show irregular color lines, forming a unique decorative effect of batik products (as shown in Picture 4.65).

Multicolor batik is realized by "over dyeing". For example, to over dye deep blue and light blue, it's necessary to dye the cloth in light blue, cover the part where the color is to be retained with wax, and then dye the cloth in deep blue. Madder, arnebia euchroma, balsamine, bayberry juice, or sometimes ox blood is used for dyeing red. Gardenia, Chinese ash leaves or buddleja officinalis is used for dyeing yellow.

Tie dyeing, *jiaoxie*, is also called *cuoxie* or *zhaxie* in China, and there are five methods. One method is to sew and tie the fabric first, and then dye it. The pattern thus resembles grids and flowers. The second is to clamp the folded fabric between pairs of small plates, making the patterns similar to the shape of the plates. The third method is called knotting, which means folding, knotting, and tightening the fabric, making the pattern lace shaped. The fourth is to pinch up and bind the fabric, which makes the pattern look like hollow squares. If the fabric is dyed red or purple with white flowers, it is like stripes of the deer, thus called "deer's embryo dyeing". The

Picture 4.62   "Picture of a
Hundred Sons", clamp dyed

Picture 4.63   Wax knives

Picture 4.64 Drawing patterns on the fabric with a wax knife

Picture 4.65 Finished batik product

fifth is called bandaging, which means wrapping beans and stones in the fabric and tying them (as shown in Picture 4.66).

The tied fabric should be dipped in clear water for more than 20 min before dyeing, and the thick fabric for 30 min. If it is directly put into the dye solution, which will penetrate the tied place in large quantities, no patterns will be produced

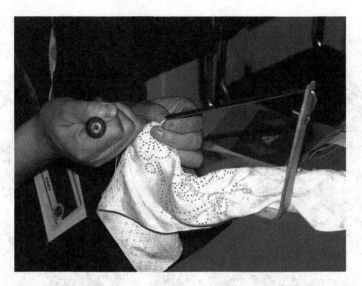

Picture 4.66 The method of sewing and tying

in the tied parts. On the contrary, patterns with insufficient coloring will show up after dip dyeing due to the lack of dye infiltration or insufficient infiltration into the tied parts. Because of the infiltration of dye solution on the edge of the tied parts, color halo from deep to shallow is naturally formed, which makes the fabric look layered and has the artistic effect of shading and blurring (as shown in Picture 4.67). This kind of color halo is difficult to achieve with other methods, so the effect of tie dyeing is non-copyable.

4.3.3   Relief Printing

Relief printing means engraving relief patterns on flat and smooth wooden boards, then brushing the patterns and finally pressing the boards onto fabric to get patterns. This kind of technique reached a higher level in the Western Han dynasty in China, and the raised-and-painted gauze and gold and silver printed gauze unearthed in Mawangdui tombs in Changsha city were made by combining the technique of relief printing with painting.

The techniques of wooden stamp printing and wooden roller printing, which are still used by Uygur people in Xinjiang, are quite distinctive. The former is similar to stamping, and the engraving patterns of the wooden stamp is extremely deep, which can reach 7–10 cm. It's also feasible to print the frame with one stamp first, and then print the flowers with another to form a two-color pattern (as shown in Pictures 4.68 and 4.69). The latter means printing patterns on fabrics using a cylindrical roller, the

Picture 4.67 Finished tie dyeing product

surface of which is engraved with patterns. It is similar to modern roller-printing, which can print infinitely continuous patterns.

4.3.4 Products of Dyeing and Finishing

4.3.4.1 Blue Calico

Blue calico was called "medicine spot cloth" or "*jiaohua* cloth" in ancient China, which was made with physical resist dyeing technology, that is, to prevent dye solution with some slurry, and make a blue background with white patterns or a white background with blue patterns. It was very popular among the people in the Ming and Qing dynasties, with an annual output of 600,000 bots in the late Qing. It was mainly produced in Jiangsu, Zhejiang, Hunan, Shandong, Henan, Hebei, Sichuan, Shanxi, Shaanxi provinces and Northeast China, which had their own unique local styles, especially Jiangsu and Zhejiang (as shown in Pictures 4.70 and 4.71). Wu Yuanxin (as shown in Picture 4.72), a national inheritor of blue calico technology and a master of Chinese arts and crafts, has devoted himself to the inheritance and protection of blue calico since his youth. After years of hard work, he founded Nantong Blue Calico Museum, which was well received by people.

Picture 4.68 Wooden stamps used for printing fabrics in Xinjiang

Hollowed pattern templates are used for gluing of blue calico (as shown in Picture 4.73). They are mainly made of paper. Fibrous skin paper or *xuan* paper is pasted into multilayers and then coated with tung oil or persimmon oil to enhance its waterproof performance. The method of cutting patterns is similar to paper cutting. The patterns must be connected as a whole, and the hollowed-out parts should be reduced as much as possible, so there are inevitable many short lines and dots, and these basic shapes form the various patterns. The best work is one named "Mandarin Ducks Playing with the Lotus", which is formed by small dots only. The template of this pattern is made by chiseling dots on paper boards with a special tool called "*yuezi*", and it is ground flat with pebbles after the patterns are made. Afterwards, it is brushed with raw tung oil and then cooked in tung oil or Chinese lacquer. In the old days, there were artisans who made a living by engraving pattern templates in Shandong and Jiangsu provinces.

Before dyeing, cover the pattern template on a white piece of cloth, and glue it with the slurry made of lime, soybean powder and water (as shown in Picture 4.74). Then dip dye it in the indigo solution 6–8 times or even more than 10 times. Dry it after dyeing, scrape the slurry on the cloth with a knife, and the fabric will show blue and white patterns (as shown in Picture 4.75).

Picture 4.69 Wooden stamp printing

4.3.4.2 Color Calico

Color calico is made by the technique of overprinting with pattern templates of different colors and then being dyed into various colors (as shown in Picture 4.76). The color calico made in Junan, Linyi and Cangshan counties in Shandong province, Xi'an in Shaanxi province, and Weixian county in Hebei province shows rich local flavors. The patterns are mostly centripetal; the outline and structural lines are usually created with blank lines; the cloth is overprinted with bright colors such as bright red, green and orange. The production of pattern templates is similar to those of blue calico, with one template for each color.

In the past, many farmers liked to wrap gifts and cover dowries with color calico, which increased the festive atmosphere. Such cloth was also widely used as daily necessities such as door curtains, belly wraps, aprons and ticking. Zhang Mingjian of Linyi county, Shandong province is the inheritor of local color calico, and his designs are fashionable and well received by the people (as shown in Picture 4.77).

4.3.4.3 Watered Gauze

Watered gauze, also known as gambiered Canton gauze or cloud gauze, is calls "roasted silk" in Beijing. It is also called "*xiangyun* gauze (rustling cloud gauze)"

Picture 4.70 Calico of blue
background with white
patterns in Zhejiang province

because it rustles when people walk while wearing clothes made of the material, and later was called "*xiangyun* gauze (fragrant cloud gauze)" with homophonic pronunciation. To make watered gauze, dip dye the silk fabric repeatedly in the juice of *shoulang* yam to get the light brown semi-finished product, then smear the black pond mud rich in iron on one side of the silk and expose it to the hot sun. After the iron ions and other biochemical components in the mud fully react with tannic acid in the *shoulang* yam juice to generate black ferrous tannate, shake off the pond mud, wash the silk clean, and the watered gauze with black and glossy face and light brown back is thus made (as shown in Picture 4.78). The gloss on its surface is made by the combination of the colloidal polypeptide bond in the fibroin and tannin-based hydrogen bond of the *shoulang* yam.

Shoulang yam is a kind of food as well as Chinese herbal medicine, whose juice has the functions of cooling, dehumidifying, detoxifying, and clearing heat. Therefore, watered gauze, soaked with *shoulang* yam juice, is also of the above-mentioned health care functions, and has a crisp and smooth hand feeling, which is especially suitable for wearing in hot and humid summer.

Picture 4.71 Calico of white background with blue patterns in Jiangsu province

4.4 Weave and Variety of Fabric

4.4.1 Fabric Weave

Fabric weave refers to the order of interlacement of warp and weft threads at certain angles, and it determines the variety, physical properties and appearance of a certain woven fabric.

The simplest weaves are the three foundational weaves, including plain weave, twill weave and satin weave, and they are also the basis of various fabric weaves (as shown in Picture 4.79). They are characterized by: (1) The step number of warp and weft interlacing is a constant; (2) There is only one warp (weft) interlacing on each

Picture 4.72 Wu Yuanxin at work

warp (weft) yarn, and the others are weft (warp) interlacing; (3) The number of warp yarns and that of weft yarns in a weave repeat are equal.

Plain weave is the simplest type of fabric weave, which is formed by the interweaving of warp and weft alternatively, and every two warp and weft threads repeat.

Twill weave means that continuous warp (weft) interlacing on adjacent warp (weft) yarns are arranged in diagonal lines, and the fabric surface presents continuous patterns of diagonal lines. There are at least three threads in a weave repeat for twill weave.

The distance between the warp and weft interlacing of satin weave is relatively far. The interlacing is independent and discontinuous, and is arranged in a certain order. There are at least five warp and weft threads in a complete weave repeat, with five and eight the most common. Generally speaking, warp satin is represented by warp step number, while weft satin by weft step number. Step number refers to the number of threads separated by two adjacent threads or two adjacent warp and weft interlacing, and the step number of satin is not less than two.

Most complex weaves are formed by the change or combination of the above three foundational weaves. It can be concluded from the unearthed cultural relics that plain weave and skein weave were already invented in primitive society in China. During the Shang and Zhou dynasties, there were twill weave, variant forms of plain and twill weave, joint weave and skein weave, as well as some complex weaves such as warp

Picture 4.73 Pattern template of blue calico

backed weave, weft backed weave and thread drawing weave. In the Tang dynasty, satin weave, the most characteristic and complex type of the three foundational weaves, was invented in the form of six-end irregular satin weave, which ushered in the future development of fabric varieties. The great diversity of traditional textiles was based on the skillful use of various fabric weaves.

4.4.2 Variety of Woven Fabric

There are many kinds of traditional Chinese silk, hemp, wool and cotton fabrics, the most representative of which are *juan* (tabby silk), *luo* (gauze), *ling* (thin silk),

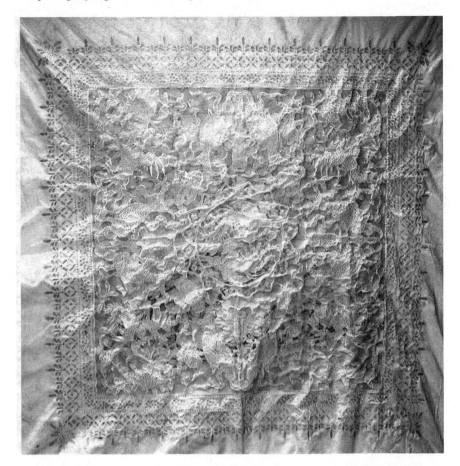

Picture 4.74 Fabric with the pattern of "Auspicious Double Fishes" after the slurry is scraped

satin, brocade, velvet, *kesi* (Chinese tapestry) and blanket, with many varieties in each category.

4.4.2.1 *Juan* (Tabby Silk)

All the plain silk fabrics with plain weave, such as *sha* (plain gauze), *hu* (crepe), *chou* (plain silk), *su* (plain silk), *jian* (fine tabby silk), *wan* (pure white fine silk), *gao* (white silk) and *lian* (scoured silk), can be classified as tabby silk. The main differences between them lie in the different thickness, density and twist of warp and weft, or whether they have been scoured or dyed or not. Take *sha* (plain gauze) for example. It is made of quite thin threads, and the warp and weft density is very small, which makes the fabric light and thin. Look at *hu* (crepe) next. The reason there are wrinkles on this type of silk fabric is inseparable from its yarn twist. It is woven from

Picture 4.75 Blue calico with the pattern of "Auspicious Double Fishes"

raw silk with strengthened twist, and then scoured to make the silk threads untwist, shrink and bend under its internal stress, thus forming wrinkles on the fabric surface.

4.4.2.2 *Shaluo* (Leno)

This type of fabric is made of the leno weave, in which the warp and the weft threads of the regular yarn and "doup yarn" are twisted with one another, and the eyelets thus formed are called skein eyelets. The weave with uniform distribution of skein eyelets and no strip shapes is called *sha* weave, while the weave with skein holes arranged along warp or weft are called *luo* (gauze) weave. Because of its light texture and good air permeability, *shaluo* (leno) is especially suitable for underwear and summer clothes and has been a major category of silk fabrics since the Spring and Autumn period and Warring States period in China.

Picture 4.76 Color calico named "May There Be Surpluses Every Year", made by Zhang Mingjian

4.4.2.3   Satin

Satin is a general term for all kinds of patterned and plain fabrics based on satin weave. Satin fabrics first appeared in the Tang dynasty. Since the Song and Yuan dynasties, with the widespread application of five-end, eight-end satin weave and various irregular satin weaves, satin has become a major category of silk fabrics along with *luo* (gauze), brocade, *ling* (thin silk) and *sha* (plain gauze).

Satin weave is developed based on twill and characterized by the uniform distribution and discontinuity of the individual interlacing on two adjacent warp or weft yarns. Because the individual interlacing is often covered by the float of adjacent warp or weft yarns, the fabric surface is smooth, even, and lustrous, and the patterns are three-dimensional. Therefore, satin weave is most suitable for weaving patterns with complex colors. The combination of satin weave and the *jin* (brocade) weave results in the making of brocade, the most gorgeous silk fabric.

Picture 4.77 Color calico named "To Be the Champion" to welcome the Beijing Olympic Games by Zhang Mingjian

4.4.2.4 *Ling* (Thin Silk)

Ling (thin silk) is a silk fabric with twill patterns on twill base. The texture of the ice *ling* is similar to that of pointed twill and lustrous, so it was also called "ice" before the Han dynasty. During the Three Kingdoms period, Ma Jun simplified the loom for *ling* weave, which greatly helped increase the output of such fabric and make more complex patterns. The production of *ling* reached its peak in the Tang dynasty, when a "*ling* workshop" was set up under the Weaving and Dyeing Department and it was stipulated that different levels of official robes should be made of *ling* with

Picture 4.78 Watered gauze

Picture 4.79 Schematic diagram of three foundational weaves

special colors and patterns. After the Song dynasty, *ling* began to be widely used in the mounting of paintings, calligraphic works, and scriptures.

If all kinds of silk fabrics are ranked according to their exquisiteness and preciousness, *ling* will rank second only to brocade, since its utilization of natural silk is quite successful, and both plain and patterned weave can fully reflect the excellent characteristics of the material. The plain *ling* is smooth, soft, and clear while the patterned *ling* has good clarity and reveals three-dimensional effect.

4.4.2.5 *Jin* (Brocade)

Brocade is a colorful jacquard fabric with heavy warp or heavy weft woven by joint weave or complex weave. It was said in ancient China that brocade is like gold. It takes a lot of work to make and is as valuable as gold, so only the nobles can afford it. The invention of brocade has far-reaching influence on the development of textile technology. Cloud brocade, Shu brocade and Song brocade are the three famous types of brocade in China, which represent the most outstanding achievements of brocade weaving. On the other hand, some ethnic brocade such as Tujia brocade, Li brocade, Zhuang brocade and Dai brocade demonstrates strong local characteristics and distinct ethnical aesthetic taste.

Shu brocade in its early form was mainly warp brocade, which is made by the interweaving of warp and weft color stripes with warp stripes as the basis, thus creating a unique style with rich colors (as shown in Picture 4.80). Two types of Shu brocade came into being after the Song and Yuan dynasties, namely, warp brocade and weft brocade, but patterns were still woven based on warp color stripes.

To make Shu brocade, there are many processes including designing, finalizing the draft, choosing artisans, making figure designs for weaving, installing the figure design onto the loom, and weaving, as well as some special techniques such as knotting, thread pulling, shuttle picking and ends jointing. For example, the warping method of weaving Yuehua brocade is quite unique. The different colors of the stripes and the warp should be numbered first, and spools are arranged according to the sequence and width of the color stripes as well as the color shade of the warp thread. Every time a thread is warped, some of the spools must be rearranged, which is called "hand-to-hand". The *hualou* drawloom for weaving Shu brocade is 1.8 *zhang* (6 m) and the warp width is more than 1.2 *zhang* (4 m).

Song brocade is produced in the Jiangnan area centered on Suzhou and Hangzhou. It's said that after the Song court moved south of the Yangtze River, the Weaving Department in charge of Song brocade production was set up in Suzhou to meet the needs of imperial clothing and mounting scrolled paintings and calligraphic works. Robes for civil and military officials were also made of Song brocade, and the patterns on the robes were specified according to their positions. During the Ming and Qing dynasties, most of Song brocade was produced in Suzhou, and it was recorded that the sound of thousands of drawlooms was heard in the northeast part of the city.

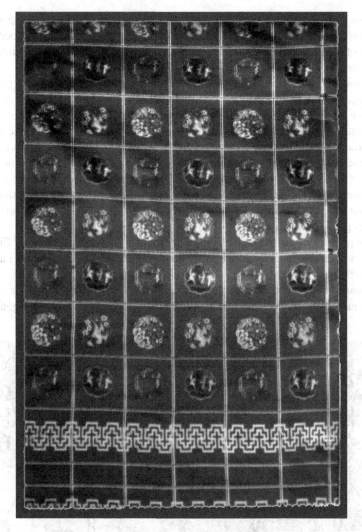

Picture 4.80 Quilt cover with square patterns, modern Shu brocade

Song brocade is a weft-brocade with color weft, which falls into four categories, namely, heavy brocade, fine brocade, box brocade and small brocade. As the most valuable of Song brocade, heavy brocade is best known for its delicate texture and rich colors, and its weaving technology is characterized by the incorporation of gold and silver threads, as well as the weaving process of long, short and partial shuttle throwing with multi-strand silk threads (as shown in Picture 4.81). Fine brocade, the most representative of Song brocade, is widely used in the decoration of clothing, precious gifts and high-grade paintings and calligraphic works, with geometric patterns as skeleton filled with flowers and auspicious grass. Qian

Xiaoping, a senior engineer, has made great efforts to rescue the weaving skills of Song brocade, which was about to disappear. Song brocade weaving has been listed in *List of the National Intangible Cultural Heritage*, and Qian was recognized as its inheritor (see Picture 4.82).

Cloud brocade, or **Yunjin**, is the representative of brocade and is produced in Nanjing, Jiangsu province. It is made by the *hualou* drawloom, with one craftsman drawing the figure design on the *hualou*, or figure tower, and another one throwing the shuttle for weft insertion and beating-up on the loom. There are three kinds of Yun brocade, namely, tinseled brocade, satin brocade and *zhuanghua* brocade.

Tinseled brocade, named so because it was stored in the "Satin Library" of Internal Affairs Office of the Qing dynasty, is the inheritance and development of *nashishi* in the Yuan dynasty. It is featured with patterns woven with gold or silver threads on the satin base. And "multicolor tinseled brocade" is woven with colored weft threads besides gold and silver threads.

The patterns of satin brocade are mainly rounded floral ones, such as "Five Bats Surrounding the Character Longevity" (implying blessing and longevity) and "All the Best".

Zhuanghua is the most complex and exquisite type of Cloud brocade weaving technique and is divided into *zhuanghua* satin, gauze, velvet, and plain gauze. The

Picture 4.81 Song brocade with *Badayun* patterns on blue ground of the Qing dynasty

Picture 4.82   Qian Xiaoping, inheritor of Song brocade weaving

technique of swivel weaving is employed to weave patterns on *zhuanghua*, which is a kind of multi-weft and multi-color sateen. Weft cops with different colors in the weft direction can help to weave weft threads of different colors in sections according to the patterns, thus creating the technique of maintaining the warp and breaking the weft. There are no knots on the back of colored weft, with long floating fluff. This unique weaving method cannot be replaced by modern looms (as shown in Picture 4.83).

One type of *zhuanghua* satin incorporates gold thread (as shown in Picture 4.84), which is divided into two kinds, namely, round and flat gold thread. The weft rod, a special weft insertion tool should be used to ensure that the thread does not twist during weaving. In 2010, Cloud brocade was added to the *Representative List of Oral and Intangible Heritage of Humanity* at UNESCO. Jin Wen, a master of Chinese arts and crafts, is a representative inheritor of Cloud brocade. He wrote the part about

Picture 4.83 *Zhuanghua*
satin

Cloud brocade in two monographs including *History of Science and Technology in China: Textile* and *Complete Works of Traditional Chinese Handicrafts: Weaving, Dyeing and Embroidery*.

Tujia brocade is called "Xilankapu" in the Tujia language, which means "brocade woven by the girl named Xilan" or "quilt cover with flower patterns" (*huada pugai* in Chinese). This type of brocade is mainly used as quilt cover, and three pieces of brocade are connected into a whole quilt cover (as shown in Picture 4.85). It is mainly produced in Xiangxi Tujia and Miao autonomous prefecture of Hunan province and Enshi Tujia and Miao autonomous prefecture of Hubei province, especially Longshan, Huayuan and Yongshun counties in the above two prefectures.

Tujia brocade is woven of cotton, woolen, or silk thread on the backstrap loom, with a width of less than two *chi* (0.67 m). Weft threads A and B are inserted in turn, with shuttle A throwing thin threads in the method of long throwing and shuttle B creating patterns, changing colors in sections, thus maintaining the warp, and breaking the weft. Most of the patterns are everyday objects and natural products.

Picture 4.84 *Zhuanghua* satin with gold thread woven in contemporary China

As the local proverb goes, "It is better not to raise a daughter if she doesn't weave brocade". Hard-working and intelligent Tujia women learn to weave at an early age and become quite proficient in Tujia brocade when they get married (as shown in Picture 4.86).

Li brocade of Hainan has a long history and was sold to the mainland in the Song dynasty. It is recorded in *Well-balanced Records of Guihai* by Fan Chengda of the Southern Song dynasty that "many people in Guilin buy Tujia brocade as their bedding". Even today, it is still a representative handicraft of the Li nationality.

It is woven with the primitive backstrap loom. The weaver sits on the ground and picks in different weft color threads according to the design patterns with a figure-picking knife. The usual geometric patterns include square, triangle, and diamond, which are used as the decorations of figure, animal, or plant patterns with some artistic

Picture 4.85   Tujia Xilankapu

techniques such as deformation and abstraction. There are 40 kinds of commonly seen patterns including various figures, peacocks, rice flowers and local palm trees. Brown and black are the main colors, alternating with cyan, red, white, blue, and yellow, showing an elegant and quiet artistic effect (as shown in Picture 4.87).

4.4.2.6   *Kesi* (Chinese Tapestry)

Kesi is a high-grade patterned fabric woven with natural silk as warp thread and boiled-off silk of various colors as weft thread. The unique knot weaving technique of "maintaining the warp and breaking the weft" means interweaving the weft parts of various colors with warp yarns by using the small shuttle to knot, penetrate, hook, prop and long and short shuttling (as shown in Picture 4.88). According to *Chicken Rib Compilation* by Zhuang Chuo of the Southern Song dynasty, "Craftsmen in Dingzhou, Hebei province don't use big looms to weave *kesi*; they design flowers, grass and animals skillfully by using boiled-off silk threads of various colors. When

Picture 4.86 Tujia brocade
of 12 zodiac

weaving the weft with the small shuttle, weave the parts of the same colors and leave
those of different colors out, thus forming various colorful patterns, which look as if
discontinuous. If looked at without any background, it seems like that the patterns
are hollowed out by knives, hence the name *kesi*, meaning 'cut silk' in Chinese".

The efficiency of silk reeling is very low, but it is honored as "the king of the
weaving fabrics" because of its excellent production, simplicity, and elegance, and
it wins the reputation of the artistic fabric that won't wear out "even after a thousand
years" because it can stand touching, rubbing, and washing (as shown in Pictures 4.89
and 4.90).

Picture 4.87 Tube skirts of
Li brocade

4.4.2.7 Rugs

The history of the Chinese rug can be traced back to more than 3,000 years ago, and it was created by Uyghurian, Mongolian and Tibetan people in plateau pastoral areas.

The surface of traditional woven rug is densely covered with plush, which is of solid texture and good elasticity. The 8-shaped knotting method is adopted in weaving rugs (as shown in Picture 4.91), which means tying a knot with the plush yarn on the warp head of two adjacent warp yarns and break it with a knife, which is called "*shuantou*" (tying the head). The whole piece of the rug is woven like this. Knots are tied from one end to the other along the weft direction, after which a thick weft thread is woven and smashed flat with a milling rake. Then a thin curved weft is woven along the outer edges of the front and back warps and smashed solid. Finally, the thread

Picture 4.88 *Kesi* weaving

end is cut flat and even with a pair of scissors. Rugs produced in different regions have different styles. Some well-known types include the Beijing rug, Ningxia rug, Xinjiang rug and the Tibetan rug.

In the Qing dynasty, the Beijing rug reached its peak with exquisite weaving technology and elegant and magnificent patterns, which were influenced by the Mongolian rug's decorative techniques of silk weaving patterns.

Various types and styles of rugs in Ningxia are determined by their functions, as well as the living habits and religious beliefs of the makers. There are worship rugs for temples, and bed rugs and wall rugs for daily use.

The patterns of Xinjiang rugs are mostly composed of geometric skeletons, and the rugs are covered with pomegranate flowers and box patterns, which are rich in symbolic meaning, with distinct layers and gorgeous colors. The silk rug with a golden background and flower patterns made by Uygur people during the Qianlong period (1736–1796) was 3.27 m long and 1.64 m wide. It had rich colors and was a representative of Xinjiang rugs with cotton warp and weft, background woven by gold and silver threads, and knots made of natural silk threads.

Tibetan rugs are the necessities of Tibetan daily life, and are widely used in monasteries, houses, tents and on horsebacks. Such kind of rugs is developed from the primitive form of "*liu*" to the mature form of "*chibujie*" in later period. "*Liu*" is a type of woolen cloth interwoven with crosswise warp and weft threads, which are

Picture 4.89 Product of
contemporary *kesi*

quite thick, thus forming a certain thickness. Later, *"chibujie"* breaks through the
plain weaving technique, and adopts the unique process of U-shaped knotting.

4.4.2.8 Pulu

Pulu is the transliteration of Tibetan wool fabric, which became a high-grade fabric
as early as the Tang dynasty. According to legend, Princess Wencheng of the Tang
dynasty wore clothes made of *pulu* when she went to Tibet and got married to
Songtsen Gampo. The name of *"pulu"* first appeared in the literature of the Song and
Liao dynasties, and it was once a tribute of Tibet to the Song court.

"*Bangdian*" and "*kadian*" are the two representative products of *pulu*. *Bangdian*,
meaning "apron" in Tibetan, is woven with various colored yarns, with horizontal
color patterns on the surface. Its weaving width is one *chi* (33 cm), and the warp
and weft density is 10–14 yarns per centimeter. The best type of *bangdian* is called
"*xiema*", which is made of 14–20 kinds of dyed wool yarns. It can be woven by a

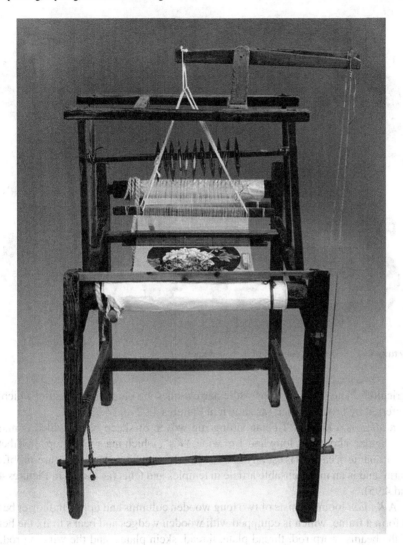

Picture 4.90 *Kesi* loom

four-heddle loom since its fabric weave is cassimere twill. The whole process is as follows:

Selecting wool—combing wool—twisting thread—dyeing—drying—warping and weft making—weaving on the loom—sewing—finished product.

Bangdian is characterized by its bold color matching, some with strong contrasting color bars, some in combination of similar colors, and some with bright primary colors interspersed with many secondary color bars. To present different visual effects and meet the needs of different people, there are broad and narrow color bars on *bangdian*. The broad color bars are rough and bright, which are loved by women in

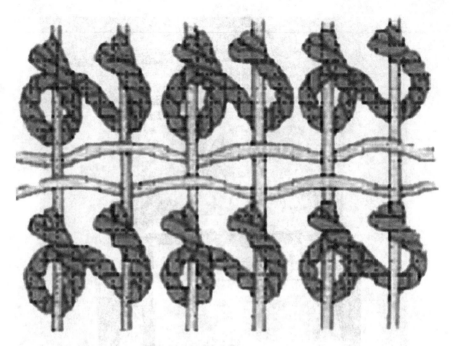

Picture 4.91 Schematic diagram of 8-shaped knotting

agricultural and pastoral areas while narrow ones are elegant and gentle, which are preferred by urban women (as shown in Pictures 4.92 and 4.93).

Kadian is a kind of Tibetan sitting rug woven of sheep wool, which is mainly rectangular, about 2 m long and 1 m wide. "*Ka*", which means "on top" in Tibetan, gets its name because it is usually laid on seats. It is soft in texture, moisture-proof, and warm, and is an indispensable article in temples and tents (as shown in Pictures 4.94 and 4.95).

A *Kadian* loom consists of two long wooden columns and upper and lower beams to form a frame, which is equipped with wooden wedges and beaks to fix the height of the beams, warp rod, thread plate, thread, skein plates, and the winding rod. To weave *kadian*, first put the loom frame against the wall. The weaver, sitting in front of the loom, first lifts the thread out and makes skeins, inserts the skein plate to fix the cross, and then makes knots with wool threads on the winding rod. The knots are smashed against the surface with a mallet. Next, the thick and thin weft threads are passed from both ends of the warp respectively and smashed tight. Then the wool threads are cut off on the winding rod with a blade. The rod is drawn out to complete the weaving of one line. The whole rug is woven by repeating the above process.

Picture 4.92 *Bangdian* with broad bars

4.4.2.9 Grass Cloth

Grass cloth is a plain cloth woven with ramie, and the clothing made of it is cool and comfortable to wear and easy to wash and dry, with good air permeability. The process of making grass cloth includes raw hemp bleaching, tearing, spinning winding, twisting, winding, and weaving. There are four steps of weaving, i.e. brushing, loading, sizing, and weaving. Brushing means separating the yarns wound into spindles, straightening and lengthening them, and then hanging on the warp frame by one end. The other end is rolled into a big knot, which is pulled onto a wooden stick, and then pressed with heavy objects. How much warp yarn should be hung depends on what kind of grass cloth is produced. Then the rice flour paste is brushed back and forth evenly on the yarn, which is loaded on the loom and woven after the sizing gets dry.

Picture 4.93 *Bangdian* with narrow bars

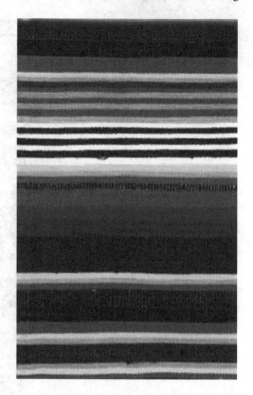

4.5 Embroidery Techniques and Varieties

Embroidery, commonly known as "flower embroidery", is a decorative fabric with needles stitching patterns on it with colorful threads.

4.5.1 Preparation Before Embroidery

Suitable embroidery fabric, thread, needles, scissors, and the embroidery frame should be selected and prepared according to the needs of embroidery beforehand.

There are distinct requirements for thread, needlework, and patterns for different kinds of fabrics, so fabrics should be chosen according to the style of the embroidery, thus making it beautiful and dazzling.

Picture 4.94 *Kadian* with
geometrical patterns

There are many kinds of embroidery thread, such as pure cotton thin embroidery thread, pure cotton thick embroidery thread, twine, linen thread, silk thread, machine embroidery thread, wool, gold, and silver thread, with pure cotton and silk embroidery thread as the most popular.

Pay special attention to the eyes and tips when choosing embroidery needles. The eye should be oval; if it is rectangular or pointed round, it is easy to "bite" the thread. For the tip, the thinner and longer, the better.

You should also be careful about choosing scissors. For example, the tips of those for cutting thread ends should be upturned to avoid cutting the embroidery, while the tips of those for cutwork embroidery and silk-spinning should be pointed and sharp.

There are square and round embroidery frames, which should be chosen according to the size of the pattern. If the pattern is too large, it should be embroidered in several batches. The area with patterns generally covers half of the embroidery frame, which should be tightened and flat, so that the embroidered patterns are flat and not out of shape.

Picture 4.95 *Kadian* with
auspicious grass patterns

4.5.2 Stitches

According to the composition characteristics of embroidery, stitches can be divided
into six categories, namely, flat stitch, stripe stitch, knot stitch, plaited braid stitch,
cross stitch, and auxiliary stitch. The three stitches that are most commonly used are
described in the following part.

4.5.2.1 Flat Stitch

It includes flat stitch, encroaching satin stitch, inlaying stitch and long and short
stitch.

Flat stitch, also known as "straight stitch" or "even stitch", is common in some
famous types of embroidery such as Suzhou embroidery, Xiang embroidery and Shu
embroidery. To do a flat stitch, the embroidery thread is arranged straight, and the
pulling out and inserting of each stitch are at the edge of the pattern, which is formed
by the change of stitch length.

Layered stitch is used to embroider in batches with short straight stitches, and
the previous ones are followed by the later ones, which looks like they are "layered"
together in batches. The more the batches, the softer the color conversion, with the
artistic effect similar to relief, which is very decorative (as shown in Picture 4.96).

Picture 4.96 Layered stitch

Inlay stitch is a typical technique of Suzhou embroidery, especially in its double-sided embroidery. The practice is to overlay the embroidery threads in batches. According to *Principles and Stitching of Chinese Embroidery* by Shen Shou (1874–1921), a famous embroidery master, "Inlaying means later batches of stitches covering the previous ones, which looks like the fangs of dogs" (as shown in Picture 4.97).

Long and short stitch is also known as "interpolating stitch", which means that the long and short stitches are mixed with each other, with the later ones being pulled out from the middle of the previous ones and arranged radially from inside to outside, so that the stitches are flexible, and the colors are soft. This kind of stitch is the main technique and characteristic of Xiang embroidery.

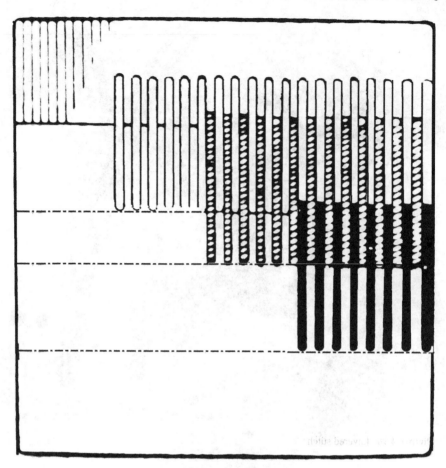

Picture 4.97 Schematic diagram of inlay stitch

4.5.2.2 Stripe Embroidery

The techniques of stripe embroidery include chain stitch, back stitch, double needle stitch, couching stitch, and round couching stitch with gold thread.

There are two basic types of chain stitch, namely, open, and locked chain stitch. When making the open chain stitch, the needle and thread are pulled out from the back of the fabric and held down with the thumb on the material to form a loop. The needle is then inserted into the right side of the spot where the thread first emerges and is pulled out again from the left side of the loop. Tighten the thread while pressing the loop with the thumb. The second stitch is made by inserting the needle into the right side of the spot where the thread first emerges and making a loop. Then the needle is pulled out from the left side. Tighten the thread while pressing the loop. And the stitch is repeated, creating the effect of a stripe. The locked type is like the

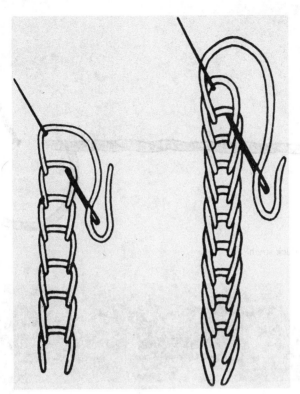

Picture 4.98 Schematic diagram of chain stitch

open type, with the only difference lying in that the needle of the former is pulled out and inserted at the same place, thus forming narrower and stronger loops than the latter (as shown in Picture 4.98).

The method of back stitch is to pull out the needle and thread from the back of the fabric, insert them backward, and pull them out in front of the spot where the thread first emerges. The second stitch is still inserted backward at the spot where the thread first emerges. The stitches are closely connected. The smaller the stitches, the better, so there is a common saying that "three stitches make one sesame" and "three stitches make an ant" (as shown in Picture 4.99).

Double needle stitch is done by two needles. The bigger needle and thread are completely pulled out of the fabric while only the haft of the smaller needle is pulled out; the bigger needle surrounds the smaller one counterclockwise with the thread to make a small ring. Pull out the small needle and thread, insert them on the left side of the small ring, and half of the small needle is pulled out from the right side of the ring. The bigger needle and thread are pulled out and make another small ring around the small needle, which is pulled out and inserted into the spot of the previous stitch to fix the ring. The process is repeated likewise (as shown in Picture 4.100).

Picture 4.99 Back stitch

Picture 4.100 Double needle stitch

Couching stitch refers to nailing embroidery thread on the fabric to form patterns, which is simple but varied.

The technique of round couching stitch with gold thread is the same as couching stitch, with the only difference being that the laid thread is gold thread, hence the name. If the gold thread is laid out on the fabric, it is called parallel couching stitch with gold thread. In Canton embroidery, the above two techniques are often used (as shown in Picture 4.101).

4.5.2.3 Crossing Stitch

Crossing stitch includes counted cross stitch and cross stitch.

The former is a general name for embroidery with leno fabric as the ground material according to the weave pattern. At the end of 1970s, Nantong Embroidery Research Institute of Jiangsu created "colored satin embroidery" by using counted cross stitch in the embroidery of decorative paintings (as shown in Picture 4.102).

Picture 4.101 Round couching stitch with gold thread

Picture 4.102 Counted cross stitch

Picture 4.103 Huayao cross stitch

Cross stitch is often made on plain cloth with crosses at warp and weft interlacing. It is extremely popular among the populace with simple stitches and durable patterns, which are made up of countless crosses (as shown in Pictures 4.103 and 4.104).

4.5.3 Varieties of Embroidery

Some famous embroidery varieties with regional characteristics are Gu embroidery, Suzhou embroidery, Xiang embroidery, Shu embroidery, Cantonese embroidery, and Bian embroidery. Miao embroidery is also highly popular because of its unique ethnic style.

Picture 4.104 Product of
cross stitch

4.5.3.1   Gu Embroidery

Gu embroidery, which was named after the family of Gu, was created by Han Ximeng, the grandson-in-law of Gu Mingshi, the founder of Luxiang Garden, one of the three famous gardens in Shanghai, during the Jiajing period (1522–1566) of the Ming dynasty. Gu's family was in close contact with Dong Qichang (1555–1636), a well-known calligrapher and painter, and Gu Shouqian, husband of Han Ximeng, was one of his disciples. Influenced by the literati painting school represented by Dong, Gu embroidery, characterized by its bookishness, focuses on embroidering ancient paintings of the Song and Yuan dynasties while being free of the stereotype and creating its own style. Its stitching techniques are complex and changeable, including more than ten kinds, such as straight stitch, satin stitch, knot stitch, split stitch, couching stitch, single inlay stitch and scale carving stitch; the thread used is also particularly stressed. Before embroidery, the silk thread must be turned into

Picture 4.105 Gu embroidery work of the *West Lake Albums*

single strands, which are dyed separately and then embroidered in turn according to their colors. Gu embroidery works, imitation of famous landscape, flower and bird, and figure paintings, are exquisite and life-like, which is quite unique. They were popular all over the country during the Ming and Qing dynasties and were once called "painting embroidery" (as shown in Picture 4.105). Suzhou embroidery, Xiang embroidery and Shu embroidery, which developed later, all benefited from Gu embroidery.

4.5.3.2 Suzhou Embroidery

Suzhou embroidery is the embroidery centered around the city of Suzhou, Jiangsu province, and is mainly characterized by its "tidiness, smoothness, straightness, uniformness, thinness, softness and denseness" summarized in *Embroidery* by Ding Pei of the Qing dynasty.

Suzhou has a long tradition of embroidery. In the Southern Song dynasty, there were some places named "Xiuxianxiang" (Embroidery Thread Lane), "Gunxiuxiang" (Embroidery Clothes Lane), "Jinxiufang" (Brocade Embroidery Lane) and "Xiuhuanong" (Embroidery Lane). There was the embroidering workshop affiliated with the Suzhou Weaving and Dyeing Bureau in the Ming dynasty. In the Qing dynasty, there were more than 150 private embroidery workshops. At the beginning of the twentieth century, Shen Shou (1874–1921), a famous Suzhou embroidery artist, created "life-like embroidery", which displayed the perspective and light and shadow effects of western oil paintings. Yang Shouyu of Danyang Zhengze Women's Vocational School in Jiangsu province created random stitch embroidery by learning from Western painting, which further strengthened the expressiveness of Suzhou embroidery. After the 1950s, the Embroidery Research Institute and the Embroidery Factory were set up in Suzhou, and some famous embroidery artisans such as Zhu Feng and Gu Wenxia were not only focused on creating excellent works but also laid great emphasis on research on technological improvement, which had a profound impact on embroidery technology in many places in China.

Suzhou embroidery is famous for its exquisite embroidery skills and beautiful patterns (as shown in Picture 4.106). Since the 1950s, double-sided embroidery has become the representative of Suzhou embroidery. It means embroidering patterns with identical stitches and colors on both sides of the ground fabric, with the common themes of kittens and goldfish, etc. There are nine categories and more than 40 types of Suzhou embroidery stitches. Embroidered daily necessities include quilt covers, cushions, clothes, shoes, and accessories. and embroidered ornamental products are screens, table screens, hanging screens, and albums.

Picture 4.106 Suzhou embroidery

4.5.3.3 Xiang Embroidery

Xiang embroidery, taking the city of Changsha, Hunan province as the center, develops its own regional characteristics by absorbing the features of Suzhou and Cantonese embroidery based on folk embroidery in Hunan. In the middle and late Qing dynasty, women embroidery workers in Changcha gradually gained fame. From the eighth year of Xianfeng period (1858) to the end of Qing dynasty (1912), there were about 40 embroidery workshops in the city, among which the one set up by Hu Lianxian in the 24th year of Guangxu period (1898) became famous for its exquisite workmanship and distinctive style. The theme of Xiang embroidery is represented by lions and tigers, and the life-like fur is expressed by "fluffy hair stitch". This type of embroidery won four gold medals at the Panama Pacific International Exposition in 1915. At contemporary era, double-sided disparate embroidery, with completely different images, colors, and stitches on both sides of the fabric, is a token of its new development. The embroidery is mainly used for daily necessities such as costumes, skirts, chair cover, table skirts, etc., and there are also ornamental products including central scroll, strip screens and screens (as shown in Picture 4.107).

Picture 4.107 Xiang embroidery

4.5.3.4   Cantonese Embroidery

The city of Guangzhou in Guangdong province is the production center of Cantonese embroidery, which has a long history and was sold to Portugal, Britain, and France during the Zhengde period (1506–1521) of the Ming dynasty. According to *Records of Silk Embroidery in Cunsutang* by Zhu Qiqian (1872–1964), palace of the Qing rulers stored the Cantonese embroidery antique folding screen of the Ming dynasty, which is described as "stitches are thinner than the hair with great care". In the 58th year of the Qianlong period (1793), "Jinxiu Hang", the guild of embroidery, was established in Guangzhou. The embroidery works "Birthday Celebration of Guo Ziyi" and "Su Wu Herds Sheep" by Lin Xinquan, Wang Bingnan from Chaozhou won awards at the Nanyang Industrial Exposition in 1910. Cantonese embroidery, taking flowers, birds, dragons, and phoenixes with festive colors as its main themes, are characterized by full composition, neat shape, and exaggerating expression (as shown in Picture 4.108). Commonly used stitches include back stitch, blanket and layered stitches, satin stitch, fastening stitch, web stitch and so on.

Picture 4.108 Cantonese embroidery

4.5.3.5 Shu Embroidery

Shu embroidery is an embroidery centered on Chengdu, Sichuan province. Yang Xiong, a writer of the Western Han dynasty, describes it in his *Ode to the Capital of Shu*, "If you cover the brocade embroidery on the land, it seems that it is boundless". A representative work was named "Guan Shang Jia Guan" (depicting a rooster and cockscomb flowers, which are homophone to "official" in Chinese) made during the Shaoxi period (1190–1194) of the Southern Song dynasty. During the Daoguang period (1821–1850) of the Qing dynasty, workshops specializing in embroidery was set up and the number reached 75 during the Xuantong period (1909–1910). In 1915, Shu Embroidery won an award at the Panama Pacific International Exposition in San Francisco, USA. Since the 1980s, new technological improvement has been made, such as double-sided embroidery with different colors or shapes. This type of embroidery is characterized by the combination of both thick and thin silk thread and even stitches; and soft satin quilt cover are the main daily necessities made of it (as shown in Picture 4.109).

4.5.3.6 Bian Embroidery

Bian embroidery is centered in the city of Kaifeng, or Bianjing, the capital of the Northern Song dynasty. The government of the Song dynasty set up the Embroidery Institute in Bianjing, the capital, where there were over 300 embroiderers working. The street where most of the folk embroidery workers lived was named "Xiuxiang" (Embroidery Lane). Since the Qing dynasty, the theme of Bian embroidery has been focused on words of good wishes and characters of stories, especially the imitation of Chinese paintings and calligraphic works, with delicate style (as shown in Picture 4.110).

4.5.3.7 Miao Embroidery

Miao embroidery is of a unique artistic style by expressing the genesis myth and legends of the Miao nationality with a lot of deformation and exaggeration. The common patterns are mainly composed of geometric patterns, such as square, spiral, cross and so on. Miao women seldom make manuscripts when embroidering; instead, they rely on their inspiration and imagination to make plump embroidery works by skillfully combining the graphics with good proficiency and extraordinary memory (as shown in Picture 4.111).

Picture 4.109 Shu embroidery

4.6 Social and Cultural Attributes of Weaving, Dyeing, and Embroidery

Culture is the human material and intellectual achievement regarded collectively. Traditional weaving, dyeing, sewing and embroidery skills and their products, as the manifestation of Chinese wisdom and creativity, display multiple features of technology, history, and reality with rich cultural symbolism and meaning, influencing all aspects of Chinese people's daily lives.

Picture 4.110 *Along the River During the Qingming Festival* of Bian embroidery

Picture 4.111 Miao embroidery

4.6.1　Social Division of Labor Between Men and Women

It is recorded in the verse *July* in *Lessons of Bin State* in *Book of Songs*, "Yellowbirds are singing merrily in the warm spring sunshine. A girl, walking along the path with a bamboo basket, is picking tender mulberry leaves".

Treatise of Geography in *History of Han* also records that men cultivate grain and rice while women rear silkworms and weave. The natural economy of China is based on the social division of labor in which men farm and women weave, hence the saying that "people will suffer from hunger if men do not farm and from cold if women do not weave". Women have made great contribution to the formation and development of the Chinese nation.

4.6.2   Twelve Symbols of Sovereignty

Since China became a civilized society in the Xia dynasty, rites and laws have been in place to distinguish social hierarchy, with the system of official dresses as one of the important aspects. It is said that Emperor Shun stipulated the 12 symbols on the basis of the ancestors' costumes as the patterns exclusively on the emperors' costumes. The symbols are the sun, the moon, stars, mountains, dragons, pheasants, *zongyi* (tiger and monkey), algae, fire, rice, *fu* (黼, white and black axes), and *fu* (黻, blue and black patterns), each with special symbolic meaning. The 12 symbols have remained unchanged (as shown in Picture 4.112) since then. In the Western Zhou dynasty, rites and music were emphasized, and the shape and pattern of clothes and crowns were different according to the wearer's hierarchy, which was described as "the clothes of the noble and commoners are hierarchical…The world can judge people's social status according to their clothes". and "the noble men wear clothes that are decorated with pictures of mountains and dragons while low-status people wear coarse clothes". Colors are also used to show etiquette and enlightenment. Blue, vermilion, yellow, white, and black are considered as pure colors and green, red, bluish green, purple and dark yellow as secondary colors. Confucius once said, "Gentlemen don't wear clothes embroidered with reddish blue or reddish black, nor wear household apparel of red and purple color". Such kind of grading system of clothing is a symbol of autocratic cultural tradition.

4.6.3   Characteristics of Clothing Fashion in Different Times

Different times have different dress culture. During the Shang and Zhou dynasties, people's hair was bound up, and the left lapel of the garment overlapped the right one. People usually wore the upper outer garment and petticoats. During the Warring States period, King Wuling of Zhao carried out the policy of "shooting on horsebacks in Hu dresses", which changed the Chinese costume style of loose garments with a large girdle. Meanwhile, because Duke Huan of Qi liked purple very much, the Qi people took the color as fashion, and five plain clothes are not worth one purple dress. In the Qin and Han dynasties, it was the trend to dress in black clothes with purple silk ornaments. During the Wei, Jin, Southern and Northern dynasties, when the ethnic minorities invaded the Central Plains from the north, the symbolic elements

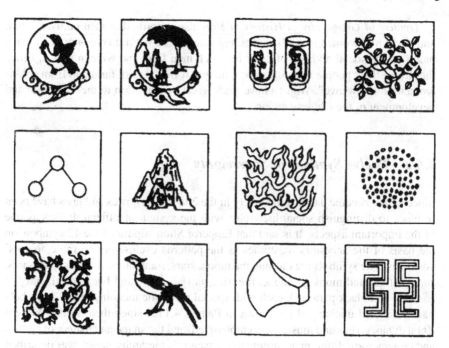

Picture 4.112 Patterns of 12 symbols of sovereignty

of their dresses such as narrow sleeves, close-fitting shape, round collar, and slits in garments were integrated into people's daily clothing. In the prosperous period of the Tang dynasty when Chinese and foreign cultures flourished, women's clothing was unprecedentedly luxurious and stylish. Influenced by Neo-Confucianism in the Song dynasty, the culture was relatively convergent, and the dominant clothes were simple and quietly elegant instead of being more elaborate as those in the previous dynasty. At the beginning of the Qing dynasty, Manchu women wore robes, while Han women still wore garments and petticoats. However, soon after, Manchu followed the trend of the Han nationality by wearing the Han clothes and replacing long robes with short clothes.

4.6.4 Production and Living Customs

Worship of the silkworm goddess and the practice of sericulture ceremony have been in existence since ancient times, which are recorded in many monographs and ancient books mentioned earlier. They are usually recorded at the beginning of the agricultural books in past dynasties. For example, *Book of Agriculture* by Wang Zhen (1271–1368) records that "the sacrifice to silkworms is like that of wine and food, which is to pay tribute to their inventors". There are special temples for the silkworm goddess in many places, and some sericulturists build shrines at home to worship the statue of silkworm goddess. Some of the customs related to sericulture, which were formed over the course of thousands of years, are for the healthy growth of young silkworms, such as "closing the silkworm gate", which is described in *Annals of Huzhou Prefecture during the Tongzhi Period* as "there are many taboos during the silkworm season. For example, neighbors don't pay visits to each other during that month. Fan Chengda of the Southern Song dynasty says in his poem, 'People don't meet each other during the mulberry picking season', from which we can see that the custom has a long history. During that time, even the government officials don't collect taxes". It is also a taboo to say words that are pronounced "*shu*", "*jiang*", "*pa*" and "*chong*", because they are homophones of some ominous words related to silkworm rearing.

There are various customs with special characteristics about festivals and marriage formed around silkworms and weaving and dyeing in places where silkworms are reared.

Silkworm prayer festival

In Hanshan county, the northwest part of Hangjiahu Plain in Zhejiang province, sericulturists, especially women, gather here to pray for a good harvest, commonly known as "silkworm prayer festival".

Touching the breasts of female sericulturists

When people gather for the silkworm prayer festival, unmarried men are allowed to touch the breasts of unmarried women, and those who are touched will be qualified to raise silkworms and her family's silkworms will surely have a good harvest that year.

Inviting silkworm babies

People in Yuhang, Zhejiang province call silkworms "babies". Before the Qingming Festival every year, silkworm egg sellers will give a picture of a goose, a chicken, and a horse to silkworm buyers, indicating that fortune and treasure are right at their doorstep. After the silkworm eggs come in the house, every household will paste their windows with peach blossom paper, stick red paper cuttings such as silkworm cats, treasure basins and cash cows, and insert peach branchlets onto doors and windows to express good wishes and ward off evil spirits.

Offering thanks to silkworms

After the busy season of silkworm rearing and silk reeling around the Dragon Boat Festival, sericulturists will worship the silkworm goddess to celebrate the harvest and thank her for her blessings, which is called "offering thanks to silkworms".

Sending silkworm eggs and looking after silkworms

When a boy and a girl get engaged, the girl's side often sends a sheet of silkworm eggs as a token to the boy's family, and the future mother-in-law must pick them up while wearing a red silk cotton-padded jacket, which is called "picking up silkworm eggs". In the first year of marriage to her husband's family, the bride should raise a sheet of silkworm eggs independently, which is called "looking after silkworms".

Sacrifice to the Gods of the Vat on the 9th day of the 9th month of the Lunar Calendar

The dyeing workshops worship Mei Fu and Ge Hong as the founders of the industry. They are collectively called "two Saints of Mei and Ge" or "two Immortals of Mei and Ge' and are honored as "Gods of the cloth dyeing vat" by dye makers. Mei Fu of the late Western Han dynasty was famous for practicing alchemy, and Ge Hong was a famous Taoist priest and alchemist in the Eastern Jin dynasty. According to folklore, the dyeing industry came into being because Mei and Ge taught people the dyeing skills. The craftsmen in the industry hold a sacrifice ceremony every year on the birthday of the "two Saints of Mei and Ge", that is, the 9th day of the 9th month of the Lunar Calendar (as shown in Picture 4.113).

4.6.5 Weaving, Dyeing, and Embroidery in Words and Literature

Textile, an important industry and means of livelihood, is surely reflected in words and literature. There are more than 100 Chinese characters with "糸" (meaning "silk") as the radical in oracle bone scripts in the Shang dynasty, and 267 ones in *Discussing Writing and Explaining Characters*. In *Jade Book*, a dictionary in the Southern and Northern dynasties, over 400 words with the radicals of 糸, 丝, 素, 索, etc. are included; in *Kangxi Dictionary* of the Qing dynasty, there are 830 characters with "糸" as the radical. In addition, there are also a large number of words and idioms related to sericulture, silk weaving, dyeing, sewing and embroidery, such as *sang yu* (with the literal meaning of "mulberry and elm trees" and extended meaning of "twilight years of people"), *sang shu weng you* (with the literal meaning of "pivot made of mulberry trees and windows made of urns" and extended meaning of "extremely poor"), *zuo jian zi fu* (with the literal meaning of "cocoon oneself like silkworms" and extended meaning of "be caught in one's own trap"), *qian si wan lu* (with the literal meaning of "thousands of silk threads" and extended meaning of "inextricable links with sb. or sth".), and *nie er bu zi* (with the literal meaning of "be dyed in a dark liquid without becoming black" and extended meaning of "be noble without being affected by bad environment"). Besides, some specialized words about textile are now common ones, such as *zong he* (meaning "synthesis", *ji gou* (meaning "institution") and *zu zhi* (meaning "organization"). Descriptions of spinning, weaving and clothing in literature are also very common, such as the stories about Luofu (a beautiful girl at the end of the Han dynasty) picking mulberry leaves and Xishi (a beautiful girl of the Yue State during the Spring and Autumn period) washing silk at the riverbank, and some well-known verse lines, including "Alas oh alas! Alas oh alas! Mulan is weaving cloth of topmost class", "The thread in the hand

Picture 4.113 Gods of the cloth dyeing vat

of the loving mother, is woven into the roving son's garments", and "She dares to boast fantastic needlework with ten fingers…making wedding gowns for other people but not herself". The poem *Crimson Silk Threaded Carpets* by Bai Juyi of the Tang dynasty, which depicts the author's sympathy of the hardships of sericulture, is one of the famous literary works that have been handed down through the ages. It goes like this, "Crimson silk threaded carpets, hands that marveled, Cocoon, selected and boiled, silk spools unraveled. Softened, whitened, red and blue dyed, fibers handled. When dyed red, the silk threads were red etched in blue, weaved silk into carpet in Chambers for dance retinue. More than thirty meters, the palatial area by measure, these crimson silk threads were hand sewn for full coverage. Multi-colored and fluffy, with whisk, fragrance exuded, soft threading, the virtual embroidery and real eluded. When beauties sing and dance on this velvety carpet, silk stocking and embroidered shoes sank with every step. …Was the prefect of Xuan Cheng aware of such hardships? Less than a meter of carpet required a thousand silk strips. The floor, ignorant of the cold, that the living required, don't steal human clothing for the ground to be attired". In the old days, officials didn't necessarily care about the hard work of cocoon selecting and silk reeling. However, we now know that one strand of silk doesn't come easily and therefore respect and value the creative work and achievements of silkworm raisers and weavers from generation to generation.

It is a routine for the evolution of production technology to replace the old with the new. Even though textile has developed into a highly mechanized and automated industry in contemporary times, traditional textile handicrafts including sericulture, silk weaving, sewing and embroidery still have their own value and space for existence and development due to the influence of cultural traditions and customs as well as the imbalance of social and economic development. At present, there are more than 70 weaving, dyeing and embroidery skills listed in *Lists of the National Intangible Cultural Heritage*, which will be protected by the government according to law. If combined with school education, vocational training, folk tourism, and cultural industry, they are expected to be better inherited, developed and revitalized.

4.7 Appendix: The Weaving, Dyeing, and Embroidery Skills Listed in *Lists of the National Intangible Cultural Heritage*

Comprehensive category (including silk, hemp, wool, and cotton)

Silk weaving in Hangzhou and Huzhou cities
Spinning, dyeing, weaving and embroidery of the Li nationality
Traditional cotton weaving in Hebei province and Xinjiang Uygur autonomous region
Grass cloth weaving in Jiangxi province and Chongqing municipality
Wool weaving and felt making in Sichuan and Gansu provinces
Bangdian and *Kadian* weaving
Tibetan blanket weaving in Yajia village
Uygur patterned felt and printed cloth
Etles silk weaving and dyeing
Rug making in Beijing, Inner Mongolia autonomous region and Xinjiang Uygur autonomous region

Brocade and *kesi* (Chinese tapestry)

Shu brocade
Cloud brocade
Song brocade
kesi (Chinese tapestry)
Lu brocade
Tujia brocade

Zhuang brocade
Dong brocade
Miao brocade
Dai brocade

Printing and dyeing

Nantong blue calico
Miao batik
Bai tie dyeing
Watered gauze
Liquidambar printing and dyeing

Embroidery, barbola and cross stitch

Gu embroidery
Suzhou embroidery
Xiang embroidery
Cantonese embroidery
Shu embroidery
Miao embroidery
Shui horsetail embroidery
Tu disc embroidery
Ou embroidery
Bian embroidery
Han embroidery
Qiang embroidery
Folk embroidery in the cities of Gaoping and Guangyuan
Yi (Sani) embroidery
Uyghur embroidery
Manchu embroidery
Mongolian embroidery
Kirgiz embroidery
Kazakh felt embroidery and cloth embroidery
Shangdang heap kam
Huangzhong barbola.
Qingyang sachet embroidery
Cross stitch (in Hubei, Hunan and Anhui provinces).

Chapter 5
Ceramics

Hua Jueming and Qiu Gengyu

Ceramics, porcelain, glass, and glassware are all silicate products, the production of which is called silicate engineering in modern times.

As early as 10,000 years ago, Chinese people were able to make ceramics. As the saying goes: "Shennong (the legendary Divine Husbandman) farms and makes ceramics". The emergence of ceramics was an extraordinary feat of the Neolithic Age. It was closely related to the agricultural revolution and the expansion of people's living needs at that time.

There are pieces of proto-porcelain found that date back to the Shang dynasty. The Yue kilns of the late Eastern Han dynasty could burn celadon, which is closer to porcelain in a real sense. The word "china" refers to porcelain, and when the initial letter "c" is capitalized, it becomes the English name for the country of China, which shows how far-reaching and significant China is as the mother of porcelain. In ancient China, ceramics was referred to as "*taoshan*" (陶埏). *Heavenly Creations* reads: "through the interaction of water and fire, clay can be burnt into ceramics", "ceramics are widely used among the population" and "ceramics are a concrete sign of civilized life". Ceramics is a technique as well as an art that uses water, fire, and earth. It passes on civilization, revitalizes the economy, enriches people's daily lives, and cultivates their temperament. The lives of Chinese people are closely entwined with pots, jars, cups, bottles, and plates.

H. Jueming
The Institute for the History of Natural Sciences, Chinese Academy of Sciences, Beijing, China
e-mail: huajueming@163.com

Q. Gengyu (✉)
Academy of Arts and Design, Department of Ceramic Design, Tsinghua University, Beijing, China
e-mail: qgy@mail.tsinghua.edu.cn

© Elephant Press Co., Ltd 2022
H. Jueming et al. (eds.), *Chinese Handicrafts*,
https://doi.org/10.1007/978-981-19-5379-8_5

5.1 Development

At early Neolithic sites such as the Nanzhuangtou site in Xushui, Hebei province, and the Xianren Cave in Wannian, Jiangxi province, coarse sand pottery of approximately 10,000 years old was unearthed. This pottery shows evidence of having been burned on flat ground at a firing temperature of about 700 °C. At those sites such as Peiligang in Xinzheng, Henan province, Cishan in Wu'an, Hebei province, and Hemudu in Yuyao, Zhejiang province, gray, black, and yellow sand pottery, clay pottery, and charcoal pottery were found from a later period, but with more shapes, larger bodies and more diverse decorations. Painted pottery from the late Hemudu period was also found, as well as a horizontal cave kiln with a firing temperature of about 800 °C to 900 °C that had been used at the Peiligang site.

There are many pottery relics from the middle and late Neolithic Age that were found all over Yangshao, Dawenkou, Longshan, Majiayao, Qijia, Daxi, Qujialing, Majiabin, and Liangzhu's cultural sites (as shown in Pictures 5.1, 5.2, and 5.3). First, the right pottery clay was selected. After which, plant ash, shells, or sand were added as auxiliary materials. The shaping methods included kneading, coiling, and molding. The introduction of the slow wheel and the fast wheel was of incredible significance for these methods. Decorative techniques such as smoothing, polishing, carving, printing, lacquering, and colored painting were pervasive, and the painted pottery patterns were vivid and realistic, fully showing the creative talents and aesthetic ideas of potters. All kinds of unearthed cooking utensils, storage containers, and food containers were made to meet the daily needs of our ancestors, but there were also ceramic devices that were not, such as net pendants, whorls and pellets. The structure of the kiln chamber and fire chamber steadily became more logical and well thought out, with more flame passages and fire holes, which made it possible for the firing temperature to be raised to 900 °C–1,000 °C.

White pottery can be traced back to the Shang dynasty. Its firing temperature was above 1,200 °C (as shown in Picture 5.4). Stamped hard pottery and proto-porcelain also emerged in southern China during the Shang dynasty. Stamped hard pottery was mostly made of china stone clay because the iron oxide (Fe_2O_3) content of clay

Picture 5.1 Early painted pottery basin decorated with dancer patterns, Majiayao culture, unearthed in Zongkou, Tongde county, Qinghai province

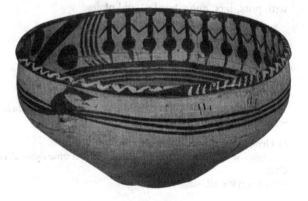

Picture 5.2 Black pottery
stem cup, Longshan culture

is low. It was fired in small dragon kilns, large fire chambers, small kiln chambers, and chamber kilns with chimneys. The firing temperature could reach 1,200 °C, which made the texture denser and harder (as shown in Picture 5.5). The surface of proto-porcelain is coated with glass glaze, using calcium oxide (CaO) as the flux (i.e. calcium glaze), and the surface is colored blue with gray, yellow with gray, or brown (as shown in Picture 5.6). During this period, ceramics were also used in buildings, such as three-way earthenware water pipes. By the beginning of the Western Zhou dynasty, tube tiles and flat tiles appeared (as shown in Picture 5.7). In the Warring States period, bricks were used on floors and walls, since these materials were generally readily available. The Terracotta Army at Qin Shi Huang's mausoleum in Xi'an really proves the scale and skill level of pottery making at that time, with its large size, large number, beautiful decoration, and fine firing techniques (as shown in Picture 5.8).

The celadon wares of southern China represent the use of Yue kilns in the late Eastern Han dynasty (as shown in Picture 5.9), which was a great leap forward in

Picture 5.3 Painted pottery, Lower Xiajiadian culture

Picture 5.4 White pottery
lei (a kind of wine vessel),
Shang dynasty

ceramic firing technology. China became the mother of porcelain, and for a long time it was the only country in the world that produced and used porcelain.

The differences between porcelain and pottery are as follows: (1) Pottery is made of fusible clay, while porcelain is made of china stone and kaolin; (2) Porcelain is fired in a dragon kiln at a temperature as high as 1,300 °C, containing more glass phase and mullite, creating a solid and dense texture that has a high strength; (3) The

Picture 5.5 Impressed pottery with trellis patterns, Zhou dynasty

porosity and water absorption rate of porcelain is very low, its surface has a glassy luster and the thin top layer is slightly transparent; (4) The color is white or slightly gray, as stated in *Heavenly Creations*: "Some of the pottery is as thin as paper, some as white as snow and as smooth as jade".

Celadon is a high-silicon and low-aluminum porcelain. Early production areas were mostly situated in the south, such as the Yuezhou kiln (越州), Wuzhou kiln, Yuezhou (岳州) kiln, Taizhou kiln, and Hongzhou kiln. In the Tang dynasty, and possibly even before that, celadon was also fired at the Wenzhou Ou kiln, Yixing Junshan kiln, Guanchong kiln in Xinhui, Guangdong province, and Qionglai kiln in Sichuan province. Objects at the Yuezhou kiln and Hongzhou kiln were fired in saggar as early as the Sui dynasty, which played an important role in ensuring the quality of porcelain and increasing the output.

The firing skills of celadon in the Tang dynasty and the Five Dynasties period became more advanced, and the shape, decoration and glazing more refined. The secret color (*mise*) porcelain made at the Yue kiln was loved by the people and was used to pay tribute to the royal family. It is considered a world-class treasure (as shown in Picture 5.10). At the Ou kiln, potters created overglaze and underglaze colors. The phase-separated opalescent glaze of the Wuzhou kiln was much earlier than the Jun wares in the Tang dynasty.

The successful firing of northern white-glazed porcelain in the Sui and Tang dynasties was another major technological breakthrough, which ignited a fierce competition between the two systems of "southern blue" and "northern white", and even

Picture 5.6 Proto-porcelain, Zhou dynasty

Picture 5.7 Eave tile of the
Warring States period,
Yanxiadu site

created prerequisites for the subsequent development of colored glazed porcelain and painted porcelain. Low-temperature lead-glazed pottery developed in leaps and bounds in the Tang dynasty. Its most representative works are the Tang tri-color glazed ceramics (as shown in Picture 5.11).

The white porcelain made at the Xing kiln, Gong kiln, and Ding kiln all used secondary sedimentary clay or kaolin and feldspar containing more kaolinite. This kind of high-aluminum low-silicon porcelain was the predecessor of modern kaolin-quartz-feldspar ternary porcelain. The content of iron oxide (Fe_2O_3) and titanium oxide (TiO_2) in this kind of porcelain was very low, so the base texture was white, and feldspar was added to the glaze, which greatly increased the potassium oxide

Picture 5.8 Terracotta
warrior, Qin dynasty

Picture 5.9 Celadon sheep
made in the Yue Kiln, the
Three Kingdoms period

Picture 5.10 Octagonal
vase, secret color
(*mise*) porcelain of the Yue
kiln, unearthed in Famen
Temple in Fufeng county,
Shaanxi province, the Tang
dynasty

(K_2O) content. By doing this, the traditional calcium glaze was transformed into a calcium alkali glaze and alkali calcium glaze. Thereby, it greatly improved the quality of the glaze.

The firing temperature of white porcelain exceeded 1,300 °C, as seen at the Xing kiln and Gong kiln, which reached 1,370 °C and 1,380 °C, respectively. The small updraft dome kiln was composed of a combustion chamber, a small kiln chamber, and double chimneys. It used wood as fuel. The improvement of the loading and firing process mainly involved selecting different styles of sagger cylinders depending on the shape of the vessel. The upside-down firing process created at the Ding kiln in the Song dynasty played a very great role in reducing the deformation of large utensils.

Early white porcelain was dominated by daily use utensils such as bowls, cups, plates, and jars, which were rarely decorated, but their rim, bottom, and base underwent great changes. In the later period, the level of carving and imprinting at the Ding kiln was very high, which had a profound influence on ceramic decoration practices (as shown in Picture 5.12).

From the Song and Yuan dynasties to the Ming and Qing dynasties, the Imperial kiln, Ge kiln, Ru kiln, Ding kiln, Jun kiln, Yaozhou kiln, Cizhou kiln, Jizhou kiln, Longquan kiln, Dehua kiln, and Jingdezhen kiln were all well known all over the

Picture 5.11 Ceramic horse, Tang tri-color glazed ceramics

Picture 5.12 White porcelain baby-shaped pillow, Ding ware, the Northern Song dynasty

world for their color-glazed porcelain, painted porcelain and porcelain molding. This enabled China's ceramic firing skills to reach its glorious heights (as shown in Pictures 5.13 and 5.14).

During the Southern Song dynasty, the Hangzhou Imperial kiln and Longquan Ge wares were known for their black base celadon glaze. The raw materials they used were mixed with purple gold soil, high in iron oxide. When they were fired in a kiln, the oxygen was reduced, turning the base black. A floating glaze with a strong jade texture was used and the color ranged from a light greenish blue to beige, and due to the fact that the potassium oxide content was high, it could maintain a thick glaze

Picture 5.13 Ge kiln's
five-footed writing-brush
washer, Northern Song
dynasty

Picture 5.14 Blue and
white porcelain vase painted
with the narrative scene of
Xiao He chasing Han Xin in
the moonlight, Jingdezhen,
Yuan dynasty

layer at high temperatures. The thin and thick glazed products that were made this
way were seen as their own separate category of porcelain. Another characteristic
was that the bottom, base and rim show the true color of the bottom layer, which
was commonly called "purple rim and iron-colored base". Most Imperial kiln and
Ge kiln porcelain showed crackles after firing because of the different expansion
coefficients of the base and the glaze. However, due to the skillful operation and
coloring techniques of potters, it became a unique way of decorating which proved

Picture 5.15 Three-footed case, Ru ware, Northern Song dynasty

very popular among the people, who commonly called it "gold-and-iron wire" or "gold-and-silver wire".

The white base celadon wares of the Longquan kiln were made of relatively pure china stone clay, which had a low iron oxide and titanium oxide content, and a white base color. The glaze was a transparent calcium alkali glaze colored with iron oxide, and the finished products were as green as jade.

The Ru kiln and Yaozhou kiln in the north also produced celadon wares colored with iron oxide. The former had a shorter firing time and only a few products remain in existence, which are now coveted (as shown in Picture 5.15). The latter was mainly known for its mold impressing and carving patterns (as shown in Picture 5.16).

Black glaze porcelain developed greatly in the Song dynasty. Two of its most famous pieces are the hare fur patterned porcelain cup from Jianyang, Fujian province, and the black glaze porcelain cup from Jizhou, Jiangxi province. After crystallization, phase separation, and recrystallization, the glaze of the hare fur could change into varying colors of gold and silver, depending on the production circumstances. When looking at the fur pattern from different angles, depending on the light's refraction, seven different colors could be seen.

This was because of a combination of the glaze phase separation and optical film interference, which was not only scientifically and technologically impressive, but also artistically.

The glazing techniques of the Jizhou black glaze cups were unique. First, the base surface was coated with black glaze, and then white glaze was applied through sprinkling, spraying, dripping, paper-cutting decals, and attaching tree leaves. After firing, various glaze layers, patterns, characters, and leaf patterns became visible (as shown in Picture 5.17). If you changed the concentration of white glaze and the size of the glaze drops, you could also get dot-like, spot-like, and strip-like patterns. As tea sets, these two kinds of porcelain have played a great role in what was fashionable in tea drinking.

Picture 5.16 Vase with interlaced peony design, Yaozhou ware, Northern Song dynasty

Picture 5.17 Black glaze
bowl with leaf pattern,
Jizhou ware

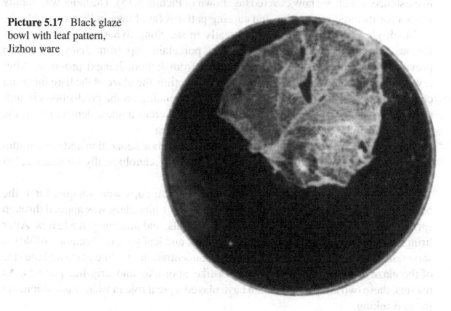

Picture 5.18 Begonia-shaped writing-brush washer, Jun ware, Song dynasty

At the Jun kiln in Yuzhou, Henan province, they applied red glaze with a copper compound as colorant. After firing, red and purple patches and strips in different sizes were distributed on the blue opalescent glaze in different colors, to fantastic and enviable artistic effect. The phase separation of this glaze was a very complex physical and chemical process, which was difficult to control. It was commonly called "kiln transmutation" (as shown in Picture 5.18).

From the Five Dynasties period to the Song dynasty, Jingdezhen in Jiangxi province was known for firing porcelain. With its unique resources and the wisdom and skills of its potters, the blue-white glazed porcelain it produced won a world-wide reputation for high quality and quantity, which also made it the largest and most famous kiln. During the Yuan and Ming dynasties, the white glaze porcelain of Shufu and the sweet white porcelain of Yongle in Jingdezhen were both superior in quality and appearance, which laid the technical foundation for subsequent colored glaze porcelain and painted porcelain. Since the Yuan dynasty, this kiln has been firing underglaze blue, underglazed red, and underglaze blue and red. In particular, blue-and-white porcelain has always been the largest, most distinctive, and most prestigious product of Jingdezhen. Yellow glaze porcelain, blue glaze porcelain, red glaze porcelain, and black glaze porcelain colored with transition metal oxides such as iron, cobalt, copper, and manganese, constituted a series of colored glaze porcelain. Low-temperature colored glaze porcelain that used lead oxide as flux was also one of its important products. In the middle of the Ming dynasty, Jingdezhen potters created the "clashing color", which was a combination of underglaze blue and overglaze. Chenghua clashing color porcelain became the most representative type of painted porcelain of this period, in particular due to its rich and gorgeous colors, its white and wet-looking glaze, and its vivid and diverse patterns (as shown in Picture 5.19). The low-temperature overglaze fired in the Qing dynasty was commonly called the "Kangxi Five Colors". The addition of arsenic oxide (As_2O_3) as an opacifier in the coloring material to make the tone light and soft, and the addition of golden red and antimony yellow to make the color richer, was commonly called "Yongzheng famille rose". Three emperors in the Qing dynasty, namely Kangxi, Yongzheng, and Qian-long, all liked porcelain, especially Qianlong. If the royal family liked something, it would become popular with civilians as well. The preference of the imperial family had a great influence on the prosperity of the porcelain industry and the improvement of skills (as shown in Pictures 5.20, 5.21, and 5.22). Similar to this is the firing of colored glaze tiles and their application in imperial religious buildings (as shown in Picture 5.23).

China's ceramic industry always had a difference between the Imperial kilns and the common kilns. With the power and strength of the royal family, the Imperial kilns could use all the assets of the national treasury and assemble excellent craftsmen and good materials from all over the country to fire fine products that were used by the royal family, nobles, and bureaucrats regardless of cost. The quality of the products was usually better than that made by common kilns, and there was innovation in both technology and art. However, this kind of production system often depended on the emperor's personal hobby, and the art was often complicated but not innovative in itself. The low cost of labor and material resources was not something that could be emulated at common kilns. Whereas, common kiln had the widest distribution of

Picture 5.19 Clashing color pot with sea mammal pattern, from the Ming dynasty's Chenghua period

Picture 5.20 Red glaze
vase with phoenix pattern,
from the Qing dynasty's
Kangxi period

products, the largest number of kilns, and the most vitality. Most of the major innovations came from these common kilns, and sometimes they were introduced to the
imperial house and reintroduced to the civilians, to be renewed and improved again.
This is how Imperial kilns and common kilns were interactive and complementary
to each other. However, from a historical point of view, common kilns have been in
the leading position.

Since the Ming and Qing dynasties, of all the famous kilns, only the Jingdezhen
kiln produced large quantities of colored glaze porcelain and painted porcelain of
various colors and was able to establish itself as the world-famous porcelain capital.
Later, Dehua kiln's white porcelain dominated, known for its exquisite porcelain
molding. It was also an influential kiln (as shown in Picture 5.24).

Ancient China attached less attention to science and technology than to statecraft
and literature, which means there were only a handful of descriptions of ceramic firing
techniques written. The earliest known article is *Pottery Records* written by Jiang Qi
in the Southern Song dynasty, which was later included in *Annals of Fuliang County*
during the Kangxi period (1662–1722). In the 10th year of the Chongzhen period
(1637) in the Ming dynasty, Song Yingxing's book *Heavenly Creations* explained
the pottery making technique of tiles, bricks, jars, and urns with a special chapter:
Ceramics. In view of porcelain making, the kilns of Dingzhou and Pingding in the
north, Dehua and Wuyuan in the south are mentioned, and the porcelain making skills

Picture 5.21 Famille rose vase with peach pattern, from the Qing dynasty's Yongzheng period

of Jingdezhen, kiln transmutation, and Mohammedan blue are thoroughly introduced. Japan, France, and Germany reproduced, and abridged these books in 1771, 1869, and 1882, respectively, which had an impact on the porcelain industry in these countries.

Porcelain, silk, and medicine made up the bulk of foreign trade in ancient China. As early as the Tang dynasty and the Five Dynasties period, porcelain produced at the Changsha kiln, Yue kiln, and Xing kiln in China was sold in large quantities via the Silk Road and the trading ports of Guangzhou and Ningbo throughout Central Asia, Iran, Korea, Japan, Southeast Asia, the Indian subcontinent, and Arab countries (as shown in Picture 5.25).

During the Song and Yuan dynasties, the sales volume of porcelain increased further and wider, expanding to West Asia and East Africa. The Jingdezhen kiln and Longquan kiln were important in the production of porcelain for export, especially the most popular blue and white porcelain. In the early Ming dynasty, the ban on maritime trade was strictly enforced, but the export of porcelain was never interrupted. During the Zhengde period (1506–1521), porcelain specially designed to supply overseas markets began to be fired (as shown in Picture 5.26), especially daily utensils such as tableware, which resulted in special names such as "Shantou Porcelain" and "Kraak Porcelain". During the Ming and Qing dynasties, millions of porcelain pieces were shipped to Europe every year, and blue and white porcelain

Picture 5.22 Glazed vases
in various colors, from the
Qing dynasty's Qianlong
period

spread all over Asia, Africa, Europe, and America. The large demand for porcelain caused the further development of the Jingdezhen kiln and Dehua kiln. After being shipped to Guangzhou, the white porcelain produced in Jingdezhen was painted using western painting methods, which was well received by Europeans. It was commonly called "Kwon-glazed porcelain", while Dehua porcelain enjoyed the laudatory title "Chinese White" because of its white glaze.

With the export of porcelain, China's porcelain-making methods also spread abroad. For example, Korea started making blue and white porcelain in the fifteenth century, and Persia began to fire porcelain with the help of Chinese craftsmen in the sixteenth century, which in turn affected the surrounding countries. Arabs spread Chinese porcelain-making technology to Italy and the Netherlands, which greatly improved European porcelain production. At the same time, Chinese porcelain workers also imported raw materials, techniques, and modeling utensils from abroad. For example, the earliest blue and white porcelain was fired with the technology of coloring with cobalt oxide from Western Asia, the Ming dynasty's monk's-cap jug and octagonal candlestick were modeled after similar ones from Western Asia and the Middle East, and the techniques of enamel painted porcelain and carmine colored porcelain in the Qing dynasty were also inspired by those in Western countries.

Picture 5.23 Colored glaze decoration on a building in Jiexiu, Shanxi province, Ming dynasty

Ancient people often praised the excellent quality of porcelain with the words: "as blue as the sky, as bright as the mirror, as thin as paper, and the sound as the chime stone." Among silicate products, porcelain dominated for a long time, while pottery was mostly used to make household utensils such as pots and jars. This situation changed in the Southern Song dynasty, which saw the rise and dominance of purple clay (*zisha*).

Yixing, Jiangsu province, called Yangxian in ancient times, made pottery as early as the Neolithic Age. It had been one of the pottery-making centers of the south during the Eastern Han dynasty. In the Tang dynasty, celadon wares were fired with dragon kilns and saggar, but due to a lack of high-quality kaolin, everyday pottery such as vats, jars, basins, and pots were produced the most. A lot of purple clay can be found in Yixing, which, because of its rarity, was referred to as "a place blessed with rich soil". This lump of rock, once separated from the ore bed, was milled, sieved, mixed, aged, and tempered. It had excellent plasticity and was most suitable for clay slab modeling, which entailed the characteristic forming technique of patting and pressing the pieces together, inlaying and looping, and filling.

Most of the purple clay utensils were purple clay teapots, and their rise and dominance were closely linked to the evolution of tea-drinking customs. In the Song dynasty, purple clay was only used for tea cups and kettles. The period from the

Picture 5.24 Dehua
porcelain molding, Ming
dynasty

Picture 5.25 Arabic
porcelain bowl, Changsha
ware, Tang dynasty

Picture 5.26 Blue and
white patterned ewer,
customized for the
Portuguese royal family,
Jingdezhen ware

Southern Song dynasty to the middle of the Ming dynasty was the pioneering stage of purple clay pottery. From the Ming dynasty's Jiajing (1522–1566) to Wanli period (1573–1620), tea making was changed from decocting to infusing, making a teapot essential. Hence, the purple clay teapot only came into being and even popular because of the affection and participation of tea-loving literati. From the late Ming dynasty to the present, these teapots have flourished. The art of making them has only become more and more refined, with a large number of masters coming forth, such as Gong Chun, Shi Dabin, Chen Mingyuan, Chen Mansheng, Gu Jingzhou, Jiang Rong, Wang Yinxian, and Lv Yaochen (as shown in Picture 5.27).

Picture 5.27 Gnarled
teapot, inscribed with "Gong
Chun"

Another wonderful type of traditional Chinese ceramic art is Shiwan ceramics molding. Shiwan is famous for making ceramics. It is said that "Shiwan tiles are the best in the world" and "Shiwan ceramics have taken over Guangdong and Guangxi, and conquered the markets overseas". People from the Lingnan area call the ceramic figures molded onto the surface of a wall tile or ridge tile a "figurine" (*gongzai*). Shiwan figures are mostly statues of gods or the Buddha, as well as figures molded on the surface of upright tiles, so they are also called "Shiwan *gongzai*". These high-grade products are molded by hand. Famous artisans of this art, from the Ming dynasty until now, include Su Kesong, Huang Bing, Chen Weiyan, Pan Yushu, Liu Chuan, and Zhuang Jia. Shiwan ceramic sculpture comes in a wide range of themes, most of which are popular story characters, animals, and ridge tile figurines. While they focus on glaze color matching, they often also use exaggeration and generalization techniques to make the sculpted figures have both form and spirit, such as making a "master of the pen" tall and a "master of the sword" short, removing the neck for heroes and removing the shoulders for beauties. Its local and common nature is distinctive, which makes it well-liked by the people.

The ceramic production techniques on the *List of National Intangible Cultural Heritage* include Yixing purple clay pottery, Jieshou painted pottery, Shiwan ceramics, Li primitive pottery, Dai slow-wheel pottery manufacturing skills, Uygur molded earthen ware, Qinzhou Nixing pottery, Tibetan black pottery, Yazhou pottery, Jianshui purple pottery, Xingjing sand ware, Jingdezhen handmade porcelain, Yaozhou ware, Longquan celadon ware, Cizhou ware, Dehua porcelain, glass (Mentougou of Beijing, and Shanxi province), Linqing tribute bricks, Ding porcelain, Jun porcelain, Tang tri-color glazed ceramics, Liling underglaze multicolored porcelain, Kwon-glazed porcelain, and Fengxi porcelain. With the joint efforts of ceramics artisans, communities, enterprises, and the government, ceramic-making skills can be better protected, developed, and passed on.

5.2 Manufacturing Skills

5.2.1 Raw Material Selection and Preparation

5.2.1.1 Clay

The clay used in pottery is mostly secondary clay. Its main component is hydrated aluminum silicate, which is classified into three types: brown, tan, and cinerous. It has strong plasticity but has many impurities.

In the early days, pottery was made from local materials. Later, people gradually learned to choose high-quality pottery clay as their base material, instead of just using the clay that was readily available. The collected clay must be air-dried and weathered. The weathering time can range from a few days, to half a year, to a year,

Picture 5.28 Hydraulic tilt hammer, Jingdezhen

but there are also some kiln clays that do not need to be weathered and can be used after being slightly treated.

The weathered clay is relatively loose and can be crushed with a mallet or a hydraulic tilt hammer (as shown in Picture 5.28). Sometimes it is necessary to sieve the clay to remove impurities or to separate the coarse materials from the fine materials and filter the impurities through elutriation (as shown in Picture 5.29).

After being crushed, the clay is soaked in water, blended into mud, and then put in a wooden trough or pit to preserve its moisture and have it age (as shown in Picture 5.30). The longer the mud is aged, the better the plasticity will be. In some kilns, they also mix mud through hand kneading, foot treading, stick hammering, and shovel turning (as shown in Picture 5.31).

Different regions and different types of mud have different properties, so potters often choose mud according to specific requirements. For example, the application of engobe is intended to close up the pores on the surface of the base and smoothen the rough texture and color of the base body. After that, through practice and craftsmanship, the potter makes full use of the engobe's covering and color-changing ability to portray the effect of different shades after glazing. This process actually results in a unique decorative technique.

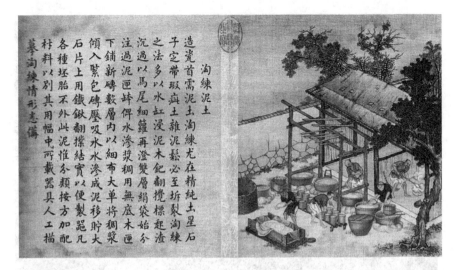

淘練泥土

造瓷首需泥土淘練尤在精純土星石
子定帶瑕疵土雜泥鬆必至坼裂淘練
之法多以水缸浸泥木钯翻攪標起渣
沉過以馬尾細籮再澄雙層絹袋始分
注過泥匣缽俬用無底木匣
下鋪新磚數層內以細布大單將稠漿
傾入緊包磚壓吸水水滲成泥移貯大
石片上用鐵鍬翻撥結實以便製器凡
各種坯胎不外此泥惟各類按方加配
材料以別其用幅中所載器具人工描
摹淘練情形志備

Picture 5.29 Sieving and filtering clay, cited from *Illustrated Handbook of Pottery*

Picture 5.30 Preserving moisture and aging

Picture 5.31 Cattle stepping on mud, Jianchuan, Yunnan province

5.2.1.2 Mixing Materials

Sand, clinker, plant ash, and charcoal crumb are lean materials. When mixed into mud, they can reduce shrinkage, prevent cracking and deformation, increase rigidity, and enhance severe heat resistance. Sometimes, different types of pottery clay can be mixed together and the same effect can also be achieved by increasing the sand content of the mud.

Sand, clinker, shells, charcoal, and coal slag must be crushed before they can be used as mixing materials. Plant ash, on the other hand, must be burned, either from wheat straw or rice husks. When burning it, the temperature needs to be tightly controlled since it is not usable if it is burnt to white ash.

5.2.1.3 Porcelain Clay

The selection and preparation of porcelain clay are more particular than that of pottery clay. For details, please refer to the relevant discussion on *Porcelain Firing* in this chapter.

5.2.1.4 Glaze

The commonly used earthen glaze is actually the same as the original ore glaze. The Dai people make lead glaze by stir-frying lead blocks and plant ash in a pottery pot to make a gray-green powder. To use the powder, water is added to turn it into a paste, which is then brushed on the surface of the base body. This paste becomes a translucent yellow-green after firing (as shown in Picture 5.32).

The glass glaze made by the Uyghurs in Kuche County, Xinjiang Uygur autonomous region, was made by smashing glass, grinding it into powder, and then mixing it with clay to form a slurry. Sometimes alkali is added to increase the viscosity of the glaze, which prevents the precipitation of the glass powder and enhances the adhesion of the glaze slurry. Green glaze uses copper oxide as a coloring agent.

For the glaze preparation at the Dehua kiln, please refer to *Dehua Porcelain Molding* in this chapter.

Glazing methods include dipping, pouring, swinging, coating, and blowing.

5.2.2 Shaping Methods

There are two types of methods for shaping ceramics: manual and mechanical. Here we expand on manual forming methods.

Picture 5.32 Firing lead glaze, Yingjisha county, Xinjiang

5.2.2.1 Kneading

Kneading is the most primitive and authentic forming method. Because they are kneaded by hand, these products are all small utensils and pottery mold, as well as pottery parts and accessories, with plain and simple shapes. In practice, kneading is often combined with patting, using tools such as wooden paddles and stone balls (used to mold the inner walls).

5.2.2.2 Clay Strip Coiling

Clay strip boiling refers to the shaping method of using a hand wheel or a pallet to stack clay strips layer by layer and builds a piece of pottery while rotating it. This method can be applied in various ways, such as clay circle stacking, spiral coiling, and spiral coiling combined with patting. The products made this way are mostly round. Take the Dai pottery made by Yu Wen, a potter in Mandou village, Jinghong, Xishuangbanna, Yunnan province, for example: The utensil she makes is small water tank. First, she dusts the slow wheel with ash to make it easier to separate the blank from it. She then uses a wooden paddle to pat a clay cake flat and thin to serve as the tank bottom (as shown in Picture 5.33). Turning the wheel with her toes, she sticks clay strips around the bottom of the tank one by one, and adds fine clay strips between the rings for reinforcement (as shown in Picture 5.34). After reaching the required height, it is trimmed with a bamboo knife (as shown in Picture 5.35). Then, the inner wall and rim are shaped with a wooden chip, so that the tank belly bulges inside and outside, and the rim folds (as shown in Picture 5.36). After drying slightly, a stone ball is placed against the inner wall and the outside of the wall is patted with a wooden paddle. Once the belly of the tank bulges outwards, the pattern is patted into it (as shown in Picture 5.37). The rim is wiped with a wet cloth, and the bottom and feet are trimmed with a wooden knife (as shown in Picture 5.38). After the base has dried, it is fired.

The above is just the general process of forming water pots using clay strips. The fine details of each process are not explored. The main purpose of this brief introduction is to show that the craft is rational, and that even the most primitive forms of pottery must conform to scientific principles as well as follow certain steps and specifications. It cannot be made by just anyone.

5.2.2.3 Patting

The pottery paddle is used to shape pottery, and there are different methods, such as (1) patting with a mud ring tube and (2) using the paddle while moving the wheel to make the rim and then the body (as shown in Picture 5.39).

Picture 5.33   Flattening clay to make the tank bottom

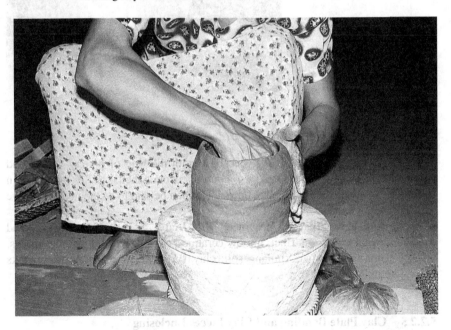

Picture 5.34   Clay strips as reinforcement

Picture 5.35 Trimming the edge

5.2.2.4   Molding

Modeling is the method of shaping the base with the help of pottery, wood, and clay molds. Taiwan's Yami people use gourds, coconut shells, and even their own knees and elbows as molds to make bowls. This can be said to be the most primitive molding method.

In Yixing city, Jiangsu province, Zibo city, Shandong province, and Maizhokunggar county in Tibet, hollow ceramic helmets are used as inner molds, while in Kuqa and Kashgar, Xinjiang, solid clay models are used to make water tanks (as shown in Picture 5.40).

5.2.2.5   Clay Plate Bonding and Clay Pieces Enclosing

Clay plate bonding is mostly used for making square utensils, with templates that can be used in mass production, while clay piece enclosing is used by the Tibetans in Zhongdian, Yunnan, to make a shoe-shaped teapot.

Picture 5.36 Shaping the
belly of the pot

Yixing potters are the best at this seemingly simple and primitive pottery craft.
Their top-quality purple clay pots are made by enclosing clay pieces.

5.2.2.6   Wheeling

Fast-wheel shaping is the most advanced technique in hand-made pottery, and it
is also the main shaping method used at most traditional kiln sites (as shown in
Pictures 5.41 and 5.42). The operation sequence is generally as follows: Spin the
wheel with your foot (as shown in Picture 5.43), and then use the stirrer to drive it,
so that it will continue to run through inertia. Use your hands to squeeze the clay
tuft into a column, then insert your thumb and push forward to form the bottom of
the pot. Then, lift the clay up with one hand on the inner and the other on the outer
sides to form the wall, which is called "throwing" (as shown in Picture 5.44). When
trimming the base, the fast wheel must rotate slowly, which is called "fast wheel slow
rotation". A wooden tool called an "angle plate" is used, which is operated by hand
and requires experience. Use a damp cloth or your fingers to smoothen the rim. The

Picture 5.37   Using pebbles and a wooden paddle to make the sides bulge

Picture 5.38   Trimming the bottom and feet

Picture 5.39 Shaping a large vessel through patting

base is cut from the wheel with linen thread or iron wire, which is called "cutting the base". Due to the manual operation and the rotating movement of the wheel, spiral fingerprints and lines will be left on the bottom and the walls of the pottery (as shown in Picture 5.45).

In actual production, various shaping methods are often used together, such as the combination of kneading and wheeling, or the sequence of kneading first and then wheeling.

5.2.3 Decoration Skills

5.2.3.1 Imprinting

Stamping, impressing, and mold impressing are decorative imprinting techniques.

Picture 5.40   Pottery horse, mold shaping

5.2.3.2   Clay Appliqué

There are three kinds of clay appliqué: appliqué, appliqué embossing, and pattern raking. Appliqué refers to the techniques of extension, rubbing, lining, and pressing. Appliqué embossing refers to three-dimensional embossing and high embossing. Pattern raking refers to using a mold to reproduce a unit pattern and then pasting it on the base to form an embossed decoration.

5.2.3.3   Engraving and Painting

There are three types, namely, engraving, painting, and incising. It takes a lot of effort to engrave patterns, which need to be carved deeply and in layers. Less effort is needed for painting, with its smooth and vivid lines. Incising is often used in combination with engraving and painting. The parts, other than the engraved patterns, are removed from the base body, which has been coated with glaze or engobe, resulting in embossing patterns (as shown in Picture 5.46).

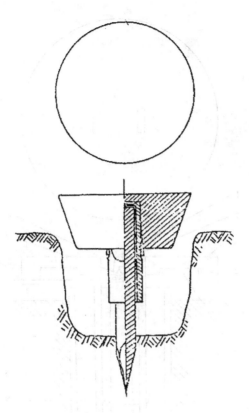

Picture 5.41 Structure of a slow wheel

5.2.3.4 Through-Carved Work

It can be divided into semi- and fully through-carved work. The wall of through-carved porcelain is transparent, which is also called "openwork carving" or "transparent patterns".

5.2.3.5 Colored Painting

It is divided into underglaze and overglaze (as shown in Picture 5.47). For details, please refer to *Jingdezhen Porcelain* in this chapter.

5.2.3.6 Inlaying

There is colored clay inlaying, porcelain inlaying, and metal inlaying (as shown in Picture 5.48).

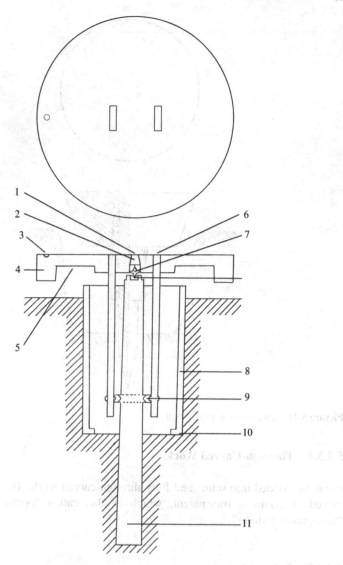

Picture 5.42 Structure of a fast wheel. 1. wheel center, 2. *luobo*, 3. wheel nest, 4. swivel, 5. rotating chest, 6. wheel eye, 7. *jiao*, 8. wheel leg, 9. *chuan*, 10. bottomless cylinder, 11. wheel shaft

Picture 5.43 Spinning a
fast wheel

5.2.4   Firing Methods and Facilities

5.2.4.1   Open Pit Bonfire Pottery Kiln

Dried pottery is piled on flat ground and burned with wood or rice husks, turf, and
cow dung. This is the most primitive firing method. The Wa people in Yunnan and
the Yami people in Taiwan still use this method (as shown in Picture 5.49).

5.2.4.2   Mud-Cake Thin-Shell Kiln

The kiln bed on ground is covered with branches and wood chips. Then the pottery
base is put on top, covered with straw, and then the mud is pasted onto the thin-shell
kiln body. A fire hole is pulled out from the bottom, and a smoke hole is opened at
the top after the fire is ignited. After about 6 h, the mud shell will begin to crack,
causing more air intake and more vigorous combustion. The cover plays the role

Picture 5.44 Throwing

of evenly spreading and preserving heat, and the mud shell can be taken out of the kiln in about one day and night. Yunnan's Dai people still use this method to make pottery.

5.2.4.3 Vertical Cave Kiln

The kiln chamber is in the shape of a vertical cave. The kiln bed with fire holes is separated from the combustion chamber by a basket. The pottery base is put into the kiln chamber through the top and it uses firewood and straw as fuel. The Xinjiang Uygur pottery kilns are square with smoke holes leading from the walls of the kiln to its roof. Stacked firing and nested firing techniques are used, with the mouth of each bowl and pot facing down while separated with soil nails. There are also similar pottery kilns in Yuanmou and Jianchuan, Yunnan province. Black pottery can be fired in these kilns by reducing flame (as shown in Picture 5.50).

Picture 5.45 Shaping tools

Picture 5.46 Engraving the glazed base and removing the blank glaze to reveal the true color of the base

Picture 5.47 Paining rust-colored patterns

Picture 5.48 Colored clay inlaying, Jianshui, Yunnan province

5.2.4.4 Round Kiln

The round kiln is also known as the steamed bun kiln or horseshoe kiln. The flames of a semi-inverted round kiln are stoked and pulled from the combustion chamber through the kiln top to the back of the kiln through a chimney (as shown in Picture 5.51).

Picture 5.49 Open pit bonfire pottery kiln

Picture 5.50 Dai people's Cave Kiln

Picture 5.51 Horseshoe kiln, Chenlu town, Tongchuan, Shaanxi province

However, in the improved fully inverted round kiln, flames are introduced into the kiln bed through a flame suction hole, resulting in a wider area being covered with a uniform heating base, which means a higher energy utilization rate, a shortened cycle, and better product quality (as shown in Picture 5.52).

5.2.4.5 Dragon Kiln

The dragon kiln is also known as the snake kiln or centipede kiln. It includes three types: (1) the full-body dragon kiln, (2) the split-chamber semi-inverted flame kiln, and (3) the horizontal-chamber connecting-room kiln, all of which are built in a long strip on hillsides (as shown in Pictures 5.53, 5.54, and 5.55).

The full-body dragon kiln can range from 10 m to nearly 100 m. It is built of refractory bricks or stones. There is a combustion chamber in the front, a number of firewood holes in the middle, and a chimney at the back. The flame jumps out

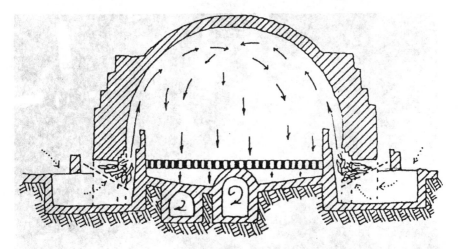

Picture 5.52 Structure of fully inverted round kiln

Picture 5.53 Dragon kiln, Longquan, Zhejiang province

Picture 5.54 Ascending kiln

Picture 5.55 Horizontal-chamber connecting-room kiln, Altay, Xinjiang

Picture 5.56 Full-body dragon kiln

along the body of the kiln, with the kiln temperature controlled by an experienced craftsman (as shown in Picture 5.56).

The split-chamber semi-inverted flame dragon kiln has several kiln chambers. Each chamber is provided with a separate kiln door, firewood hole, and smoke exhaust hole. There are fire holes under the fire barriers between chambers to form a semi-inverted flame structure. When firing one kiln, each kiln chamber is fired in sequence, which is an improvement on the full-body dragon kiln (as shown in Picture 5.57).

The kiln chamber of the horizontal-chamber connecting-room kiln is long in a horizontal direction and is equipped with multiple ignition devices. The fuels are long sticks of wood. Like the split-chamber dragon kiln, it is also fired from bottom to top.

This "fast firing" of pottery is unique. For example, in Zibo city, Shandong province, they use a rectangular kiln with multiple furnace openings, and porous pottery covers are used to buckle onto the furnace openings. After the pottery is burning red, it is taken out and covered with a nonporous pottery cover which blackens it by depriving it of oxygen (as shown in Picture 5.58).

For the assembly and firing of the base, cushions and spacers were used in the early stages. These were later changed to saggars, which prevented the fumigation of flames and smoke (as shown in Picture 5.59). In the Northern Song dynasty, people at the Ding kiln created the upside down firing process, which used combined gasket saggars. The base of each layer was placed on the gasket faced down. This way,

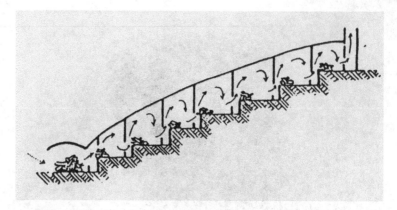

Picture 5.57 Split-chamber semi-inverted flame type dragon kiln

Picture 5.58 Pottery-covered fast-firing kiln, Zibo, Shandong province

they could make full use of the space of the kiln chamber and greatly increase the output. The technique was introduced to the southern regions of China later, making the pottery there more readily available. However, to avoid adhesion to the gasket, the mouth of the vessel was not glazed, resulting in a defect known as "*mangkou*" (unglazed rim). At present China, kilns in various places still mainly use saggars for loading and firing.

Picture 5.59 Saggar sets being loaded for firing

5.3 Ceramics Firing

Nowadays, China still has intact pieces of all kinds of ceramics ranging from its founding to when it became highly developed, which is rare and valuable. The following is an introduction to the representative art of ceramics firing.

5.3.1 Folk Ceramics

Hainan's Changjiang Li autonomous county and Sanya, a famous tourist city, still preserve their original pottery craftsmanship in its original ecology. This pottery, which is mostly made by women during the slack season, is all formed by hand or with clay sticks, piled up and fired on open pit bonfire. The products are mainly daily utensils as well as animal-shaped toys.

The original pottery techniques of the Dai people in Yunnan are spread across Mandou and Mannongfeng villages in Jinghong city, Manzha village in Menghai county, and Mangyang village in Menglian autonomous county in Xishuangbanna Dai autonomous prefecture. Take Mandou village for example. Its pottery workshop covers an area of about 90 square meters, with five slow wheels and a foot-pedaled tilt hammer to pound pottery clay. The clay is kneaded on raw cowhide and covered with burlap for storage. The slow wheel is driven by foot, and the mud cakes and the clay strips are formed and trimmed with wooden paddles, bamboo knives, and pebbles. The pottery base must be dried and preheated, and then placed on a kiln bed made of tree branches and wood chips, covered with straw, then coated with mud. Then the fire is ignited, smolders, cooled for about 10 h, and finally taken out of the kiln. The firing temperature is about 700 °C. Most of the utensils made are pots and jars. This production method enjoys a high and successful completion rate. The work at these kilns is all done by women. Famous artisans include Yu Tao, Ai Zhang, and Yu Wen (as shown in Picture 5.60). Men were not allowed to watch during firing. This is a custom inherited from ancient times and reflects the social division of labor back then.

The places where they use the above-mentioned primitive pottery-making techniques are few and far between. Other hand-made pottery in China is mostly made on fast wheels and fired in kilns.

The kiln field in Chenlu town, Tongchuan, Shaanxi province, has a long history and is enormous. It is known as the "furnace mountain that never sleeps". The earliest kiln temple and kiln god tablet *Deyinghou Stone Tablet* in China were found here. This tablet reads:

"The inhabitants make profit from pottery and rely on it for their livelihoods. Their pottery is as exquisite as gold and finer than cut jade. The clay is mixed and the wheel rotates to make the base. The size or circumference all comply with set

Picture 5.60 Clay strip coiling, by Ai Zhang, Menglian autonomous county, Yunnan

Picture 5.61 Double-wheel throwing, Chenlu town, Tongchuan, Shaanxi province

rules. Then the pottery is put into the kilns and fired. The flames are sent in, the blue smoke flies away, and after several days, the firing is done with impressive success. When you tap the completed product, it clangs. When you observe its color, it is warm and mild. The people can and do rely on it to provide for them. They owe it all to the help of the divine." Large products such as vats and bottles use double wheel to throw, which is a precise technical feature (as shown in Picture 5.61).

Yixing, a city in Jiangsu province, is known as the "Pottery Capital". Women in Yixing's Dingshu town make rough and fine ceramics with shaftless slow wheels by clay pieces enclosing and mold shaping to form the base, glaze it and then fire it in a dragon kiln. The resulting products include jars, pots, containers, and chamber pots. Because of their excellent texture, many distilleries still use the large jars made in Yixing to hold aging Baijiu.

In Cizao town, Jinjiang city, Fujian province, there are ancient kiln sites dating back to the Southern Dynasties period. Previously, more than half of the residents lived on pottery. The shaping method is a combination of windlass throwing and patting. Most products are unglazed and fired with wood in a dragon or ascending kiln. There are eight ceramics factories in Anhai town. Their main products are common earthenware, such as jars, pots, and containers. Smaller pieces are made using a windlass wheeling system, while large pieces are made by clay strip coiling and then patting. There is no other decoration used besides glazing. It is fired in a full-body dragon kiln with a firing temperature of about 1,100 °C.

Tibetan pottery is the most recognizable of the ethnic minorities. The pottery is made by men in Tangdui village, Zhongdian County, Yunnan province. At the age of 12, craftsman Sunnuo Qilin became apprentice to Nong Buen, who had a total of 14 apprentices, including his eldest son Lurong Enzhu. The clay is selected from the mountain slopes to the southwest of the village. They have two kinds of clay, (1) red clay and (2) white clay, which are mixed at a ratio of 2:1. Pottery is mostly made on the second floor of a Tibetan *tulou*, and the utensils made include teapots, yak butter teapots, braziers, hot pots, and earthen pots (as shown in Pictures 5.62 and 5.63). The shapes of these vessels are quite complicated. They are shaped by clay pieces enclosing or clay strip coiling with a shaftless slow wheel. The tools used are mallets, wooden paddles, wooden cones, pottery helmets (mold), and wooden slabs. The decoration methods include patting, stacking, pasting, carving, and porcelain inlays. The patterns include dragon patterns, flower and plant patterns, geometric patterns, and mat patterns. The bases are dried on a drying platform or in their houses, using the heat generated by the firepit and then piled on the flat ground. The fuel they use is pine wood. Tibetans like to make black pottery with gold and silver for decoration. Some pottery has copper parts, such as the lid and spout of a yak butter teapot.

Uyghurs, such as those in Mangxin town, Yingjisha county, southern Xinjiang, make ceramics as a family unit (as shown in Picture 5.64). A famous Uyghur artisan, Aili Aimaiti, learned to make ceramics at the age of 7. The locals addressed him respectfully as "Wusida" (master). Their pottery workshops have two floors, the lower floor is used to store ceramic mud and contains a small kiln. There are holes between the floors and double-wheeled windlass are equipped on the beams. The operator sits on the beam to pedal the wheel and conduct the throwing. The bases are decorated with techniques such as engraving, painting, printing, pasting, and painting. The patterns are mostly geometric, flower, and plant patterns, which are

Picture 5.62 Tibetan potter Sunnuo Qilin

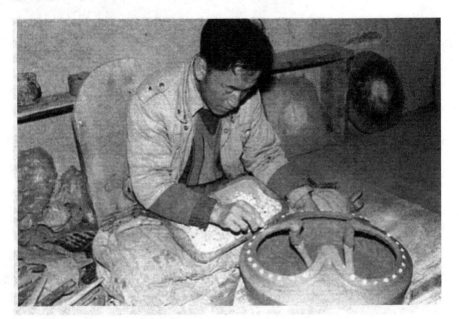

Picture 5.63 Tibetan black pottery brazier, inlaid with porcelain pieces

glazed by brushing, painting, and dipping. The glaze is a low-temperature lead glaze made in-house. Copper oxide or rust is added to form a green or a yellow glaze. The wares they make include bowls, plates, basins, jars, pots, candle holders, and vases, which are loved for their bright colors, different shapes, and strong ethnic characteristics. These products are usually sold on markets and sometimes to other vendors who come to trade goods. The *nang* (a kind of crusty pancake) pit made in Gongliu county in northwestern Xinjiang is a stove for roasting Uyghur *nang*. It looks like a water tank without a bottom. It is shaped on a hollow inner mold and fired in a horseshoe kiln, which makes it quite distinctive.

Earthen pot, most commonly known for its clay baking pots, is a kind of pressed sand earthenware. It is commonly used to cook food and decoct traditional Chinese medicine. Because of its good heat resistance, it does not crack easily. The cooked food and boiled traditional Chinese medicine have no peculiar smell either and are perfect for storage. This makes it popular and widely used (as shown in Picture 5.65).

Different from other pressed sand earthenware, the mixing material of sand ware is coal powder or coal slag, which can ignite spontaneously and quickly heat up during firing. The sand ware can be fired in 10–20 min and sometimes even within a few minutes (as shown in Picture 5.66). Important areas that make earthen pots include Sichuan's Yingjing, Guizhou's Zhijin, Guangxi's Luocheng, Shanxi's Pingding, Hebei's Handan, and Shandong's Zibo.

There are many places where they make pottery kilns. This chapter does not include all of them. For details, please refer to Qiu Gengyu's book entitled *Research on Modern Chinese Folk Ceramics*.

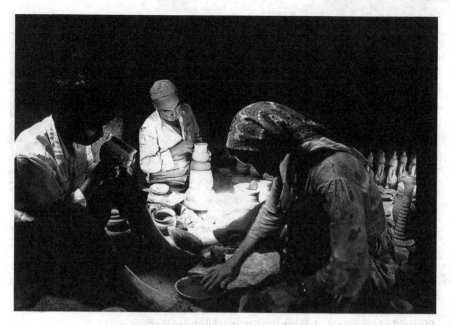

Picture 5.64 Uyghur in-house pottery making

Picture 5.65 Earthen pot

5.3.2 Low-Temperature Lead-Glazed Ceramics

Low-temperature lead-glazed ceramics use lead oxides as flux, and use iron, copper, cobalt, manganese oxides as colorants. The melting point of the glaze is about 800°C. The viscosity is limited, the fluidity is good, the glaze is clear and transparent, and the glaze surface is smooth and rich in luster. On the other hand, its chemical stability is poor, it corrodes easily, has low density, and scratches easily.

Picture 5.66 Taking out the red hot earthen pots from the furnace and place them in the thermal pit

5.3.2.1 Tang Tri-Color Glazed Ceramics

Lead-glazed ceramics were first created during Emperor Wudi's reign of the Western Han dynasty, mostly with green glaze and some with yellow glaze. The Tang dynasty was the peak of firing low-temperature lead-glazed ceramics, especially during the Kaiyuan and Tianbao periods (713–756). White clay was used to make the base and was fired at temperatures from 750°C to 1,000 °C. The items that were made include both daily utensils and building items. The most commonly made were pottery figurines, pottery horses, camels, and tomb beasts due to the popular elaborate burial traditions at the time. The female figurines wear long dresses; have their hair in a high bun and a plump face. The Hu people figurines, typically leading a camel carrying people and singing. The colorful glaze reflects the atmosphere of the prosperous Tang dynasty and the extravagant life of the aristocratic bureaucracy. It was mainly produced in Chang'an, the capital, and in Luoyang, the eastern capital. It gradually declined in the middle and late periods, and almost became extinct during the Five Dynasties period. However, it saw a revival during the Song and Yuan dynasties, when it was mostly used for creating common household utensils and bricks. There is no historical title officially named after the "Tang tri-color glazed ceramics". When the Longhai Railway was built in the early twentieth century, many lead-glazed ceramics were unearthed from the Tang tombs in Luoyang. They were mostly green, yellow, and white, so they got the name "Tang tri-color glazed ceramics" and this name simply stuck and is still used today.

5.3.2.2 Colored Glaze

Colored glaze originally refers to the colored glass imported from overseas in the Wei and Jin dynasties, which is also known as glassware. Later, lead glaze containing glass was applied to tiles, and this type of low-temperature lead-glazed ceramics became known as colored glaze. Colored glaze products have various forms, require exquisite skill, display vivid colors, and have high artistic value. Colored glaze buildings, rich in ethnic characteristics and cultural significance, are a fine product of traditional architecture and still have further development potential in the modern age.

Colored glaze was first fired in Shanxi, and then spread to Beijing, Shandong, Henan, and Nanjing. The palaces of the Northern Wei dynasty were already decorated with colored glaze, but these techniques really matured in the Tang dynasty. Wu Zetian's *Tablet of Prince Becoming Immortal* at Gou Mountain, Yanshi, Henan province, states that "the households hang colored glaze with the same color as a sunny day". The glaze used at that time was yellow, green, blue, white, and ocher. The Song dynasty even used black-colored glaze. For example, the Kaifeng Iron Pagoda was built using it. In the Yuan dynasty, kilns were set up in Dadu to fire plain white-glazed bricks and tiles. Several famous buildings, such as the Jin Memorial Temple in Taiyuan, Huayan Temple in Datong, and the Yongle Palace in Ruicheng, all use colored glaze components. In the Ming dynasty, imperial-colored glaze kiln factories were established in Beijing and Nanjing. This dynasty's great works of glazed architecture is exemplified in the Forbidden City, Temple of Heaven, Temple of Earth, Temple of the Sun Altar, and Temple of the Moon Altar, and the Beihai Nine Dragon Wall.

In the Qing dynasty, a special colored glaze kiln was set up in Mentougou, Beijing. Under the imperial dictatorship, the architectural format and materials became graded. Common people could only use gray flat tiles, noble bureaucrats gray tube tiles, princes and nobility green-glazed tiles or trimmings, and only the royal buildings and temples could use yellow-glazed tiles and trimmings. The multi-colored glaze tiles that contained purple, green, black, and blue were used in the palace and imperial gardens are called "Collection of Colors" (as shown in Picture 5.67).

Colored glaze was mostly fired in Shanxi, more specifically in Taiyuan, Yangcheng, Hejin, Jiexiu, and Yangquan (as shown in Picture 5.68). In Taiyuan's Mazhuang village, there are three families (Su, Bai, and Zhang) who have been producing colored glaze ever since the Wanli period of the Ming dynasty. Su Jie, the sixth generation descendant of the Su family, was the most influential artisan of colored glaze. His last disciple, Ge Yuanyuan (1940–), who graduated from Shanxi Agricultural University, developed a peacock blue glaze formula and obtained a national patent on it. He is the author of *Introduction to Glass Craftsmanship* and *Colored Glaze Firing Crafts*. He has a wealth of practical experience and is an academic expert on the subject.

Picture 5.67 Colored glaze screen wall of Haihui Temple, Yangcheng, Shanxi province

Picture 5.68 Firing colored glaze

It is said that colored-glaze firing in Yangcheng started in the Yuan dynasty. The Qiao family had the largest number of various colored-glaze schools in Shanxi and the longest history. There are still two ceramic factories producing colored glaze in Houzeyao village, Yangcheng today.

Hejin-colored glaze in Shanxi province is most famously produced by the Lu family. The colored glaze components at Temple of Guan Yu in Jiezhou, Nainai Temple in Yongji, Town God's Temple in Quwo, and Five Dragon Palace in Jishan were all made by the Lu family. Lu Hongjian, a descendant of the eighth generation, was part of the restoration of Confucian Temple in Yuncheng and Yongle Palace in Ruicheng. His sons Lu Yantang and Lu Yanrong set up Lu's Colored Glaze Factory and Colored Glaze Craft Factory in the early 1990s (as shown in Picture 5.69).

The major processes of making colored glaze are: material preparation, shaping, biscuiting, glazing, and glaze firing. Different from other ceramics, colored glaze must be fired twice. The first time is called biscuiting, which requires a high temperature, and the second one is glaze firing at a low temperature. The clay used is a low-aluminum crucible soil. The composition and properties of the clay are judged empirically using traditional methods such as visually inspecting, kneading, licking, scratching, and biting. For example, if a section of clay is licked with the tip of the tongue and the clay's suction is strong, it shows that its absorption capacity and plasticity are better. The shaping method depends on the type of object that needs to

Picture 5.69 The ninth-generation descendants of the Lu family skilled in Hejin-colored glaze, the Lu brothers as well as the photos of their father, grandfather, and great-grandfather

be made. Since the shape of the roof ridge and the dragon-head ridge ornament are usually vivid and full of changes, they are often made by hand. On the other hand, the shape of the tiles is uniform and often molded. Mud used for shaping should be mixed with cotton to prevent cracking.

The high-temperature biscuiting of colored glaze is about 1,100 °C–1,150 °C. The preparation of the glaze is to fry the lead ore or a lead block into powder in an iron pan under gentle heat. Then, after sieving, it is rinsed with water. Dipping, pouring, and brushing methods are used to apply the glaze, and the thickness of the glaze slurry is controlled solely through experience. The preparation of colored glaze is the most difficult and the most secret technique in this industry. There is a saying that it is "passed on within a family, from daughter-in-law rather than daughter". Even nowadays, only a few people know it and have mastered it. Low-temperature glaze firing uses wood as fuel, since the resulting ash is relatively pure and does not easily pollute the glaze surface. The firing temperature ranges from 800 °C to 950 °C. The control of the kiln temperature has a great influence on the color of the glaze. This is why kiln workers play the most important role in the kiln yard.

5.3.3 Purple Clay Pottery (Zisha)

Yixing's purple clay is mined in open air or by tunneling. The mined ore must be weathered, aired, crushed, sorted, milled, sieved, soaked and stirred into wet mud, and then crushed, cut, and aged into raw mud. It must then be moisturized, aged, and tempered before it becomes pugged mud suitable for making pottery (as shown in Picture 5.70).

Purple clay mud is a kind of native clay, which produces excellent pottery without the need of adding other raw materials. Its main components are quartz, clay, and mica with iron oxide, making it a type of clay-quartz-mica clay. Due to its excellent plasticity, it can be repeatedly processed and is suitable for clay plate shaping. In addition, due to its high strength, it only shrinks a small amount and does not easily deform during firing, which makes it perfect for producing complex shapes. The clay contains just the right amount of sand so that the product can withstand both the heat and cold, making it ideal for tea sets.

Most of the purple clay is formed by hand, requiring various tools to create various shapes. Artisans of purple clay pottery pay great attention to their tools, and often make their tools themselves. They sometimes have to prepare special tools to suit specific pottery shapes. Commonly used tools are wooden bats for patting mud slices and strips, wooden paddles for shaping pottery bodies, spouts and kettle buttons, shaftless slow wheels, bitterling-shaped knives, sharp knives, *guiche* for cutting round mud slices, *qiangche* for cutting mud strips, and ox horn slices for scraping (as shown in Picture 5.71). The shaping methods of purple clay pottery include patting the body, splicing the body, and making the base, which are collectively referred to as clay plate shaping. Taking the patting method as an example, there are six major steps: (1) Make various round clay pieces that will make up various parts of the kettle, while

Picture 5.70 Pugged purple clay

some of them are only used to support the spout during the modeling process; (2) Pat the clay pieces to make into the kettle body, enclose them, press them together tightly, shape them, and then press them onto the kettle bottom so that the spout can be retracted. In order to keep the spout from deforming, the spout is sealed with clay; (3) Press and roll the kettle body out with a paddle, trim its shape, scrape and press the kettle body with a comb to make it smooth and regular, cut out the pieces for the neck and adhere them to the kettle body; (4) Knead the kettle mouth and the kettle handle, and press them onto the kettle body; (5) Make the top opening and the lid that fit onto each other to finish the overall shape. Then, make the lid button and cut out the spout that was sealed with clay before; (6) Trim the inner wall of the pot with bamboo tools, install the lid, and let it dry in the shade (as shown in Picture 5.72).

The shape of purple clay teapots can be divided into three categories: embossment pottery, plain pottery, and longitudinal line pottery. Embossment pottery involves embossing, engraving, and kneading techniques to incorporate animal or plant patterns onto round and square pottery, such as pumpkin, pine, bamboo, and plum teapots. Plain pottery is made of geometric shapes such as spheres, cylinders, and squares. They are concise, stiff, and smooth, with clear lines and angles, and the most notable ones are the Han flat teapots and the monk's hat teapots. Longitudinal line pottery is made by embossing petals, melon ridges, and other ribs on the round pottery's base, most exemplified by the water chestnut teapots and the wind-rolled sunflower teapots. The above-mentioned patting method can be used to make round pottery, embossment pottery, and longitudinal line pottery. The splicing method is

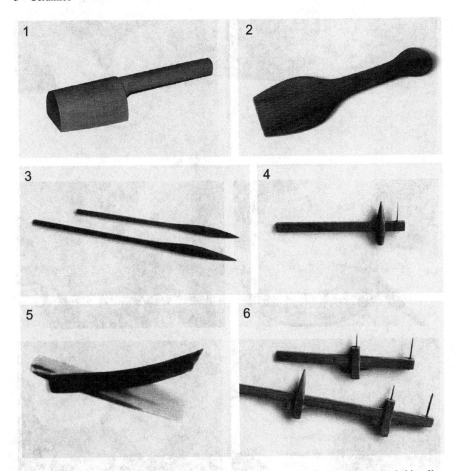

Picture 5.71 Tools for making purple clay pottery. 1. wooden bat 2. wooden paddle 3. bitterling-shaped knife 4. *guiche* 5. ox horn slice 6. *qiangche*

used to make square and polyhedral wares. The base is formed with a mold and then bonded into a whole.

The decorative methods of purple clay pottery include pottery carving, clay painting, piling, twisting, sculpting, and color painting, but the most common method is pottery carving. Potters often inscribe vessels with poems and characters and combine the shape of the pots with poems, calligraphic works, paintings, and seal cuttings to form a unique decorative effect. The participation of literati imbues purple clay pottery with a rich cultural significance and a fresh, elegant, and artistic character.

The purple clay wares were all set in pottery jars and earthen urns and were fired in a dragon kiln previously. After the mid-Ming dynasty, a special dragon kiln was built, which used hay as fuel and the purple clay ware was burned in a saggar (as shown in Picture 5.73). After the 1950s, it was fired in a reverse-flame round kiln and in a tunnel kiln.

Picture 5.72 Making round pottery. 1. patting mud pieces (about 3 mm thick), 2. using a *guiche* to cut the round bottom piece, full piece, neck piece, and enclosing piece, 3. cutting the mud cylinder with a bitterling-shaped knife, 4. inlaying the bottom piece, 5. using a bamboo comb to standardize each part of the pot and complete the base, 6. bending the hollow mud stick into a kettle mouth, 7. adhering the mouth, handle, and body to each other, 8. checking the fit between the spout and the lid

Picture 5.73 Dragon kiln in Yixing

Based on the above information, it can be seen that the production process of purple clay pottery is extremely complex and at the same time extremely standardized. There are hundreds of processes in the making of pottery, which are independently completed by a potter in strict accordance with the predetermined procedures. Although it is made by hand, it can be repeatedly produced and still achieve uniformity of product in shape and size. Although some shapes are simple and bright, the detailed description and line processing are very delicate. This shaping method seems to be primitive, and the shaftless slow wheel and wooden paddle used are still the same as the tools used at the very beginning of pottery making. However, under the careful operation of purple clay artisans, the manual skills handed down for thousands of years have been brought into full play, thus creating works with high precision, high standards, artistic value, and rich cultural significance, which can be called the best kind of craftsmanship.

Since the Ming dynasty, famous artisans have emerged from generation to generation in the purple clay pottery industry, and the exquisite pottery art has been well-known for a long time. This art is inseparable from the fact that any purple clay teapot is sturdy and practical in function, elegant and refined in aesthetics, and rigorous and standardized in production. The underlying reason is that throughout the whole development process, the overall harmonious technical concepts have always held a dominant position and have been passed down for a long time through the visionary talent-training mechanism. One example of this is when, in the 1950s and 1960s, young talents were sent to art colleges for further study. This resulted in us, now, having many people that are master inheritors. Among the traditional handicrafts, the development of the purple clay industry is stable, continuous, and keeps pace with the times, which is worth learning from.

5.3.4 Shiwan Pottery Molding

Shiwan ceramics in Foshan, Guangdong, began in the Song dynasty and became a famous ware in the Ming and Qing dynasties.

The ceramic figures are closely related to the firing of glazed ridge-tile ceramic figures. The local people call the ceramic figures on the glazed ridge tiles of buildings "*gongzai*".

Therefore, Shiwan figures are often collectively referred to as "Shiwan *gongzai*", especially the ridge-tile ceramic figures (as shown in Pictures 5.74 and 5.75).

Most of the bulk products of Shiwan ceramics are made by semi-manual methods, such as printing bases, while high-grade artworks are molded by hand. Pottery artisans find inspiration in real life and, after refining, summarizing, and exaggerating, they create works full of life. They even strive to create humanized images of Buddhist and Taoist immortals. Some characters do not conform to their true proportions, but they are still natural and more artistic. For example, ridge ceramic figures often have small heads and big feet and lean forward, which is to correct the optical illusion caused by looking at them from below and at a distance. Scattered perspective is adopted in the layout of human scenery, which makes the artistic conception rich and far-reaching.

There are four kinds of ceramic-making techniques and expression methods:

Picture 5.74 Ridge-tile ceramic figures

Picture 5.75 Shiwan ceramic figure

5.3.4.1 Pasting

The technique is mainly used for ridge ceramic figures. The works thus made are ethereal and transparent, with distinct layers and a strong focus on decoration and optical correction.

5.3.4.2 Kneading

The technique is vigorous and bold, which is similar to the freehand brushwork in traditional Chinese paintings. It is mostly used on rocks and figures.

5.3.4.3 Pressing

Like kneading, pressing is also a freehand technique mostly used to create the shapes of utensils, birds, and animals.

5.3.4.4 Knife-Molding

This technique requires the use of a blade and is mostly used for antique production.

Most of the ceramic figures are specially shaped, so they must be put into saggars for firing. The firing temperature is high and the holding time is long so that the color

Picture 5.76 Purple clay teapots. 1. antique *Jinglan* teapot, made by Yang Pengnian, inscribed by Chen Mansheng, 2. one of the 18 styles of Chen Mansheng, 3. loop-handle jade-like teapot by Gu Jingzhou, 4. purple clay teapot made Gu Jingzhou, a famous master

glaze can fully melt. This is why the kiln body is made low, why the slope is gentle, and why the temperature rises slowly (as shown in Picture 5.76).

Early famous artisans in Shiwan included Zu Tangju and Su Kesong in the Ming dynasty, and Huang Bing, Feng Zhilai, Huang Guzhen, and Liao Rong in the Qing dynasty. Huang Bing was good at paintings and calligraphic works and was especially good at animal shapes like ducks, cats, monkeys, and his "fetal hair" technique is one of the defining characteristics of Shiwan figures. Famous artisans from the late Qing dynasty to the Republic of China included Chen Weiyan and Pan Yushu. Chen studied under Huang Bing, and his Taoist immortals and Buddhas are both vivid in shape and smooth in appearance. His most representative work is "Nun". Pan's talents were outstanding and he was good at shaping historical figures. His works are of a scholar's style. His representative works include "Treading Snow to Seek Plums" and "Drunken Concubine". Contemporary masters include Liu Chuan, Zhuang Jia,

Liu Zemian, and Liao Hongbiao. They inherit the local characteristics and techniques of the Shiwan ceramic figures and, as a result, have created many innovative works of their own.

5.4 Porcelain Firing

Porcelain must be made of porcelain clay and fired at a high temperature above 1,200 °C to form a vitreous glaze layer on the surface. This has been the way that porcelain has been produced since its inception, resulting in its excellent quality and a wide variety of beautiful wares.

5.4.1 Celadon Wares

Celadon ware, also known as celadon, is a porcelain that is fired at a high temperature with green or yellow iron-containing glaze on the body. It is also the earliest "real" porcelain after it was developed based on the production of proto-porcelain.

Celadon wares were created in the Eastern Han dynasty at the Yue kiln in Shangyu, Zhejiang province and continued to expand in scale and technology in the Song dynasty for more than a thousand years. It was the most dominant type of porcelain for a long time (as shown in Picture 5.77).

The glaze of Yue wares was thin and even, showing a lake-green color. Yaozhou kiln in the north created a transparent glaze and boasted mold impressing and carving patterns. Ru kiln celadon was valued all over the world for its perfect color due to proper iron content allocation and reducing flame control, which made the glaze perfect. The quality of imperial wares in the Southern Song dynasty continued to improve, and new varieties were created, which had a great influence on the celadon at the Longquan kiln celadon wares. The two were listed as the representatives of southern celadon. At the beginning of the Ming dynasty, with the emergence of Jingdezhen blue and white porcelain and its unique position in the world, the demand for celadon started to decline. The Longquan kiln ceased firing in the early Qing dynasty and only resumed production in the late 1950s. Artists such as Mao Zhengcong and Xu Chaoxing made important contributions to the restoration of celadon.

The glass glaze of celadon was translucent, and the tone of its color was affected by the quality of the body. The imperial kiln and Ge kiln produced black body celadon by mixing the clay with purple gold soil with high iron content and firing it in an oxygen-deprived atmosphere. The Di kiln produced white body celadon, by using china stone clay with low titanium and iron content, which resulted in a whiter color and green jade-like texture. In order to obtain a mild and subtle jade effect, thick glazes were often used, often even thicker than the base. For this reason, it was necessary to carry out biscuit firing and glazing multiple times. Due to the different

Picture 5.77 Celadon lotus wine vessel, Northern and Southern dynasties

expansion coefficients of the base and the glaze, they would show a crackle texture after firing. Special artistic effects can be obtained through the artisan's skillful coloring technique, which was commonly called "crackles". "Purple-mouth iron-feet" and "cinnabar bottom" were formed due to the loss of the glaze layer along the mouth rim of the pottery wares and the lack of glaze on the bottom feet.

Celadon was usually fired in a dragon kiln. Different glaze colors could be obtained by controlling the atmosphere in the kiln. For example, in a weakly reduced atmosphere, the proportion of iron oxide (Fe_2O_3) in the glaze becomes higher and the glaze color turned green and yellow. But, in a strongly reduced atmosphere, there was more FeO, and the glaze turns a plum blue.

The Jun kiln was famous for its glaze color change and opalescent effect; purple, red, white, and other dark or light spots were scattered and changed endlessly.

Celadon was originally a monochromatic glaze, and the Jun kiln added copper oxide to the celadon to obtain an intricate Jun red glaze of azure, rose purple, and begonia red. Later, kiln discoloration glazes such as red jasper and aubergine appeared as well as the Shiwan Guang Jun glaze and Yixing Yi Jun glaze, which ended the dominance of celadon and opened up a new era for the diversified development of porcelain.

5.4.2 Black Glaze Porcelain

Black glaze porcelain generally refers to nearly black glazed porcelain such as dark brown, ochre black, and crimson. The difference between it and celadon lies in the iron content of the glaze. Glaze with iron content of 1% or less becomes shaded cyan, with 1%–3% cyan, and if the iron content is as high as 3%–9%, it becomes dark brown (as shown in Picture 5.78).

Black glaze porcelain was founded in the Shangyu area in Zhejiang province during the Eastern Han dynasty. During the Sui and Tang dynasties, the Ding kiln, Xing kiln, and Yaozhou kiln in the north produced celadon and white porcelain while also firing black glaze porcelain. From a technical point of view, black glaze porcelain and celadon belong to the same system, since black glaze porcelain can be fired by simply increasing the iron content in the celadon glaze.

Potters in the Tang dynasty used black glaze as the base glaze and then coated it with a glaze rich in copper, manganese, and titanium to create colored glaze porcelain. The black glaze presented milky white, brown, and gray-blue colored patches. The Cizhou kiln, Ding kiln, Pingding kiln, and Chen kiln in the Song dynasty all fired black glaze porcelain. The most famous wares were the hare fur porcelain cup from Jianyang, Fujian province, and the black glaze porcelain cup from Jizhou, Jiangxi province. The popularity of these two types of porcelain was related to the change in the custom of drinking tea in the Song dynasty from pressed tea cubes to powdered tea. By this time, the tea customs had developed into the tea art called *diancha* and *doucha*. The tea powder now had to be turned into a paste in the tea set, then filled with boiling water, and finally a *xian* (tea whisk) made of bamboo filament was used to stir it. After such processing, the tea mixture becomes white. As the saying goes, "a black tea cup is suitable for white tea", which caused the black glaze teacup to become popular. The hare fur pattern of the hare fur teacup came in gold, silver, yellow, and gray, but especially the milky white ones were the most precious (as shown in Picture 5.79). This kind of porcelain needed to be fired above 1,300°C to make the glaze flow and produce milliforms through complex physical and chemical changes. The Jizhou kiln was most famous for its wood leaf black glaze teacup, which was produced by soaking the leaves in an alkaline solution that left only the veins behind. The veins were attached to the black glaze, and after firing, the alkaline substances contained in the leaves formed the texture of the wood leaves.

The glaze soil for the Jian kiln was taken from a valley called "the glaze treasure house", which was high in iron and phosphorus. Plant ash needed to be added to increase the calcium content. At high temperatures, the iron in the glaze cooled

Picture 5.78 Black glaze pot, Deqing ware, Eastern Jin dynasty

slowly, some of which could not be melted in the glaze and became crystalline which interacted with other metal elements to form textures such as hare fur, oil droplets, and partridge spots due to the changes in temperature.

5.4.3 White Porcelain

White porcelain is fired at a high temperature with a transparent glaze applied to the white base. Its appearance is the result of the successful removal of the excessive iron content in porcelain clay. The northern white porcelain, represented by the Xing

Picture 5.79　Hare fur bowl, Jian ware, Song dynasty

kiln, created the two major systems of southern blue and northern white. The engobe white porcelain of the Cizhou kiln in the north was produced by imitating that of the Ding kiln (as shown in Pictures 5.80 and 5.81), but after the Song dynasty, the southern white porcelain industry was dominated by the blue and white porcelain of Jingdezhen.

The Classic of Tea by Lu Yu of the Tang dynasty describes the Xing kiln's white porcelain "snow-like" and "silver-like". Its rise in popularity was due to the local high-quality porcelain clay. It was dethroned by the Ding kiln because its resources were completely depleted in the Song dynasty. Both of the Xing and Ding kilns used secondary sedimentary clay and kaolinite or feldspar containing more kaolin. China was the first to use kaolinite and feldspar to make porcelain. These raw materials were high in aluminum, but low in iron and cobalt. The content of potassium oxide was increased by adding feldspar in the base material, so the base texture became white, which was similar to the kaolin-quartz-feldspar porcelain used in modern times. The addition of feldspar in glaze increased the potassium content, which even exceeded the calcium oxide content, leading to the transition of traditional calcium glaze to a calcium alkali glaze and an alkali calcium glaze, which greatly improved the enamel. The white glaze and white base complemented each other, forming a white porcelain like silver snow.

The Xing ware was white and elegant, while the Ding ware trumped it with its engraving, painting, and embossed printing. Other decorative techniques, such as stacking, incising, and appliqueing, were also used. Early white porcelain was fired

Picture 5.80 White porcelain jar, Sui dynasty

in a steamed bun kiln at a temperature of 1,380 °C. From the Tang dynasty onwards, it was fired in a funnel-shaped saggar. Ding porcelain craftsmanship was lost in the Yuan dynasty but was restored in the 1980s through the unremitting efforts of people such as Chen Wenzeng.

The engobe white porcelain was made of high-quality porcelain clay attached to a green base, covered with a transparent glaze, and then fired at a high temperature in an oxidizing atmosphere to produce a warm white tone. The Cizhou kilns were located in Pengcheng town, Handan city, Hebei province. The utensils made there were mostly for civilian use. Because the local clay was impure, it created the craft of using engobe clay to fire white porcelain, which developed into underglaze rust patterns, carved patterns, red-and-green overglaze, and the kiln transmuting black glaze. Their products were natural, bold, fresh, and lively, which reflected the aesthetic tastes of the common people and became historically significant in the art of porcelain decoration. The underglaze painting of the Cizhou kiln perfectly combined the ink-painting techniques with pattern structures. Flowers, figures, birds, beasts, insects, fish, and landscapes all appeared vividly, opening a new era of porcelain painting and

Picture 5.81 Carved engobe pickle jar, Rongchang, Sichuan province

influencing freehand painting in later generations. North Korea's "Painting Korean", Japan's "Painting Karatsu" and "Engraving Mishima" ceramics all developed their own patterns and styles on the basis of the decorative techniques of the Cizhou kiln. The Cizhou kiln prospered for a long time and it only started to decline at the end of the Qing dynasty. In the 1950s, the Cizhou Kiln Ceramic Research Institute was established, which has made great efforts to reproduce Song-style porcelain with famous artisans such as Ye Linxiang, Ye Guangcheng, and Wu Xingrang, as well as professionals such as Li Yunzhong and Wei Zhiyu, who joined the institute later to meet the challenge. The Cizhou Kiln Art Porcelain Factory was established in the 1970s to develop traditional products. Since the 1980s, people like Yan Baoshan and Liu Lizhong have created a large number of excellent works with both traditional characteristics and innovative charm. Zhao Lichun and Zhao Xuefeng wrote and published the *Chinese Cizhou Kiln* and *Cizhou Kiln Culture*. All of the above are important contributions to the protection and inheritance of the skills of the Cizhou kiln.

5.4.4 Jingdezhen Porcelain

Jingdezhen in Jiangxi province began to fire ceramics as early as in the Five Dynasties period. By the Song dynasty, with its unique porcelain clay resources, water and land transportation, and supply of materials, it attracted outstanding craftsmen from all over the country. They created a magnificent porcelain kiln system, which fired a blue and white porcelain with a high degree of whiteness and transparency like the verdant jade. This kind of exquisite porcelain was fashionable and made the kiln prosper throughout the Yuan, Ming, and Qing dynasties (as shown in Picture 5.82).

The advantage of the Jingdezhen kiln is that it is able to keep pace with the times and innovate constantly in absorbing the production skills and experience of other famous kilns. After the bluish-white porcelain, potters fired warm-toned egg-white porcelain in the Yuan dynasty. After that, a lot of painted porcelains were created, among them underglaze blue and white porcelain became the mainstream product. It

Picture 5.82 Panoramic view of the Jingdezhen Imperial Kiln Factory, painted on the blue and white tabletop, Qing dynasty

Picture 5.83 Porcelain-making workshop

not only occupied the domestic market but also sold well overseas. The Jingdezhen Imperial Kiln Factory was set up in Jingdezhen in the Ming dynasty, and the official kiln system was established, which forced craftsmen to naturalize and monopolized blue and white pigment and technology. Because of its "deliberate refinement", which was "regardless of cost" and labeled "one out of 100", many fine products were produced, which greatly promoted the deepening of porcelain-making skills and the accumulation of experience. At the beginning of Qing dynasty, the system of craftsmen's registration was abolished. The imperial kiln would use folk kilns to fire porcelains, which released the potential of folk kilns. This caused the government and the common people to compete, which pushed forward the overall development of the Jingdezhen porcelain industry. It was the peak period of traditional ceramics (as shown in Pictures 5.83, 5.84, 5.85, and 5.86).

5.4.4.1 White Porcelain

Bluish white porcelain gets its name because its glaze color has white in blue and blue in white. The traditional aesthetic makes this jade-textured porcelain very popular.

The glaze of egg-white porcelain contains iron, which gives it the white and blue color. It is also called "*shufu* porcelain" because it is often inscribed with the two Chinese characters *shu* and *fu*.

Before the Yuan dynasty, Jingdezhen used china stone as its only material, which was historically called the "unitary formula". After long-term exploration, the addition of kaolin shapes the "binary formula". The introduction of kaolin is a major event in the history of porcelain making.

Picture 5.84 Interior of porcelain-making workshop

In Jingdezhen, potters always pound the china stone with water, then wash it, drain the sand, precipitate it into wet mud, and make it into white bricks of the same size and weight by using wood molds (as shown in Pictures 5.87 and 5.88). However, kaolin can be used directly after washing. The ratio of china stone to kaolin is standardized. To determine the proportion of the two, porcelain trial pieces need to be made first. Tang Ying, a well-known manufacturing inspector of the Qing dynasty, stated in his *Compilation of Pottery Making Pictures*: "Porcelain making first needs clay. Washing porcelain clay and pugging especially focuses on purification [...] Washing and pugging mostly starts with immersing clay in a water tank, stirring clay with a wooden rake, and bleaching the dregs using a thin horsetail basket. Then a double-layer silk bag is used to purify it again with the saggar filled, and the water seeps through thickly and the slurry thickens. Next, several layers of new bricks are spread under a bottomless wooden box, with a fine cloth inside serving as cover, and the thick mud is poured in. Then, the bricks are tightly packed to absorb the water. As the water seeps through, the slurry becomes mud. Then it is put on the large stones, and a shovel is used to pound, to make it strong for pottery crafting." This preparation process has always been used by the Jingdezhen porcelain industry. The prepared mud must be aged in the mud room for 1–3 months and then can be used to make a base after stepping on it. The industry uses the formula of "chrysanthemum core, lotus petals, three feet stepping and two shoveling" (as shown in Picture 5.89).

Ash glaze is applied to daily use porcelain, as Zhu Yan said in *About Porcelain*: "Glaze cannot be made without mortar". Ash glaze is made from glaze stone slurry and glaze mortar. It is a kind of alkali glaze, more specifically a lime glaze. The main ingredient of glaze ash is calcium carbonate ($CaCO_3$). Its preparation method is an industry secret. The main materials are slaked lime and local false staghorn

Picture 5.85 Layout of porcelain-making workshop. 1. aging of the mud, 2. airing rack, 3. elutriating, 4. dewatering (mud-airing bucket), 5. making and printing the base, 6. repairing the base, 7. peplenishing water and glazing, 8. painting on base and engraving, 9. base storing room 10. raw material room

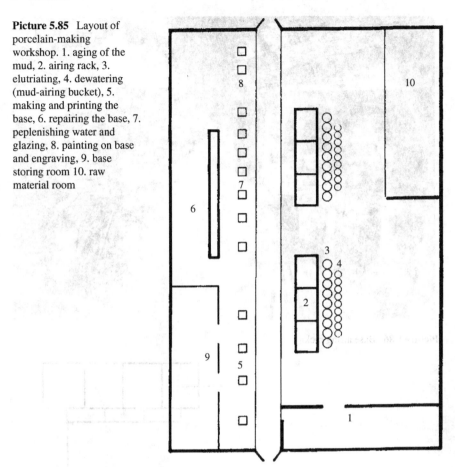

ferns. When making ash, lime is sieved on the firewood layer by layer to up to 1-m high. After three rounds of simmering, aging, homogenization, and washing, it becomes "first mortar" and "second mortar". The former is the fine slurry obtained after elutriation, which is used for glazing coarse porcelain. The latter is the "water mortar" obtained by adding the fine slag precipitated during elutriation to human urine, aging it for one or two months, and then pounding and elutriating it, which is used for glazing by sufflation or pouring fine porcelain. The ratio of the glaze depends on the product. The glaze is blue when there is more glaze ash and white when there is less glaze ash because calcium oxide influences the flux as well as the color.

Jingdezhen porcelain can be divided into three categories: round wares, curved wares, and inlaid wares. Round wares are open utensils such as bowls and plates, curved wares are utensils with deep mouths and lower abdomen, such as pots and jars, and inlaid wares are all square wares, prismatic wares, and special-shaped wares formed by inlaying. The shaping method of round wares is similar to that of curved

Picture 5.86 Base airing rack

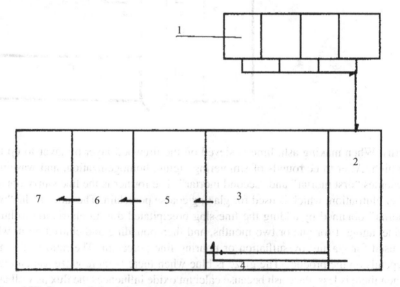

Picture 5.87 Process of elutriating china stone. 1. tilt hammer and mortar, 2. elutriating tank, 3. settling tank, 4. sand ditch, 5. tank for multiviscosity and concentration, 6. mud bed, 7. *dunzi* making

Picture 5.88 Clay mud bricks

wares. It needs to go through the processes of kneading, throwing, drying, printing, water replenishing, glazing, sharpening, water replenishing, soaking, scraping, and bottom glazing (as shown in Pictures 5.90 and 5.91). To make the product quality as perfect as possible and the technical level as professional as possible, Jingdezhen's porcelain-making division is extremely meticulous. This is embodied in the phrase: "A cup needs to go through seventy-two procedures before it is made into a finished product".

5.4.4.2 Painted Porcelain

The underglaze blue and white porcelain of the Yuan dynasty added an extremely beautiful and elegant blue to the porcelain, breaking through the limitations of monochromatic glaze and black decoration, and opening up a broad prospect for later generations of colored painting. It was from this time that the demand for porcelain-making techniques turned to the pursuit of glaze color richness and decorative diversity. Since then, Jingdezhen's painted porcelain has become the mainstream of China's porcelain production.

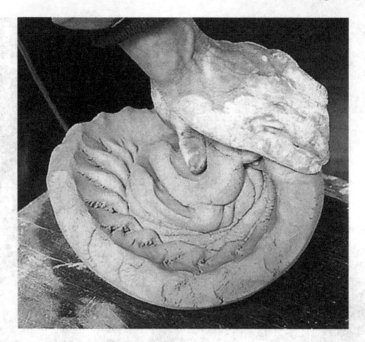

Picture 5.89 Chrysanthemum core kneading method

Picture 5.90 Molds for printing bases

Picture 5.91 Trimming
base with the knife handle
resting on the cheek to
maintain stability

(1) Underglaze

It refers to the porcelain that is decorated with colored patterns on the base surface,
applied with a transparent glaze and then fired in a kiln. It first appeared as early
as the Three Kingdoms period. The Changsha kiln of the Tang dynasty used iron
oxide and copper oxide as colorants, which was the prelude to the underglaze black
patterns (rust-colored patterns) of the Cizhou kiln in the Song dynasty. The latter
also served as the precedent of the white porcelain underglaze in Jingdezhen. In the
Yuan dynasty, the Fuliang Porcelain Bureau was set up in Jingdezhen, where many
underglaze blue wares were fired. The booming market demand attracted porcelain
craftsmen from all over the country to come to the town. From the early Ming to
the Qing dynasty, blue and white, and underglazed red pieces have always been
important in porcelain production (as shown in Pictures 5.92, 5.93, and 5.94).

The raw materials for painting blue and white porcelain patterns used in the Yuan
dynasty were mostly imported Samarra-blue pigment, a mixture of domestic and
foreign materials, or a domestic cobalt blue pigment. The pigment is a natural mineral
containing cobalt oxide. During production, the cobalt earth ore is first calcined,
rinsed several times, mixed with limestone and filler, and then ground in a bowl for
more than a hundred hours. The painting process starts with a rough draft, then the

Picture 5.92 Batchwise dyeing the feather

Picture 5.93 Filling the crane body with glass fost

draft copying, drawing and dark–light-processing. The last step refers to a technique that is unique in blue and white porcelain. The "material water" used is divided into five grades, namely (1) first thick, (2) second thick, (3) concentrated water, (4) thin water, and (5) light shadow, which are distinguished by the amount of blue and white material it leaves on the base body. The dryness and wetness of the base, the rapid

Picture 5.94 Blue and white porcelain, Ming dynasty

movement of the painting, and the length of the water accumulation all affect the shade of the water.

The coloring agent of underglazed red is copper oxide. At the end of the Yuan dynasty and the beginning of the Ming dynasty, the underglazed red in Jingdezhen reached a very high level. The color materials were made of copper crumbs, glass powder, and porcelain clay. It was fired at 1,280°C using a reduced flame. Because copper had different colors in different atmospheres and was volatile at high temperatures, it was necessary to strictly control the temperature and atmosphere of the kiln.

(2) Overglaze

Overglaze refers to the porcelain that is painted with colored materials on the existing glazed surface and fired twice (commonly known as "decoration firing"). It makes up for the shortcomings of the single underglaze color tone and makes the colors of the painted porcelain truly colorful and radiant.

When used to create a clashing color, it is first painted with blue and white materials on the base surface, covered with transparent glaze, and after being fired at a high temperature, the outline of the blue and white is filled with color. Once done, it is then fired at a low temperature for the second time. It is technically partial underglaze and partial overglaze rather than pure overglaze. The blue underglaze and the overglaze clash in beauty and fascination, hence name "clashing color". Clashing color techniques were developed greatly during the Chenghua period (1465–1487) of the Ming dynasty and were the mainstream of colored porcelain decoration at one point.

Famille verte, on the other hand, refers to porcelain that combines underglaze blue with a variety of overglaze colors, and it is not limited to five colors. The Kangxi famille verte technique was characterized with its blue, pushing the overglaze technique to its peak. The Kangxi famille verte included the black color, which was also called "antique color" or "hard color".

Famille rose was made by adding arsenic to the glass fost as an opacifier to obtain a chalky feeling so that the paint came in both light and dark. It created rich layers and formed a unique style that embodied elegance and grace. This kind of decoration technique was created during the Kangxi period (1662–1722) and was further refined during the Yongzheng period (1723–1735). It was the mainstream of Jingdezhen painted porcelain together with blue and white porcelain.

5.4.4.3 Colored Glaze Porcelain

Colored glaze porcelain is also monochromatic glaze porcelain. The Jingdezhen kiln created copper red glaze and cobalt blue glaze porcelain in the Yuan dynasty. Shiny red glaze porcelain was fired during the Xuande period (1426–1435) in the Ming dynasty. It was used to create sacrificial vessels, because the glaze was solemn, nonflowing, and free of cracks. The Qing dynasty used more than 10 kinds of color glazes, such as white glaze, blue glaze, yellow glaze, brown glaze, peacock blue glaze, green gold blue glaze, aubergine glaze, and water-melon green glaze, forming a series of high-temperature monochromatic glazes. During the Kangxi period, the red glaze developed greatly, and the Lang-kiln red was the most precious. Its glaze is jade-like and bright. Because of its high fluidity, the rim of the mouth was mostly pink and light blue, and the glaze of the overhanging parts of the body was covered more thickly with glaze and was a deeper color, forming a unique style of "blurting out, but not flowing" (as shown in Picture 5.95).

The Jingdezhen porcelain industry reached its peak during the Kangxi, Yongzheng and Qianlong periods in the Qing dynasty, which was jointly promoted by the expansion of the domestic and foreign markets and the preferences of the emperors (as shown in Picture 5.96). Although the pottery and porcelain skills made great progress during these periods, it is not advisable to learn from them due to deliberately seeking out further refinement regardless of cost, cumbersome stacking, the pursuit of luxury and the main goal of showing off wealth. This approach also made many works

Picture 5.95 Lang-kiln red *zun* (an ancient wine vessel), named after the kiln official Lang Tingji who presided over the firing, during the Qing's Kangxi period

become tacky and less about craftsmanship, which had a negative impact on later generations.

From the end of the Qing dynasty to the Republic of China, the achievements of the Jingdezhen kiln were known for its antique porcelain and fine art porcelain. The masterpieces were the works of "Eight Friends of Zhushan", which was a collective name of eight famous potters centered at Zhushan, Jingdezhen in the late Qing dynasty. In the 1950s, Jingdezhen established a ceramic research institute, scientific research units, and art colleges to restore the porcelain production and skills to their highest level in history. During this period, famous artisans such as Wang Xiliang, Yang Houxing, Qin Xilin, Zhang Songmao, Wang Enhuai, and Wang Longfu emerged. In the meantime, however, blindly following the pattern of mass-producing everyday ceramics, the replacement of the manual windlass, engraving, and printing techniques with machines resulted in some precious craftsmanship becoming invisible or even lost. This should serve as a lesson that mass production can lead to the disappearance of craftsmanship. Jingdezhen's hand-made porcelain skills were included in the *List of National Intangible Cultural Heritage* in 2006. With the joint efforts of artisans, communities, enterprises, and the government, they will be better protected and inherited.

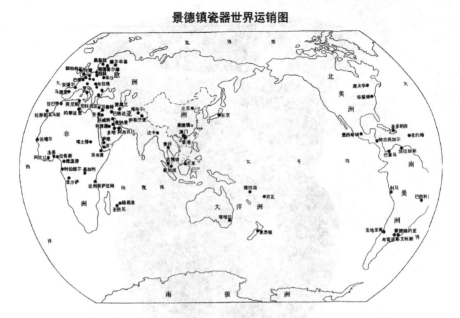

景德镇瓷器世界运销图

Picture 5.96 Map of the export area of Jingdezhen porcelain

5.4.5 Dehua Porcelain Molding

The Dehua kilns in Fujian province began to fire celadon wares as early as the Five Dynasties period. They changed to fire white porcelain and bluish white porcelain in the Song and Yuan dynasties, reaching their peak in the Ming and Qing dynasties. After the mid-Ming dynasty, the maritime trade between China and Southeast Asia and Europe developed greatly. During this period, the Dehua kiln located in the south produced a new variety of white porcelain with high-quality raw materials. The porcelain's main characteristics were that it was white, compact, soft, and crystal clear. Due to the different firing atmosphere, the porcelain colors were "ivory white", "lard white", and "baby red". The baby red was white and rosy, and it was as tender as a baby's skin. It could be made occasionally but not consistently, resulting in it rarely being handed down from ancient times. Because a large part of Dehua white porcelain was sold abroad and was regarded as a treasure of porcelain, it was awarded the reputation of "Chinese White".

The traditional Dehua porcelain glaze was made of orthoclase, limestone, and rice husk ash, which were used in low-temperature glaze and high-temperature glaze in different proportions. The round wares were formed by the "throwing" technique, the special-shaped wares mainly by mold impressing, and the porcelain molding by kneading, mold impressing or grouting. Decorative techniques included printing, engraving, appliqueing, embossing, and openwork carving. The types of kilns used were round kiln, dragon kiln, ascending kiln, and chicken cage kiln. The chicken cage kiln was a partitioned dragon kiln. It was divided into chambers by partition walls without grading. It was named after its resemblance to a row of chicken cages.

The characteristic product of the Dehua kiln was porcelain molding ware. It was created in the early Ming dynasty and remains until this day. Its biggest feature was that it was made in monochromatic colors, giving full play to the beauty of the material, adding various themes and details, thus forming a unique artistic feature that even greatly influenced Jingdezhen's and Shiwan's ceramic molding. The origin of the Dehua porcelain molding has a lot to do with the religious belief. Buddhism prevailed in Fujian during the Tang and Song dynasties. With the popularization of Zen Buddhism, the demand for small Buddhist statues greatly increased. Dehua artisans created new types of porcelain molding figures of Tathagata, Guanyin, Arhat, and Bodhidharma to get rid of the limitations of the earlier traditional rituals by creating new images. Porcelain molding figures of Taoist gods from the Jade Emperor and the Queen Mother to Magu and the God of Earth rapidly gained popularity. A great master He Chaozong's works made Dehua porcelain molding famous all over the world. In order to meet the needs of literati and commoners for indoor furnishings, porcelain molding works covered a wide range of myths and legends, historical figures, folk tales and animals, flowers and birds, and even bonsai. Western-style portraits were also made specifically for conquering European markets. Among them, the Guanyin, Bodhidharma, and Wenchang (a Taoist god) of the late Ming dynasty were exemplary due to their elegance and otherworldliness, with their warm and clean porcelain and superb craftsmanship. They were the perfect unity of art, technique, and material. They are the best of Dehua porcelain molding works (as shown in Pictures 5.97 and 5.98).

He Chaozong, also known as He Lai, was a native of Longtai village in Dehua, who had studied craft with his father ever since he was a child. The clay molding of Guanyin at Bixiangyan and Shancai (depicting a man good at playing Chinese lute) at Chengtian Temple were all made by him. His masterpieces of porcelain molding works are Guanyin statues, and many of his works are kept at the Palace Museum in Beijing, Shanghai Museum, and museums in Europe, America, and Japan.

The most reputable Dehua porcelain molding craftsmen from the late Qing dynasty to the Republic of China were Su Xuejin and Xu Youyi. Su Xuejin's porcelain molding work "Plum Blossom" won the Gold Medal at the Panama Pacific International Exposition. Xu Youyi made 500 arhats for Longji Temple in Xianyou county, Fujian province in 1930, which was an unprecedented production scale for porcelain molding. Since the Reform and Opening Up, the Dehua porcelain industry has made great progress. Famous contemporary artisans include Chen Qitai, Zhou Yage, Su Qinghe and Xu Xingtai.

Picture 5.97 Clay mold draft

5.5 The Ceramics Industry's Division of Labor, Guild Organizations, and Customs

Fang Lili did a detailed field investigation on the division of labor, guild organizations, and customs at the Jingdezhen kiln yard:

A major feature of Jingdezhen's ceramics industry is that it has a very fine and strict division of sub-industries and types of work. The craftsmen's skills are highly specialized, and they even specialize in only one specific type of work throughout their lives, which is conducive to the improvement of skills and the improvement of product quality. The three main business branches of the kiln are (1) firing, (2) shaping, and (3) painting. The firing branch is divided into the *chai* kiln (which uses pine wood as fuel) and the *cha* kiln (which uses branches as fuel).

Picture 5.98 Porcelain molding Guanyin

The shaping branch is divided into round wares and curved wares, each of which contains 11 lines of businesses, such as bodiless lacquering, large piece making, white glazing, and Ding white ware making. The painting branch produces colored porcelain and is divided into four major lines of business, such as painting porcelain and painting bodiless lacquerware. There is a detailed division of labor within each line of business. For example, the operation and management personnel of the *chai* kiln include the kiln owner, the accountant, and the firewood supply manager. The three top craftsmen are collectively known as the "*shangshanjiao*" (three top-level workers): (1) Highly skilled craftsmen who have the most authority as kiln-burning masters and are also known as fire heads. There are also (2) those who oversee carrying bases, filling the kilns with bases, firing the kilns on the midnight shifts, and (3) those who are responsible for the saggars and firing the kilns after midnight.

Those who do heavy manual work include the *shoudoujiao* (workers taking out the bottom layer of saggars), *xiaohuoshou* (workers required of one year's internship after the apprenticeship is completed), *yifuban* (first apprentice), *erfuban* (second apprentice), *sanfuban* (third apprentice), and handymen. The shaping branch's round wares department is led by the throwing workers, under whom are those in charge of printing bases, sharpening bases, brake closing bases, cutting bases, assembling bases, selecting porcelain, and other odds and ends. The workers in charge of sharpening bases and brake closing bases (responsible for blowing and dipping glaze) need more advanced skills, so they are collectively referred to as the "*shangsanjiao*" together with the throwing workers. The shaping branch's curved wares department regards the throwing workers, the sharpening workers, and the carving workers as the "*shangsanjiao*", and the rest of the craftsmen as the "*xiajiaogong*" (lower-level workers).

In addition to the above-mentioned main businesses, there are also service, sales, and supply businesses that all together form a close and complete ceramics industry production cluster, such as saggars, packaging and handling, tool making, repairs, sedans, horse handling, porcelain buying, porcelain selling, and supplying of paint, clay, glaze, mortar, firewood, and other raw materials. As stated in *About Jingdezhen Ceramics*: "There are kilns for pottery, workshops for the kilns, workers for the workshops and homes for the workers. The ceramics industry supports a variety of people".

Jingdezhen is known as the "Eighteen Provinces Terminal", where people from all directions come together over water and land. There are more outsiders than natives. Historically, different industries were monopolized by craftsmen from different regions. For example, most of the kiln owners are from Duchang county, and most of them are surnamed Feng, Yu, Jiang, and Cao. Most of the round ware craftsmen are also from Duchang. Therefore, in the traditional porcelain industry, the "Duchang Group" is the most powerful. Most of the craftsmen of curved wares are from Fuzhou and Fengcheng counties, with most of the painters and sculptors being specifically from Fengcheng. The people who do the miscellaneous and heavy tasks are mostly from Nanchang and Xinjian counties.

This way, many guilds, public offices, and other guild organizations were formed on the basis of place of origin and kinship, to coordinate conflicts of interest and maintain order within the industry.

The guild hall is a place where fellow townsmen and colleagues gather and interact. The guild hall in Jingdezhen was built in the Ming dynasty and gained importance in the Qing dynasty. There were as many as 27 guilds in the early 1950s, such as the Duchang, Fengcheng, Fuzhou, and Nanchang guilds. These guild halls are all built with temples and shrines, as well as main halls where gods were enshrined, conference halls, academies, theaters, and guest rooms. The statues in the hall each have their own regional characteristics, such as the Tianhou Empress of the Fujian Guild Hall and the Great Sage Guan Yu of the Shanxi Guild Hall. The guild has one master, the position of which is set up on a rotation schedule according to each industry and surname. Their main source of income is through fund-raising and donations but also includes rental income from land and real estate. The academy provides accommodation and study rooms for children. During the Republic of China, many guild halls transformed their academies into primary and secondary schools. In addition to business transactions, the guild also resolves disputes within the industry, organizes public welfare undertakings, and assists countrymen in need. The Meeting, which is a memorial to the gods, is the most important event of the year. Guan Yu is honored in May, and Xu Zhenjun is honored in August.

Jingdezhen's porcelain industry has three trade associations, namely, merchants, handicraft owners, and craftsmen, as well as three regional ones, namely, Huizhou, Duchang, and Zabang. The merchants are divided into 26 trade associations, and the handicraft industry is divided into 8 professions and 36 subbranches. Craftsmen are further divided into subgroups according to their type of work. Taking tool craftsmen, for example, there are 5 types and 18 subgroups, each group with their own leader who is in charge of their group's affairs. Each group protects their monopoly within the industry, safeguards the interests of the group, prevents the skills from disseminating, and has a certain degree of care and restraint for its members. After thousands of years of evolution, the porcelain industry has formed customary rules for starting, stopping, hiring, welfare, and inter-industry relations. For example, the round wares industry is the largest industry in Jingdezhen. It involves over 70% of the total number of craftsmen. The management of workers is fully controlled by the foreman. There can only be one foreman who is hired by the "boss" to complete one or multiple projects. If allowed by the boss, the foreman can hire a sub-foreman (known as "*banbande*") but no more than one. This sub-foreman can also hire assistants and apprentices. The *banbande* cannot dismiss assistants or apprentices arbitrarily. If he really wants to dismiss anyone, he must suspend the work for three days, treat all the craftsmen, explain the reason for the dismissal, and settle the salary. If the *banbande* must quit his job, he must work until the changeover period and cannot leave halfway. One year is divided into four changeover periods: Dragon Boat Festival, Chinese Zhongyuan Festival, Mid-Autumn Festival, and the 10th month of the lunar calendar. A resigning craftsman must invite the foreman in advance to drink tea and agree on the date of resignation. On the next day, the foreman must invite the craftsman for tea and persuade him before the craftsman makes the final decision. If a craftsman's performance is poor, the boss shall not interfere. Instead, the foreman will have to deal with it. If the boss violates the rules, he should treat people for tea and apologize.

If a craftsman violates the rules, he should treat people for tea and pay a fine. As the saying goes, "The Feng, Yu, Jiang and Cao families rank top, and the Zhang, Wang, Liu and Li families stand on both sides". Any possible punishment for those with these eight surnames is minor. They can only pay the tea bill without getting fined. If the apprentices have not completed the quota, they must take off their clothes and are beaten with a rope, and they will be fined with a night's work and no time off during festivals.

The work starts on the first day of the third lunar month and is closed on the 13th day of the 12th lunar month. If the working period is extended, each craftsman will have an extra 200 g of pork per day. From the Spring Equinox to the Autumnal Equinox, workhouse during the day is from 6 am to 5 pm every day, while night work is limited to 10 h. From the Autumnal Equinox to the Spring Equinox, work hours during the day are from 7 am to 6 pm every day, while night work is limited to 11 h. At noon, workers can take a lunch break. The kiln workers work in shifts and the work never stops. There are 39 holidays in the whole year. They get time off for the Water-sprinkling Festival in April, Dragon Boat Festival, Work Change Festival in mid-July (five days), Mid-Autumn Festival, Water-sprinkling Festival in September, and from the 13th of the 12th lunar month to the next 12th of the second lunar month.

The craftsman's food is provided by the boss. The throwing master receives two meals. If he does not consume it all, he can give what is left to his family or resell it. The sharpening workers get one and a half meal, and the rest enjoys one meal. They all eat meat once every 10 days, 200 g each time. Usually, they eat steamed brown rice and "kiln dishes" (pickled vegetables or green vegetables and fermented soybeans). For three festivals a year or after completing a large project, the boss will invite the craftsmen to a feast. If a craftsman gets sick, the boss shall lend them money for treatment. If, unfortunately, a craftsman dies, the boss shall buy them a coffin.

The rules vary from industry to industry. For example, according to the regulations of the curved wares industry, they eat taro at noon and drink in the afternoon on Mid-Autumn Festival. In addition, each person will be given half a jin of moon cakes. In the evening, the boss and craftsmen admire the moon together, while eating shortbread, melon seeds, peanuts, and fruits. The carving craftsmen only accept new apprentices once every 5 years, and each craftsman is limited to only one apprentice. Apprentices need to pay 20 yuan to enter the industry, and someone must vouch for the apprentice. The apprentice signs an agreement and cannot leave before the expiration date of the agreement. In the event of death or injury, the master will not be held responsible. During the first year of apprenticeship, they will be paid a salary. After 2 years, their salary will be increased by 2 yuan per year. After apprentices finish their apprenticeship, they must assist their master for 1 year, earning a salary that is half or 60% of a full salary to show gratitude to their master.

There are 10 rules in the curved wares industry, such as: You cannot steal the porcelain and materials from the base room; you shall not mess around with other people's personal effects (most of the porcelain workers' homes are in other places, and they don't have boxes, and the clothes and money are placed on the bedside. Rules were set a long time ago. Tampering with things without consent is deemed stealing); you shall not take a part-time job; no night shifts are allowed upon the start of work or during the three festivals; if the boss deducts food and welfare, you can protest and ask the master to mediate; you shall not arbitrarily reduce wages in the off-season; guild decisions should be followed. The penalties for violations range from having to pay a fine to "*duocaoxie*" (being expelled and barred from being hired again). These rules are strictly followed and supervised by the master of this group. According to local customs, women cannot enter the kiln; they can't even go to the "*yizitiao*" (the springboard in front of the kiln). Except for painting, there are no female workers in the base house and kiln.

Jingdezhen produces 50,000 metric tons of porcelain each year. Each day, workers consume 1,000 pigs and 500 metric tons of rice. However, more than half of the population is illiterate and semi-illiterate, so even the boss may not be able to keep the accounts and issue invoices. In this case, the transaction is often settled orally. Sometimes the order is written by the merchant, then the boss stamps it, and the agreed price is never broken. "It is better to go bankrupt than to break your promises". Honoring contracts has become a traditional virtue.

The Jingdezhen porcelain industry is dedicated to the business god Huaguang Tianwang (Heavenly King of Flowering Brightness). It is said that Huaguang used to be a boy serving Tathagata and was good at changing and subduing demons. Porcelain workers build temples for him and pray to him every year for blessing and fruitfulness in the firing business. Workers that work in base making, kiln firing, and the saggar industry pray to Zhao Kai, Tong Bin, and Marshal Qian (the founder of the saggar) as their respective business god (as shown in Picture 5.99).

Picture 5.99 "Worshipping God" in *Compilation of Pottery Making Pictures* by Tang Mei

Zhao Kai, with the courtesy name Shuming, once held official positions in Fujian, Zhejiang, and Jiangxi. Because of his uprightness, he was a thorn in traitorous officials' eyes. Therefore, he retired to Fuliang county, Jingdezhen city, where he introduced the Yue kiln's celadon-firing techniques, which made important contributions to the development of Jingdezhen's porcelain industry.

Tong Bin, with the courtesy name Dingxin, was born in 1576 in Li village, Fuliang county, where he studied kiln art since childhood. In the 27th year of the Wanli period (1599) in the Ming dynasty, the eunuch Pan Xiang, while serving as a Jiangxi mine envoy and Jingdezhen kiln clerk, supervised the construction of a large-scale Qinglong urn that failed for a long time. He angered the kiln workers, flogged, and even killed them. Upon seeing this, Tong Bin plunged into the kiln and successfully completed the urn. Later generations cherish the memory of this martyr, regarding him as the god of wind and fire, building temple statues and worshipping him. They name themselves *Tongqingshe* (society to honor Tong), regard the descendants of Tong Bin as their relatives, and hold a grand ceremony every 20 years to greet Tong Bin, the god. On these unprecedented occasions, people wear new clothes and make new jewelry, do lion dances and stilt-walking.

In the Qing dynasty, the master of the base-making industry also added a craftsman Zhi Si as their god, who was a local potter. Because of the kiln owner's stinginess, the craftsmen lived a difficult life. They didn't have any meat for months, so Zhi Si launched a strike, got arrested by the government and was killed. His martyrdom aroused great indignation among the potters, and they rose up to fight. The kiln owner was forced to succumb and add half a kilo of pork per month to each potter's meal. To commemorate Zhi Si, the potters put his memorial tablet in the master temple, and every time they ate meat, they would set up a stage on the drying rack, burn incense and set off crackers to worship him.

Chen Dongxing and Qiu Gengyu inspected the division of labor, guild organizations, and customs of the kilns in Chenlu town, which is in the southeast of Tongchuan city, Shaanxi province. The residence and kilns are built along the mountainside. According to a survey in 1941, there were more than 40 porcelain kilns and 121 workshops in the town, with an annual output of 8.5 million pieces of porcelain. More than 3,000 people in the town were engaged in porcelain-making as their main source of income and concurrently engaged in agriculture and other sideline businesses. This special economic model of combining handicraft industry with agriculture, self-employment, and family inheritance continued until the end of the 1950s.

Like the Jingdezhen kilns, the porcelain industry in Chenlu town also had a strict division of labor, with the concept of "four business tenants" and "three lines". The "four business tenants" were porcelain tenants (base making), kiln tenants, commercial tenants (shopkeeper and purchasers) and dealer tenants (businessmen). The "three lines" were the bowl kiln, urn kiln, and black kiln (which mainly produced small pieces such as spice boxes and cupping pots). The area of each business was relatively set. Shangjie, Yaojie, and Beitou were dominated by bowl kilns, Shui-quantou was dominated by urn kilns, and Wanli was dominated by black kilns. This was commonly known as "three lines of business not interfering with each other". Each production line made porcelain of their own kind and did not interfere with each other's business. Head of every business was responsible for its own management, supervising each kiln to abide by the rules and regulations and not go beyond. Otherwise, the offenders would be punished. The craftsmen divided the work according to the types of work, such as kneading, throwing, and trimming. Each kiln had its own division of benefits: Of the 10 shares of a bowl kiln, the kiln owner had five, kneading, throwing, and trimming each got one-third of the remaining shares; Of the 12 shares of an urn kiln, the kiln owner had six, kneading two, throwing three, and trimming one. This distribution of benefits grew organically based on the investment of financial resources, skills, and manpower.

There are many kiln temples in the porcelain area. The earliest kiln god tablet in China, *Deyinghou Stone Tablet*, was discovered at the site of the Huangbao Kiln Temple. The kiln gods include earth gods, mountain gods (overseeing pottery clay and porcelain clay), Shun (who once ploughed in Lishan and is regarded as the ancestor of pottery), Laozi (using fire for alchemy), as well as the Thunder God, the King of Cattle and the King of Horse (The King of Cattle and King of Horse are the gods of transportation). Every year on the 20th day of the first lunar month and on the Mid-Autumn Festival, sacrificial ceremonies were held at the kiln temples. Senior kiln workers would tie a red thread to the statue of the god, which was called "*dahong*" (literally means "hanging up red"). The temple fair had activities like carnival illuminations and lion dances, which served to encourage trading exchanges as well as entertainment. At the theater building opposite the kiln temple actors and actresses sang opera for three days. Besides these ceremonies, there was another one where the owner put wheatgrass as well as a hot roller in a vinegar bowl when installing a kiln, and then go around the kiln carrying the bowl. This was to pray for a smooth kiln firing so that the porcelain fired would be as good and fresh as vinegar. To keep secrets, craftsmen were not allowed to leave the town to work elsewhere, and offenders were punished severely. This rule still has an impact today. A master at Chenlu Ceramics Factory went to a Beijing school to assist in building a kiln and making pottery in the 1990s. He was fired from the factory and even his retirement was forfeited. This overly conservative approach was detrimental to the development of ceramic production.

Since the 1950s, the social, political, economic, and humanistic conditions have undergone tremendous changes, guild halls, groups, rules, and customs have subsequently declined or even disappeared. In any new historical era, impractical rules and customs that do not conform to the signs of the times will always die out. However, the most important qualities, namely, that the ceramic crafting groups perform their duties and strive for perfection, protect the rights of the industry as well as their own interests, adhere to their professional ethics, and keep trying to innovate, should still be inherited, and carried forward. Potters such as Zhi Si and Tong Bin were not afraid of violence, bravely standing against despotism and oppression. They strove for the survival and welfare of the masses at all costs, which is certainly worthy of admiration.

Since the 1930s, chemists such as Zhou Ren, Zhang Fukang, and Li Jiazhi have successively used modern science and technology to conduct in-depth analysis and research on ceramic technologies of past dynasties. Cultural relic and archeological experts such as Feng Xianming and Geng Baochang have also identified and researched unearthed and handed down ancient ceramics in detail, and systematically combed and discussed the history of ceramics in combination with literature. Scholars such as Yang Yongshan, Chen Jinhai, Qiu Gengyu have conducted in-depth and detailed investigations and analyzed on ancient ceramics and folk kilns all over the country from the perspective of ceramic processing technology (as shown in Pictures 5.100, 5.101, and 5.102). In addition, ceramic factories, kilns, and related research institutions and colleges have also done a lot of work, successfully restoring several long-lost ceramic-firing techniques. After nearly 80 years of academic research, ceramic firing has become a mature and accomplished discipline. At the same time, due to China's industrialization, its silicate manufacturing industry has been fundamentally transformed and developed into a modern, large-scale industry with a high degree of mechanization. The industrial and folk ceramics produced in Tangshan, Handan, Zibo, and Chaozhou has been continuously improved. In recent years, the modern concepts and aesthetic appeal of ceramics production have gradually been acknowledged by people, and excellent works have come into being with increased frequency (as shown in Pictures 5.103 and 5.104).

Traditional ceramic firing has also begun to receive increased attention and scrutiny. In addition to handmade porcelain in Jingdezhen and Yixing purple clay pottery, the Dai and Li people's pottery, Jieshou colored pottery, Shiwan ceramics, Uyghur clay pottery, the Yaozhou kiln, Longquan celadon, the Cizhou kiln, Dehua porcelain, colored glaze (Beijing and Shanxi), Linqing tribute bricks, Ding porcelain, Jun porcelain, Tang tri-color glazed ceramics, Liling underglaze famille verte porcelain, Fengxi porcelain, Kwon-glazed porcelain, Nixing pottery in Qinzhou, Tibetan black pottery, Yazhou pottery, Jianshui purple pottery, and Xingjing sand ware have all been added to the *List of National Intangible Cultural Heritage*. With the joint efforts of artisans, communities, enterprises, experts, scholars, and the government, these traditional skills will become better protected. In line with the country's modernization, as well as continuous exploration and innovation, the prospects for the development and revitalization of the traditional ceramic industry will certainly be broad and bright.

Picture 5.100 Professor Chen Jinhai, Dean of the Ceramics Department of Tsinghua University Academy of Arts and Design, making pottery

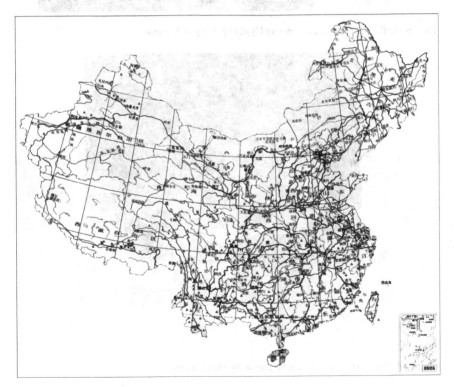

Picture 5.101 Distribution map of folk ceramics

Picture 5.102 Qiu Gengyu inspecting Uyghur pottery in Xinjiang

Picture 5.103 Sixteen-bamboo-tube teapot, by He Daohong

Picture 5.104 Polyhedron pottery box by Eva Schepova, Czech Republic

Picture 5.104 Prob.salian period. Date by Eva Scorpova Czech Republic